D1446923

A PRACTICAL GUIDE TO STATISTICAL QUALITY IMPROVEMENT

Opening up the Statistical Toolbox

Michael R. Beauregard, P.E., C.Q.E.

Raymond J. Mikulak, P.E., C.Q.E.

Barbara A. Olson, C.Q.E.

Mike Beauregard　　*Ray Mikulak*

Barbara Olson

VNR VAN NOSTRAND REINHOLD

New York

Copyright © 1992 by Van Nostrand Reinhold

Library of Congress Catalog Card Number 91-33671
ISBN 0-442-23439-2

All rights reserved. No part of this work covered by the copyright hereon may be reproduced or used in any form or by any means—graphic, electronic, or mechanical, including photocopying, recording, taping, or information storage and retrieval systems—without written permission of the publisher.

Manufactured in the United States of America

Published by Van Nostrand Reinhold
115 Fifth Avenue
New York, New York 10003

Chapman and Hall
2-6 Boundary Row
London, SE1 8HN, England

Thomas Nelson Australia
102 Dodds Street
South Melbourne 3205
Victoria, Australia

Nelson Canada
1120 Birchmount Road
Scarborough, Ontario M1K SG5, Canada

16 15 14 13 12 11 10 9 8 7 6 5 4 3 2 1

Some information contained in Chapter 1 (Sections 1.3 and 1.6), Chapter 2, and Chapter 4 (Sections 4.1, 4.2, 4.4, and 4.6) is based on material originally appearing in *SPC in Action*, copyright 1990, Resource Engineering, and *First Class Service,* copyright 1991, Resource Engineering, with the consent of the publisher, Quality Resources.

Library of Congress Cataloging-in-Publication Data

Beauregard, Michael R.
A practical guide to statistical quality improvement: opening up the statistical toolbox/Michael R. Beauregard, Raymond J. Mikulak, Barbara A. Olson.
p. cm.
Includes indexes.
ISBN 0-442-23439-2
1. Quality control—Statistical methods. I. Mikulak. Raymond J. II. Olson, Barbara A. III. Title.
TS156.B437 1991
658.5'62'015195—dc20 91-33671
CIP

Contents

Preface

A Practical Guide to Statistical Quality Improvement: Opening Up the Statistical Toolbox is designed as a reference guide for the engineer, supervisor, and manager. The intent of the text is to present conventional statistical quality improvement tools in a user-friendly form. We have worked to take some of the "mystique" out of the statistics and help others put these powerful tools to effective use in a Total Quality Manage-Management (TQM) environment.

This isn't a text on TQM. TQM has three elements (as shown in Figure i.1):

1. Creating the environment
2. The continuous improvement toolbox
3. Employee empowerment

This text focuses almost exclusively on the middle element, the continuous improvement (CI) toolbox. Further, *Opening Up the Statistical Toolbox* does not present a complete set of tools intended to "fill" the CI toolbox; only the statistical tools and some of the basic team process tools are covered.
The CI toolbox, in reality, will never get "filled". A comprehensive toolbox will include extensive team process skills and technology specific tools complimentary to the statistical tools included here.

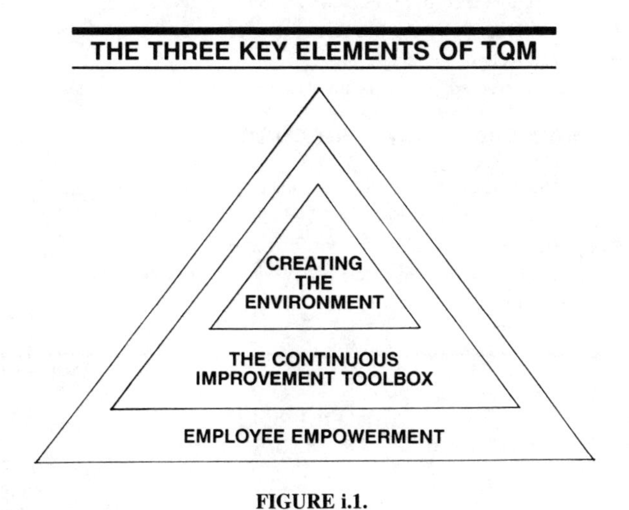

FIGURE i.1.

TOTAL QUALITY MANAGEMENT OVERVIEW

1. TQM CONCEPTS
- Traditional approaches to Productivity & Quality Improvement vs. TQM
- History of the Quality Improvement Movement
- The Concept of a Process
- Cost of Quality
- The Customer-Supplier Relationship

2. TQM: THE THREE ELEMENTS
- Creating the Environment
- Using the Continuous Improvement Toolbox
- Involving all Employees

3. CREATING THE ENVIRONMENT
- Visions that Empower Change
- Cascading the Vision through the Organization
- TQM Strategy- The Roadmap to Continuous Improvement

4. THE CONTINUOUS IMPROVEMENT TOOLBOX
- Developing a Common Language
- Establishing uniform Team Problem Solving Tools
- Setting a Base for Organization Discipline
- Establishing the "Boundaries of Freedom"
- The Basic Tools- all must know
- Fundamental Statistics- some must know
- Advanced Statistical Methods- a few must know

5. EMPLOYEE EMPOWERMENT
- Speaking the Common Language & Using the CI Toolbox
- Making the most of the Boundaries of Freedom
- Breaking Down the Barriers between Departments & Functions
- The Role of Project Teams
- Self-directed Project Teams

6. THE CONTINUOUS IMPROVEMENT CYCLE
- The PDCA Cycle
- The Kaizen Philosophy of Steady Continuous Improvement
- Benchmarking to Find the "Best of the Best"
- Stretch Goals- "Thinking outside the Box"
- The Compounding Effect of Continuous Improvement

resource engineering, inc.

FIGURE i.2.

TQM and the use of these tools are additive to the core technology of any venture. A comprehensive TQM process added to an organization with a solid core technology will significantly improve the effectiveness of that organization. Basic premises of TQM are:

- Everything can be studied as a process with inputs, value-added, and outputs.
- The overall process within an organization is really a series of internal customer–supplier relationships that need to be optimized to meet the needs of the ultimate (external) customer.
- Variation exists in everything. Reduction in variation improves the process and makes it more predictable. But, before it can be reduced, it must be understood and measured.
- Use of the team process is the best way to carry out continuous process improvements.

The overview of TQM is shown in Figure i.2. Again, *A Practical Guide to Statistical Quality Improvement: Opening Up the Statistical Toolbox* isn't a how-to text on TQM. We only address one element of TQM—the CI toolbox.

- Chapter 1 deals with some fundamental concepts of TQM focusing on the role of the engineer in a TQM environment.
- Chapter 2 covers the basic tools needed to understand and begin to measure variation.
- Chapter 3 discusses the measurement system and explores the vital role it plays in the study of process outputs.
- Chapter 4 covers statistical process control (SPC) concepts and use.
- Chapter 5 contrasts the statistical tools of improvement, designs of experiments (DOEs), to the statistical tools of control, SPC, and explains practical ways to use these powerful tools.
- Chapter 6 provides suggestions on applications in the support areas for the tools covered in Chapters 2 through 5.

A Practical Guide to Statistical Quality Improvement: Opening up the Statistical Toolbox isn't written by or for statisticians. No doubt, pure statisticians will find fault with some of the "short-cuts" we present. However, our experience has shown us that the tools, used as presented, are almost always more accurate and precise than our ability to measure the "true value" of the process outputs we are working to improve.

We are grateful to the literary executor of the late Sir Ronald A. Fisher, F.R.S. to Dr. Frank Yates, F.R.S., and the Longman Group Ltd., London for permission to reprint the chi square tables from their book *Statistical Tables for Biological, Agricultural and Medical Research* (6th ed., 1974).

Michael R. Beauregard, P.E., C.Q.E.
Raymond J. Mikulak, P.E., C.Q.E.
Barbara A. Olson, C.Q.E.

1

Fundamental Concepts of Total Quality Management

1.1 TQM—HOW DOES IT WORK?

Total Quality Management (TQM) means much more than the management of quality. TQM is a philosophy of management that drives a continuous improvement approach into all aspects of an organization. TQM does not replace the "core" technology of an organization; it is additive to the core technology, helping make it be all it can be. In a TQM environment, the culture of the organization embraces never-ending improvement (of their core technology), and change for improvement becomes a constant.

There are three key elements of TQM:

1. Creating the environment
2. The continuous improvement toolbox
3. Employee empowerment

As Figure 1.1 shows, the three elements are integrated, with "creating the environment" at the core, the first of the three sequential elements.

1.1.1 Creating the Environment

Creating the environment is the role of management. A proclamation of "quality is job 1" or adoption of a "zero defect" movement doesn't create the environment. Banners and slogans or grand targets without the means to understand, meet, and further improve the organization goals can drive the organization performance backward, not toward TQM.

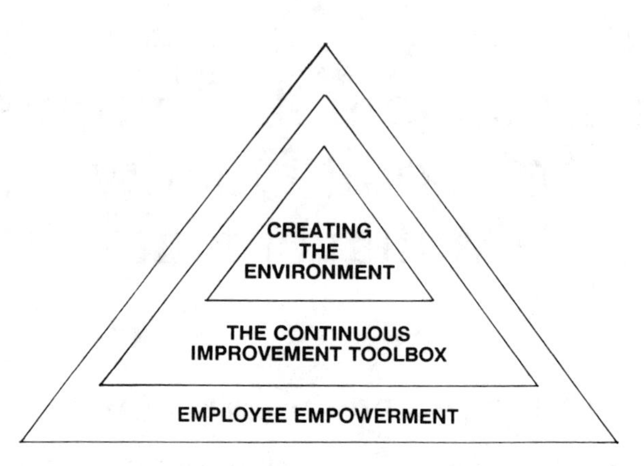

FIGURE 1.1. The three elements of TQM.

To create the environment, the job of management is to provide the leadership to:

- Develop an empowering vision: What are we all about? Where are we going?
- Communicate the vision (to everyone!).
- Develop the TQM philosophy or guiding principles.
- Reinforce the philosophy with a TQM strategy.
- Live the philosophy by action, by commitment, and with consistency. (Is it lip service or is it for real? Do we mean "it" all of the time or only when it's easy to support?)
- Drive for continuous improvement in everything, forever. Recognize and applaud all improvements, but don't stop; never stop improving.

Figures 1.2 through 1.4 show samples of TQM philosophies adopted by Xerox, Ford, and Owens Corning Fiberglas (Industrial Materials Group). Although the philosophies all have common threads of meeting customer needs, employee improvement, and continuing improvement, the philosophies are different in format, style, and emphasis, as they should be. There is no one "right" TQM philosophy, only one that is right for the organiza-

Xerox Corporation Leadership Through Quality

Xerox is a quality company. Quality is the basic business principle for Xerox. Quality means providing our external and internal customers with innovative products and services that fully satisfy their requirements. Quality improvement is the job of every Xerox employee.

Objectives Of Leadership Through Quality

- **To instill quality as the basic business principle in Xerox, and to ensure that quality improvement becomes the job of every Xerox person.**
- **To ensure that Xerox people, individually and collectively, provide our external and internal customers with innovative products and services that fully satisfy their existing and latent requirements.**
- **To establish as a way of life management and work processes that enable all Xerox people to continuously pursue quality improvement in meeting customer requirements.**

FIGURE 1.2. Xerox Corporation TQM philosophy. (*Courtesy of Xerox Corporation.*)

tion. Development (and adoption) of the philosophy statement is a tough, soul-searching assignment.

The philosophy captures the guiding principles to which the whole organization must completely and fully commit. This is the moment of truth. Don't weaken the philosophy with statements that you "wish to be"; only include what you will live by.

1.1.2 The Continuous Improvement Toolbox

Continuous improvement tools are the second element in the TQM process. Creating the environment includes communicating the vision and developing a TQM philosophy to give the organization direction. The continuous improvement (CI) toolbox arms us with the tools for continuous improvement and gives us the skills to work the philosophy. (We'll discuss shortly how the third element, employee empowerment, provides the resources to use the tools to implement the philosophy.)

MISSION

Ford Motor Company is a worldwide leader in automotive and automotive-related products and services as well as in newer industries such as aerospace, communications, and financial services. Our mission is to improve continually our products and services to meet our customers' needs, allowing us to prosper as a business and to provide a reasonable return for our stockholders, the owners of our business.

VALUES

How we accomplish our mission is as important as the mission itself. Fundamental to success for the Company are these basic values:

People — Our people are the source of our strength. They provide our corporate intelligence and determine our reputation and vitality. Involvement and teamwork are our core human values.

Products — Our products are the end result of our efforts, and they should be the best in serving customers worldwide. As our products are viewed, so are we viewed.

Profits — Profits are the ultimate measure of how efficiently we provide customers with the best products for their needs. Profits are required to survive and grow.

GUIDING PRINCIPLES

Quality comes first — To achieve customer satisfaction, the quality of our products and services must be our number one priority.

Customers are the focus of everything we do — Our work must be done with our customers in mind, providing better products and services than our competition.

Continuous improvement is essential to our success — We must strive for excellence in everything we do: in our products, in their safety and value — and in our services, our human relations, our competitiveness, and our profitability.

Employee involvement is our way of life — We are a team. We must treat each other with trust and respect.

Dealers and suppliers are our partners — The Company must maintain mutually beneficial relationships with dealers, suppliers, and our other business associates.

Integrity is never compromised — The conduct of our Company worldwide must be pursued in a manner that is socially responsible and commands respect for its integrity and for its positive contributions to society. Our doors are open to men and women alike without discrimination and without regard to ethnic origin or personal beliefs.

FIGURE 1.3. Ford's mission, values, and guiding principles. (*Courtesy of Ford Motor Company.*)

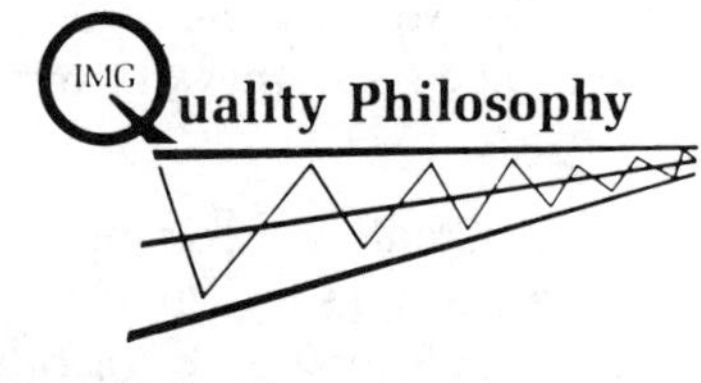

IMG QUALITY PHILOSOPHY

The Industrial Materials Group will provide products and services which meet customers' needs and expectations by establishing and maintaining an environment in which all employees pursue never ending improvement in quality and productivity.

OPERATING PRINCIPLES

1. *Management leadership and involvement are critical in meeting the objectives set forth in the operating philosophy.*
2. *Management will create and maintain an environment for our employees which eliminates fear, encourages open, two-way communications, and fosters teamwork.*
3. *Clearly defined business strategies will be communicated to all employees. Short term decisions will be consistent with long term goals.*
4. *Management will establish appropriate ongoing training necessary to improve the quality and productivity of our products and services.*
5. *Management will examine all division systems and operating principles and revise those that inhibit the concept of never-ending improvement of quality and productivity.*
6. *Management will implement systems that emphasize defect prevention rather than defect detection.*
7. *A partnership relationship will be established with all the key suppliers to the division to cause continuous improvements in the quality and productivity of goods and services which they provide.*
8. *The role of statistical methods and techniques is to assist in solving problems and identifying opportunities.*
9. *Never-ending improvement in quality and productivity is necessary to provide a competitive advantage in the world marketplace.*
10. *Our objective is to have each employee meet the needs and expectations of his/her customer.*

FIGURE 1.4. Owens Corning Fiberglas's Industrial Materials Group quality philosophy. (*Courtesy of Owens-Corning Fiberglas Corporation.*)

To work the TQM philosophy consistently throughout the organization, we need a framework or structure. The CI toolbox provides this structure by:

- Creating a common language
- Establishing uniform team problem-solving methods
- Setting a base for organization discipline (consistency of application)
- Establishing the "boundaries of freedom"

What are the continuous improvement tools? Many companies looking for a quick fix focus on one tool, usually statistical process control (SPC), equate it to the totality of TQM, think they have arrived at the "promised land," and are disappointed with their results. There isn't a quick fix. There isn't one tool used to implement TQM: You have to use a whole toolbox. Additionally, in a TQM approach, the design and use of the tools must be integrated and complimentary: using a common language, evolving a uniform team problem-solving approach, supporting consistency. Remember, TQM is a philosophy—a way of doing business that embodies a never-ending journey of improvement. As you reach milestones along the journey, additional, complimentary tools will be added to help you progress.

Let's look at three categories of statistical tools:

1. The basic team tools
2. Statistical control techniques
3. Advanced statistical methods used for statistical process improvement

The basic tools are the foundation of the common language and team problem-solving methods. They include:

- Brainstorming
- Sampling and data collection techniques
- Data display techniques (histograms, concentration charts, scatter diagrams)
- Cause and effect diagrams
- Pareto diagrams
- Process flowcharts

Everyone in a TQM organization needs to understand *and use* these basic team tools. Continuing improvement applies to cost accounting, sales (the sales process), and human resources (the hiring process) as much as it applies to a manufacturing process. Some of the basic tools are covered in Chapter 2; others are covered in Section 5.2.

In addition to the basic team tools, an understanding of the concepts and use of fundamental statistical process control (SPC) techniques is needed by line personnel (in *service functions*, too, not just manufacturing) in a TQM organization. The statistical control tools are covered in Chapter 4. The control chart is the centerpiece of this family of tools.

The third level of statistical tools, advanced statistical methods for process improvement, is discussed in Chapter 5. Typically, the technical community is the resource in an organization for application of statistical process improvement (SPI) tools. The advanced tools include the use of designs of experiments (DOE) and analysis of variance (ANOVA) to point the way for process improvements. In a TQM organization, the engineering function facilitates the use of advanced statistical tools to reach breakthrough improvements.

1.1.3 Employee Empowerment

The third element in the TQM process is employee empowerment. Empowerment is not "just" getting everyone involved, but truly unleashing the collective power, skill, talents, and creativity of all the people in the organization. Most of us have seen lots of talk with little action on the employee involvement front:

- Participative management—where only management participates
- Quality circles—who end up getting themselves dizzy and then discouraged, by going in circles, for want of training and tools to organize their problem-solving and improvement efforts
- Suggestion systems—that are really idea rejection systems

What's gone wrong? How can so many good ideas for getting people involved backfire? It's because up-front planning and preparation weren't done. In the TQM process, employee empowerment is the most powerful, the most rewarding aspect. But it's not the first, not the second, but the third "arena" visited. Management must first commit, by action, to create the environment. Then the use of continuous improvement tools must be adopted. Steps 1 and 2 provide the "boundaries of freedom" that "allow" the organization to unleash the tidal wave of employee empowerment.

Are we saying that the TQM environment, once set, is fixed, never to be changed, and that all of the CI tools need to be in place before employee involvement kicks in? Of course not. Remember, TQM is based on continuous improvement: it is a dynamic process. Management must, however, take the first steps. It is management's role to set the vision, establish the guiding principles, bring in the common team problem-solving language and approach, and then, with employee involvement, support continuous improvement.

1.1.4 Boundaries of Freedom

At first, setting boundaries may seem like a contradiction to employee empowerment. After all, aren't boundaries things that keep people "away"? Yes, boundaries do keep people away from some things, but they can also provide a focused freedom for all within the defined limits. By clearly defining employee boundaries and by providing training and tools, management can empower employees to take ownership of their work, thereby creating boundaries of freedom.

What are the boundaries of freedom? The boundaries start first with the TQM vision and guiding principles. They set the basic framework for the whole organization. The basic tools flesh out that framework by creating a common language and team problem-solving approach for the full organization to use. The statistical control techniques in the CI toolbox provide the basis for organizational discipline, assuring improvements are made based on fact (statistics), not raw emotion or opinion.

The level of employee involvement evolves in a TQM organization. Empowerment is an advanced stage of involvement. A formal, controlled project or continuous improvement team approach will evolve to self-directed project teams. The power of teamwork, the compounding of resources found when people work together (first in a structured way to learn the skills, then in an unstructured manner) will reinforce the team process.

Evolution of the team process to self-directed continuous improvement teams requires an investment in training beyond problem-solving skills. Team building skills, including training in facilitating/coaching, time management, and conflict resolution are needed to make this employee empowerment evolution. Moving to a self-directed work group continues the gradual process of transition of responsibility. This step is a big one and it should be taken as a series of smaller, integrated steps.

One of the major obstacles organizations have moving from self-directed continuous improvement teams to a self-directed work force is to transition the supervisor from an enforcer role to a facilitator–coach role. In traditional organizations, supervisors assign tasks, enforce rules, control information, direct problem-solving (or really fire fighting), and have responsibility for meeting goals. In TQM organizations that have evolved to the self-directed work force level, the supervisor's position still exists. However, the role has changed to that of a resource person. The job is one of communication, facilitation, integration of (other) resources, training, and recognition (see Figure 1.5).

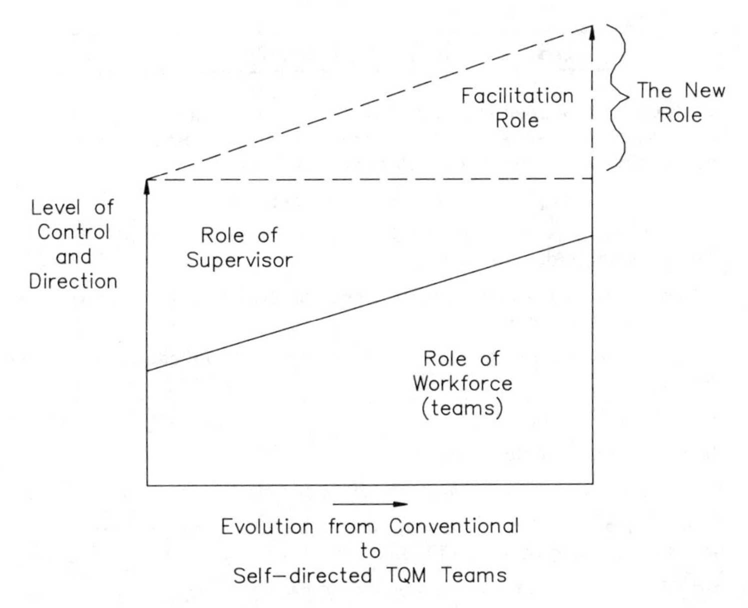

FIGURE 1.5. The changing role of the supervisor.

A TQM self-directed work force may be able to generate an "order of magnitude" (continuous improvement) advantage over a traditional organization. How can this be? With a self-directed work force, usually 10 times as many people are using their minds (not just muscle) to drive the improvement process. Training and then empowering the work force unleashes multiples of creative minds. But to just say "go to it" without the boundaries of freedom in place creates nothing short of chaos, and that will lead (back) to a dictatorial management style. As with any process, planning (before you start) is needed. The adoption of TQM is a process. Creating the environment is the first planning step. Setting the continuous improvement toolbox in place establishes the structure, the statistical tools in that CI toolbox set the discipline, and employee empowerment is where the payoff starts.

1.2 TQM LEADERS

The esteemed group of Deming, Juran, Crosby, and Taguchi are today recognized as TQM leaders. They have achieved "guru" status, and rightfully so. Dr. W. Edwards Deming emphasizes the use of statistical tools within his 14 points for management TQM model (see Figure 1.6).

DR. DEMING'S 14 POINTS *

1. Create and publish to all employees a statement of the aims and purposes of the company or other organization. The management must demonstrate constantly their commitment to this statement.

2. Learn the new philosophy, top management and everybody.

3. Understand the purpose of inspection, for improvement of processes and reduction of cost.

4. End the practice of awarding business on the basis of price tag alone.

5. Improve constantly and forever the system of production and service.

6. Institute training.

7. Teach and institute leadership.

8. Drive out fear. Create trust. Create a climate for innovation.

9. Optimize toward the aims and purposes of the company the efforts of teams, groups, staff areas.

10. Eliminate exhortations for the work force.

11.a. Eliminate numerical quotas for production. Instead, learn and institute methods for improvement.

11.b. Eliminate M.B.O. Instead, learn the capabilities of processes, and how to improve them.

12. Remove barriers that rob people of pride of workmanship.

13. Encourage education and self-improvement for everyone.

14. Take action to accomplish the transformation.

***as revised by Dr. Deming on 10 January 1990.**

FIGURE 1.6. Dr. Deming's 14 points. (*Reprinted from Out of the Crisis by W. Edwards Deming by permission of MIT and W. Edwards Deming. Published by MIT, Center for Advanced Engineering Study, Cambridge, MA 02139. Copyright 1986 by W. Edwards Deming.*)

Dr. Joseph M. Juran focuses on the project-by-project improvement approach with the Juran trilogy: quality planning, quality control, and quality improvement (see Figure 1.7). Philip B. Crosby also has a 14-step quality approach that he couples with his four quality absolutes (see Figure 1.8). Dr. Genichi Taguchi has introduced the important concept of the *quality loss function* (see Figure 1.9).

In their TQM approach, the "quality gurus" do have differences in organizational entry levels, in presentation methods, and in emphasis on (different) improvement tools. But the differences are not nearly as important as the similarities. The core message of all is commitment. Commit-

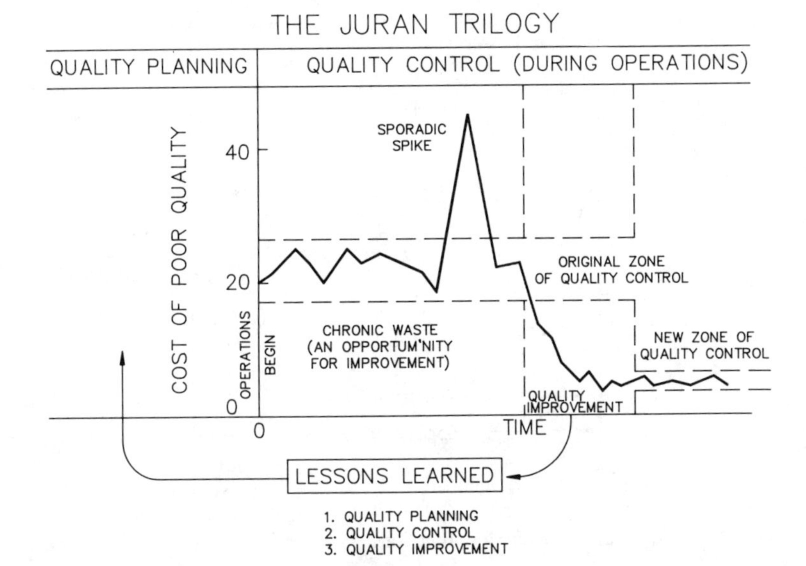

FIGURE 1.7. The Juran trilogy. (*Reprinted with permission of The Free Press, a Division of Macmillan, Inc., from Juran on leadership for quality: An Executive Handbook by Joseph M. Juran. Copyright 1989 by Juran Institute, Inc.*)

PHILIP CROSBY'S ABSOLUTES OF QUALITY MANAGEMENT

1. The First Absolute:

Quality has to be defined as conformance to requirements, not as goodness.

2. The Second Absolute:

The system for causing quality is prevention, not appraisal.

3. The Third Absolute:

The performance standard must be zero defects, not "that's close enough".

4. The Fourth Absolute:

The measurement of quality is the price of nonconformance, not indexes.

FIGURE 1.8. Crosby's quality absolutes. (*This material is reproduced with permission of the publisher, McGraw-Hill, Inc. from Quality Without Tears: The Art of Hassle-Free Management by Philip Crosby, copyright 1984.*)

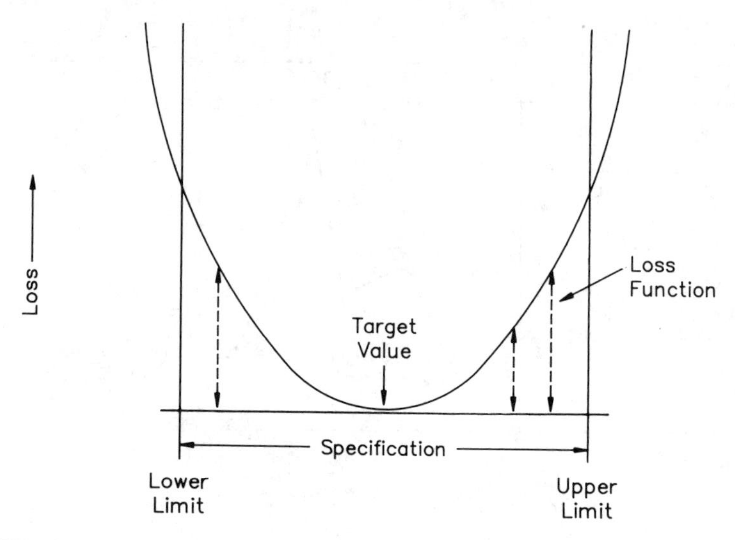

FIGURE 1.9. The Taguchi quality loss function.

ment to:

- Continuous (quality) improvement
- Involving the entire organization
- Work on systemic improvements. (Improvement efforts are focused on the system or process, not the individual. Work on all five components of the process including methods, materials, machines (or equipment), environment, and manpower, not just the "people" (or manpower) component.)
- Identify the customers (internal as well as external)
- Work to satisfy the needs of the customer(s)
- Reduce waste
- Instill pride
- Build teamwork
- Create an environment that supports all the above items

Dr. Deming's 14 points provide a complete model for TQM—a model difficult to improve upon. We can take the 14 points and group them into one of the three sequential TQM elements:

TQM Element 1. Top management commitment to creating the environment

Point 1: Create constancy of purpose (*vision*).

Point 2: Learn the *philosophy*.

Point 8: *Drive out fear.*
Point 10: *Eliminate exhortations* for the work force.
Point 11: *Eliminate numerical quotas*; instead, learn and institute methods for improvement.

TQM Element 2. The continuous improvement toolbox
Point 3: Understand the purpose of inspection.
Point 4: End the practice of awarding business on the *basis of price* alone.
Point 5: *Improve constantly* and forever *the system* of production and service.
Point 6: Institute *training*.

TQM Element 3. Employee empowerment
Point 7: Teach leadership.
Point 9: Optimize toward the aims and purposes of the company the efforts of teams, groups, staff areas.
Point 12: Remove barriers that rob employees of their *pride of workmanship*.
Point 14: Take action to accomplish the transformation.

1.3 THE CONTINUOUS IMPROVEMENT CYCLE

When we talk about a *process*, most people think "a production line." Why don't we think of cost accounting as a process? And purchasing? And order processing? How about the employment process and the sales process? If we readjust traditional thinking and look at a process as a sequence of events with an input, value-added, and an output (see Figure 1.10), we can then move to process improvement concepts.

Understanding that processes are sequences of events, it is helpful to break down the big process into a series of smaller process components. It's usually overwhelming to work on improvement of the process as a

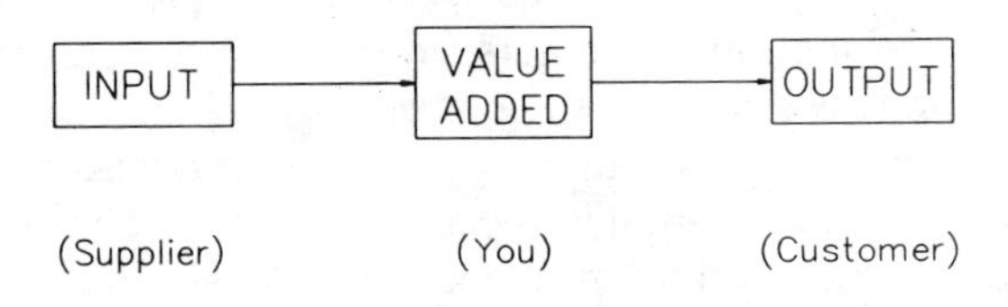

FIGURE 1.10. The process sequence.

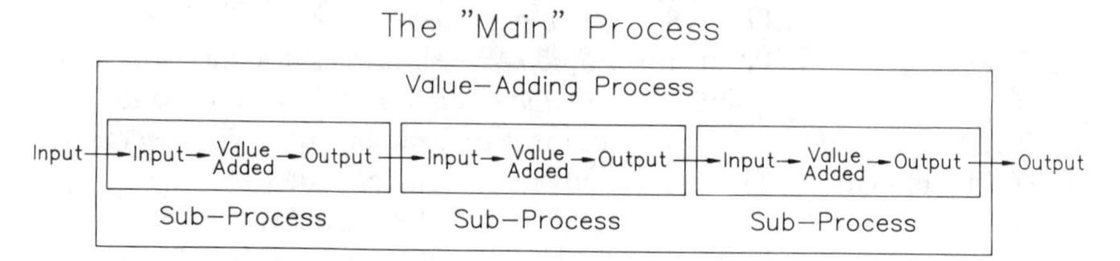

FIGURE 1.11. The main process as a series of subprocesses.

whole. Breaking down the process allows us to focus on the details, which is where the process is successful or fails. This focus makes the job ahead more doable, more manageable. Flowcharting the process is an effective technique for showing the relationship of the sequential steps in the process in a pictorial way. Flowcharting techniques are discussed in Section 2.2.

Just like the main process, each of the sequential process steps also has an input, value-added, and an output. We can look at each of the linked sequential events in a process as a subprocess in its own right as shown in Figure 1.11. The output of one step in the process becomes the input of the next step.

1.3.1 Customer—Supplier Relationships

Let's put the concept of a process into the language of quality and look at a process and its relationship to the customer. An input is provided by a supplier. An output is delivered to a customer. In between the supplier and customer, "it," the product, is processed or transformed somehow to add value for the customer. A process is then adding value to the input "raw material" (raw material can be a part, a chemical, data, or even an idea) from a supplier and delivering it to a customer.

Who is the customer? We usually think of the customer as someone external to our organization who buys our process outputs. But there are subprocesses within the whole process, and each subprocess has an output too. We can then identify customers (internal customers) internal to the process. The process stream is a sequence of suppliers to customers who are then suppliers to their customers. Everyone is a customer–supplier.

The customer always initiates the next step in the process. Processes are improved by modifying and adapting each subprocess step, and the focus of improvement is the customer and his/her needs. The continuous improvement cycle is triggered by asking three sequential questions.

1. Who is our customer(s)?
2. What are their needs?
3. How can we best meet those needs?

1.3.2 Working on the Common Causes of Variation

All processes are functions of five elements: material, methods (or procedures), machines (or equipment), the environment (our process works in), and people. All five elements make up the inputs. The interaction of the five elements is the value-added that yields the output (see Figure 1.12). Variation exists in each element and combines to create variation in the process output.

Variation can broadly be grouped into two types: common causes and special causes. Common causes of variation refer to chronic variation—the variation that seems to be "built-in" to the process. It's "always" been there and it will always be there unless we change the process. Special causes of variation are due to acute or short-term influences that are not normally a part of the process as it was designed or intended to operate. The basic tools for problem-solving in Chapter 2 are used to find variation due to both special and common causes. But when we talk about continuous improvement, we usually mean the reduction of variation due to

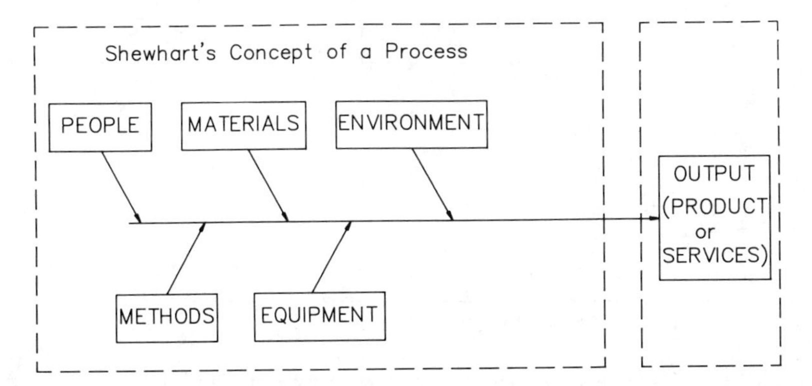

FIGURE 1.12. The elements of a process.

common causes, or reducing the historical, chronic level of variation. Finding and eliminating special cause variation is often called "fire fighting." Continuous improvement should, over time, end the need for most fire-fighting brigades by providing solutions with permanence.

1.3.3 Evolutionary Improvement

Most of us are impatient. We tend to look for big, fast improvements—for quantum leaps. We are great innovators, and, with innovation, we embrace revolutionary change. But we too often forget, or ignore, the value of slow, continuous, evolutionary change. Slow but steady change is compounded, and over time, can do as much or more than intermittent revolutionary change alone.

Should we ignore innovation? Of course not. Innovative, breakthrough change creates tremendous competitive advantages. But relying on innovation alone is not enough. Figure 1.13 conceptually shows the difference in the value-added output using an improvement process based solely on innovation versus one using innovation plus evolutionary improvement.

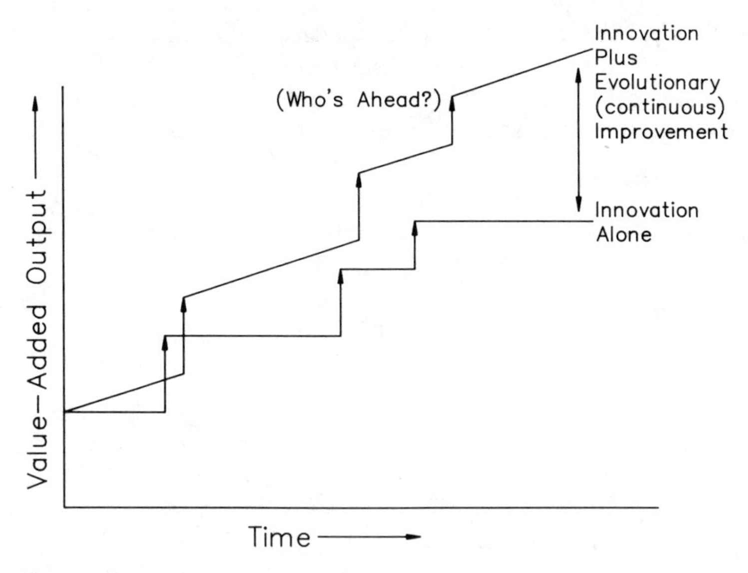

FIGURE 1.13. Evolutionary continuous improvement with innovation beats innovation alone.

Kaizen, a term used by the Japanese to explain evolutionary continuous improvement, is discussed further in Section 1.9.

1.3.4 The PDCA Cycle

The PDCA cycle—plan, do, check, act (to adjust)—embodies continuous improvement. The PDCA cycle is often called the Deming cycle; Deming calls it the Shewhart cycle. Whatever you call it, when you use it, you start a cycle of never-ending improvement (Figure 1.14). The steps of the PDCA cycle are:

P for plan	Choose an area for improvement
	Collect data
	Establish an action plan for change
D for do	Execute the action plan/the change
C for check	Check (and study) the results
	Action/change implemented
A for act	Evaluate the results
	Determine what was learned
	Identify the next area for improvement AND RECYCLE BACK TO P

The PDCA cycle is a framework for the continuous improvement of any process (or subprocess). By focusing improvement plans on customer needs, with a gradual, but constant approach, and integrating those periodic innovations or breakthroughs, you will be using the continuous improvement cycle to create a never-ending competitive advantage.

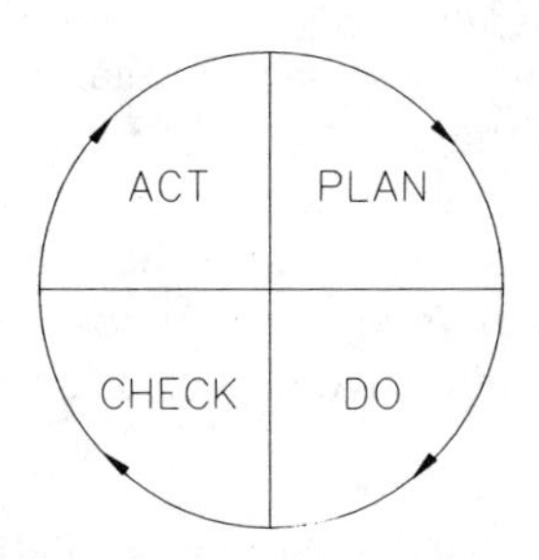

FIGURE 1.14. The plan, do, check, act cycle.

1.4 QUALITY—WHAT IS IT ANYWAY?

Quality. We all talk about it, work to improve it, and make judgments based on our perceptions of it. But what is "it"? How do we define quality?

In his 1931 landmark book, *Economic Control of Quality of Manufactured Product*, Dr. Walter A. Shewhart provides us with a good start. Shewhart acknowledges that the concept of quality is tough to pin down. He focuses on two aspects of quality—the objective side and the subjective side:

> If we are to talk intelligently about the quality of a thing or the quality of a product, we must have in mind a clear picture of what we mean by quality. Enough has been said to indicate that there are two common aspects of quality. One of these has to do with the consideration of the quality of a thing as an objective reality independent of the existence of man. The other has to do with what we think, feel, or sense as a result of the objective reality. In other words, there is a subjective side of quality. For example, we are dealing with the subjective concept of quality when we attempt to measure the goodness of a thing, for it is impossible to think of a thing as having goodness independent of some human want. In fact, this subjective concept of quality is closely tied up with the utility or value of the objective physical properties of the thing itself.
>
> For the most part we may think of the objective quality characteristics of a thing as being constant and measurable in the sense that physical laws are quantitatively expressible and independent of time. When we consider a quality from the subjective viewpoint, comparatively serious difficulties arise.

Shewhart provides us with the basis for a broad definition of quality. Some of the "modern day" attempts to define quality are good, but incomplete. Maybe they are an outgrowth of our fast food culture, or as Dr. Deming would say, we're looking for "instant pudding," the quick fix.

Juran and Gryna, in *Quality Planning and Analysis* (1980), state that "the popular term for fitness of use is quality and our basic definition becomes quality means fitness for use." "Fitness for use": it's not a bad definition for quality, but it's not a complete one either in today's world. Dr. Juran's definition deals heavily with the subjective aspect of quality.

Conversely, in *Quality is Free*, Philip Crosby (1979) offers a definition focusing primarily on the objective aspect of quality. Crosby states that:

> We must define quality as 'conformance to requirements'. Requirements must be clearly stated so that they cannot be misunderstood. Measurements are then taken continually to determine conformance to those requirements. The non-

conformance detected is the absence of quality. Quality problems become nonconformance problems, and quality becomes definable.

Both Crosby and Dr. Juran offer important contributions to the definition of quality. However, both stop short of the concept of quality suggested by Shewhart. In his fable about continual quality improvement, *I Know It When I See It*, John Guaspari (1985) helps fill in the gaps:

> You often hear that "Quality is everybody's job" and that's true. But it must start with management. Management's job is to lead people toward a goal. And, quality is the only goal that matters.
>
> Lead people from an inspection mind-set to a Prevention mind-set. That's an important start. But you've got to further. You've got to manage Quality as a whole—an integrated whole that is much, much more than the sum of its parts.
>
> Above all, listen to what your customers are telling you about Quality.
>
> Your customers are in a perfect position to tell you about Quality, because that's all they're really buying. They're not buying a product. They're buying your assurances that their expectations for that product will be met.
>
> And you haven't really got anything else to sell them but those assurances. You haven't really got anything else to sell them but Quality.

In his fable, Guaspari couples the concepts of quality, customer perception, and continuing improvement. He acknowledges that quality is many faceted, and when integrated, is larger than the sum of its parts. Additionally, he brings the customer solidly into the quality concept. As Dr. Deming states in his book, *Out of the Crisis* (1986):

> The consumer is the most important part of the production line. Quality should be aimed at the needs of the consumer, present and future.

Dr. Deming adds an important element by bringing in the future. Quality cannot be thought of as static, or defined today to meet tomorrow's needs. Quality, by definition, must become a dynamic target, adapting to future needs we may not even be able to guess at today.

So, where are we? We have chunks of a comprehensive definition of quality:

Dr. Shewhart's: Objective and subjective aspects of quality
Dr. Juran's: Fitness for use
Philip Crosby's: Conformance to requirements
John Guaspari's: Customers' perception
Dr. Deming's: The present *and future* needs of the customer

A comprehensive definition of quality has to address all of the components above. QUALITY MEANS DELIGHTING THE CUSTOMER.

1.5 QUALITY DRIVES PRODUCTIVITY

In their book, *A Two Minute Warning*, Grayson and O'Dell (1988) study the effects of productivity on world economies. They conclude:

> Productivity, more than any other factor determines a nation's standard of living and is the best single indicator of its economic performance over the long run. In the end, it determines the rank of nations. History shows that the nation that is the productivity leader eventually becomes the dominant world leader—economically, militarily, and politically.

All of the industrialized nations were included in their study. An alarming finding of Grayson and O'Dell (1988) is that:

> U.S. productivity growth is dead last. (See Figure 1.15.) The U.S. standard of living has all but halted. The U.S. share of world exports has declined in industry after industry. . . . All this has not happened overnight. Contrary to popular belief, U.S. productivity growth and competitiveness did not drop suddenly in the late 1970s or early 1980s. The decline has been under way for almost twenty years.

It doesn't take a big difference in the productivity growth rate to effect a

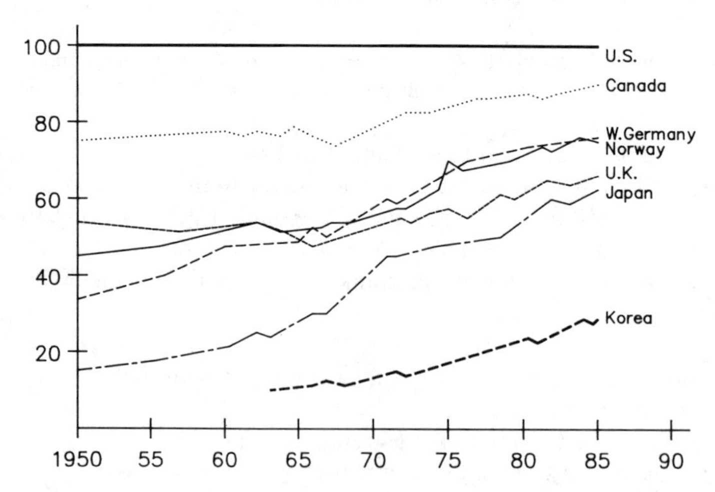

FIGURE 1.15. Productivity growth trends.

large and lasting change. Let's look back in time about 100 years ago to the 20-year period, 1870–1890. In 1870, England, not the United States, was the world productivity leader. Then, the United States (not Japan) had the fastest productivity improvement rate. Relative to today, the improvement rate wasn't large. U.S. productivity was growing at 2.1% per year, England's at 1.3% per year: "only" an 0.8% difference. But in that 20-year period, the U.S. productivity level grew from 88% of England's (in 1870) to 96% (in 1890), and passed England before 1900.

In 1990, the United States was *still* the world productivity leader. Japan's productivity level was approximately 75% that of the U.S. *But, the United States was dead last in productivity improvement!* You may ask, with our lead still so large, does the productivity improvement rate really matter? You bet it does! As recently as 1960, Japan's productivity was only 15% of the U.S. level. In those 30 years, the United States has maintained a positive productivity growth, but Japan's growth rate was bigger—and the compound effects of the growth rate difference tell the story.

By 1990, Japan had climbed to approximately 75% of the productivity rate of the United States. If current productivity improvement trends continue, Japan's productivity rate will surpass that of the United States by the year 2000. Can the trend lines be changed? Can the rate of productivity improvement in the United States equal or again surpass the improvement rate in Japan (and other countries)? It most assuredly can, by focusing on continuous quality improvement.

We don't need new plants, a new work force, or even a new generation of products to change the productivity trends. We have a "hidden factory or hidden office," right in the middle of every manufacturing plant, every insurance office, and every hospital; that is the key to raising the productivity improvement rate. The following quote is from a 1988 task force report on competitiveness from the United States House Republican Research Committee:

> Quality experts estimate that the total cost of poor quality, or the cost of not doing the right things right the first time, is 20% of the gross sales for manufacturing companies, 30% for service industries. Total [annual] U.S. production of goods and services is an estimated $3.7 trillion, so our quality "target," the potential "savings" from quality, is a staggering $920 billion that can be saved or redirected for better use. These losses, due to waste and inefficiency, hurt business on the margin, and it's on the margin where they rise and fall. . . . It's on the margin where nations rise and fall.

Think of it: 20% of every manufacturing plant is in place to make poor quality, unsalable, and unusable product, "the hidden factory" (see Figure

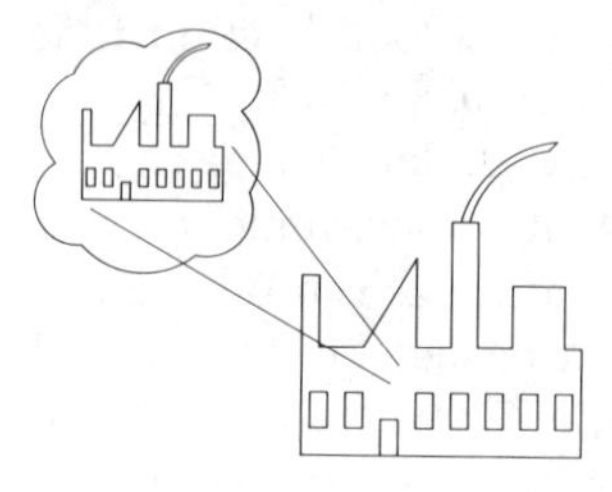

FIGURE 1.16. The hidden factory.

1.16). One out of every five employees is hard at work, doing the best job they can, making poor quality. What a shock! (What a shame!) If we can implement a quality improvement process attacking root cause chronic problems (not just fire fighting), look at the productivity gains to be had!

We have a 20% productivity improvement opportunity in manufacturing and a 30% opportunity in the services sector, and quality is the driving force. With a continuous improvement mentality, with a commitment to quality in all aspects of the business, productivity gains will compound themselves and can vault almost any organization into a world class leader.

1.6 FROM PRODUCT CONTROL TO PROCESS CONTROL

We've all been told that it is better to build quality in than to inspect bad parts (or service) out. Prevention instead of detection: It's easy to say, but how do we do it?

1.6.1 The Limitations of Mass Inspection

The output quality of most conventional operations is measured at the end of the process by checking the product quality. Sometimes all (100%) of the product is checked for conformance to product quality standards. Alternatively, a sampling plan may be used to check a prescribed fraction of the product at the end of the process. Either way, it's product control. Checking product quality at the end of the process does nothing to help make it right the first time. Product control or inspection does nothing to reduce the size of the hidden plant (or office) we discussed in Section 1.5. What does product control do for us?

- *It gives a false sense of security:* With product control, we think are separating the "bad" from the "good," and only sending the good to our customers. In fact, 100% inspection is well less than 100% effective: some bad will get through; some good will get rejected.
- *It adds to the total costs:* One result of product control is (the identification of) product rework or scrap. In some organizations, product control provides input for the "healing bin"—that magical part of the plant or warehouse where rejected product or parts are stored in, until the last day of the month. Then, with the pressures to meet the month's billing, the product is magically declared "healed" or "good enough" to ship.
- *It implies that defects (bad quality) are expected:* As Dr. Deming states [W. Edwards Deming, *Quality*, *Productivity and Competitive Position* (1982)]: "Routine 100% inspection is the same thing as planning for defects, acknowledgment that the process cannot make the product correctly, or that specifications made no sense in the first place."

"Planning for defects," none of us wants to encourage that. How do we change the focus from defect detection to defect prevention? By shifting our process monitoring approach *from* inspecting the product (at the end) *to* controlling the process that makes the product, we open the way for real and lasting process improvement and defect prevention.

1.6.2 Add Filters or Fix the Process?

Let's look at a simple, conceptual example of product inspection, trace the shortcomings, and contrast product control to process control. Our example process [from a lecture by Dr. Al Rickmers (1984)] is an oil pipeline drum-filling process (see Figure 1.17). (*Note:* This example is not intended to be a "real life" example; its role is to help us contrast process control from product control.)

In our process, oil flows through the pipe, into a drum, and is sold. The oil drums are *periodically inspected* (sampled via a sampling plan) for contamination. All is well until one day (Figure 1.18), *dirt* is found in our final product. After much discussion, consternation, and evaluation, our engineering staff decides to add a *filter* at the end of the oil pipe (Figure 1.19) to *filter out the dirt*. Again, all is well. Our process is producing drums of clean, uncontaminated oil. We are again making dollars.

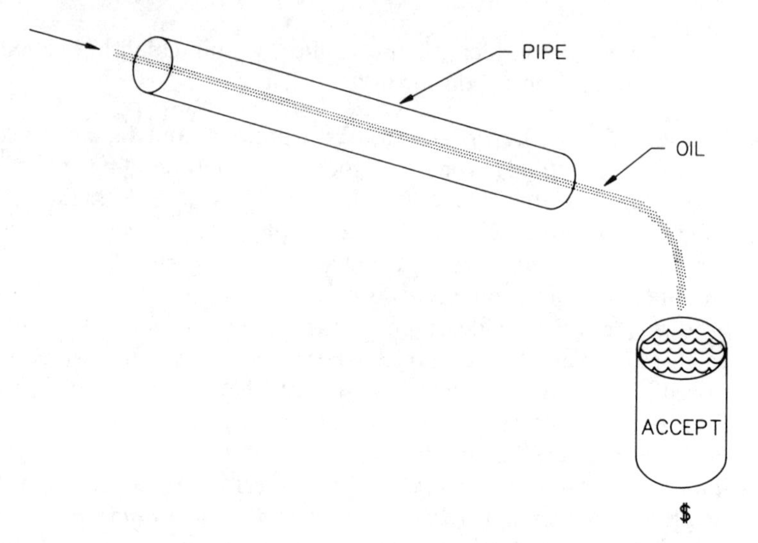

FIGURE 1.17. An oil pipeline process. (*Reprinted from First Class Service, copyright 1991, Resource Engineering, with the permission of the publisher, Quality Resources, White Plains, New York.*)

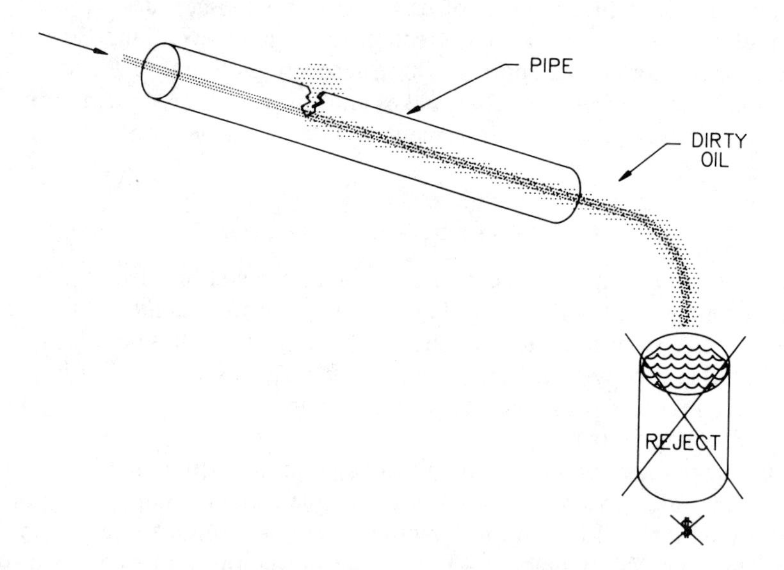

FIGURE 1.18. Pipe cracks allowing dirt into the oil. (*Reprinted from First Class Service, copyright 1991, Resource Engineering, with the permission of the publisher, Quality Resources, White Plains, New York.*)

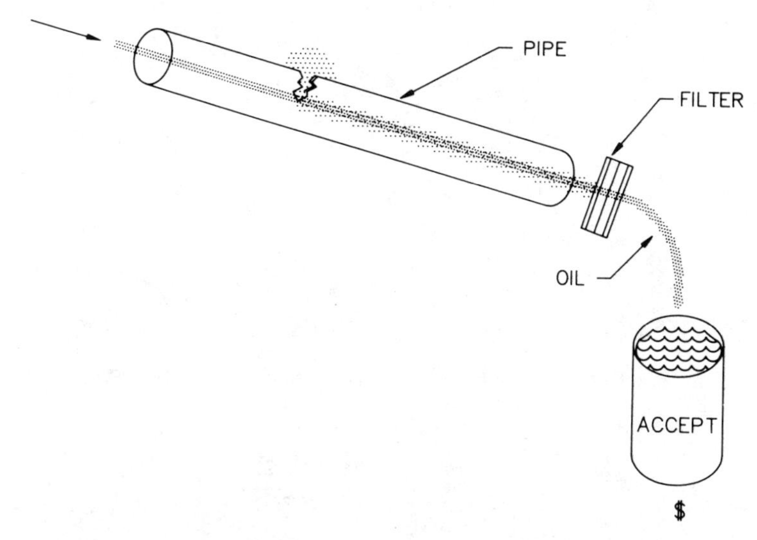

FIGURE 1.19. Filter installed to remove the dirt. (*Reprinted from First Class Service, copyright 1991, Resource Engineering, with the permission of the publisher, Quality Resources, White Plains, New York.*)

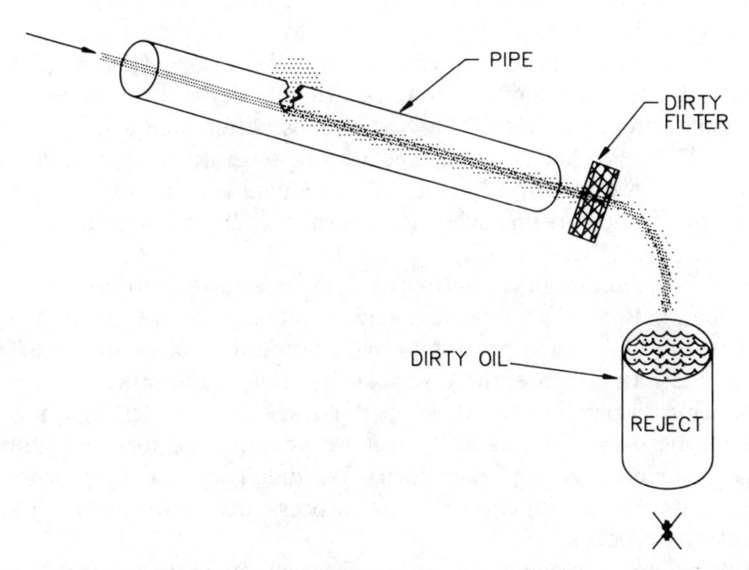

FIGURE 1.20. Filters clog, producing dirty oil again. (*Reprinted from First Class Service, copyright 1991, Resource Engineering, with the permission of the publisher, Quality Resources, White Plains, New York.*)

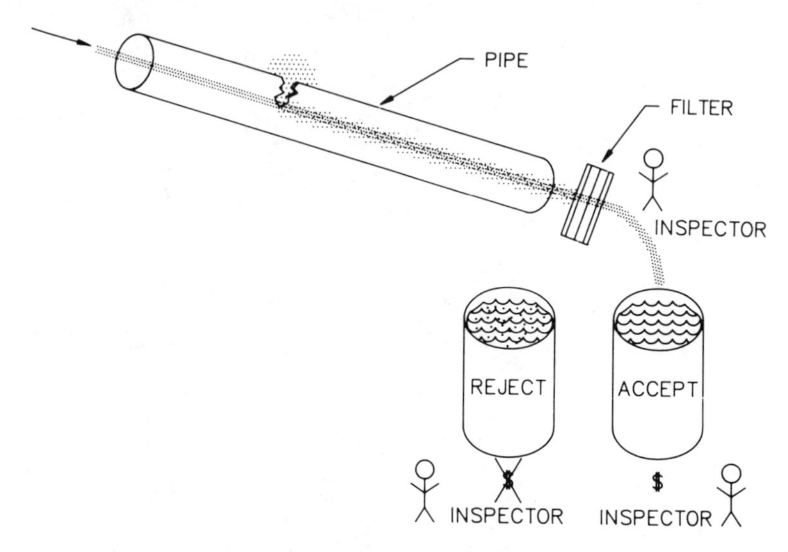

FIGURE 1.21. Inspectors added to the process to inspect out dirty oil. (*Reprinted from First Class Service, copyright 1991, Resource Engineering, with the permission of the publisher, Quality Resources, White Plains, New York.*)

Alas, one day, our final inspection once again finds dirt in our oil! How can that be? We spent money on a filter and still we have product rejects. Our maintenance department discovers *the filter is dirty* (Figure 1.20) and oil with dirt is bypassing the filter. Our quality department decides, after this repeat problem, that *100% inspection* is better than a sampling plan (Figure 1.21). We had to reject and rework several drums, and we don't even know if we caught all of the bad product. Yes, it must be better to add quality inspectors on each shift to inspect each and every drum. But is it?

By focusing on product control, we forgot about the process. We would have been better off when we first saw defective product (dirt in the oil in our case) to look at the process to make sure our process was intact (see Figure 1.22). In our case, there was a disruption in the process—a special cause. Our oil pipeline had developed a crack that was letting dirt get in (and maybe oil get out as well). But we added inspectors and filters to support our product inspection focus. We didn't look at the process. We weren't concerned with control of the process, just control of the process output, the product.

With a quality management approach based on process control instead of product control, we would have detected a problem (a special cause) within the process. We could have identified it and corrected it without the use of "filters." Think of process control as getting inside the process

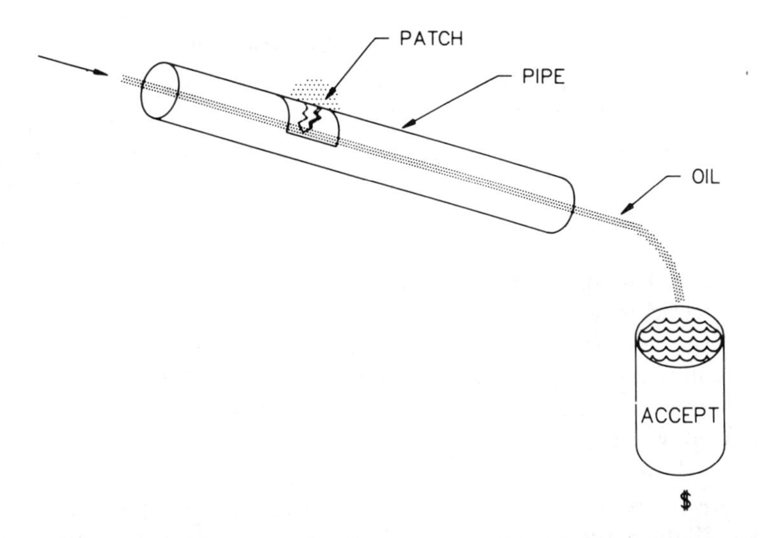

FIGURE 1.22. Or . . . fix the process, eliminate the root cause. (*Reprinted from First Class Service, copyright 1991, Resource Engineering, with the permission of the publisher, Quality Resources, White Plains, New York.*)

and learning how it runs, so we can see when a special cause enters the process. With process control we receive a warning to take action before the process produces (lots of) defective product, and, with that warning, we're able to make corrections with permanence by getting to the root cause, not just adding "filters."

1.6.3 Product Inspection Becomes a Diagnostic Role

Moving to a process control based quality management approach does not eliminate the use or the need for product inspection. But, with process control, the extent of product inspection is greatly reduced as its role is changed from a quality control to a diagnostic quality improvement tool.

The success of process control is based on the premise that, with valid process characterization, control of appropriate process parameters will assure a process output (the product) with product properties in control. A statistically based process control model (SPC) allows us to test a *few* process parameters *early in the process*, control the process based on those process parameters, and maintain our process in control. Periodic product testing (sampling and inspection) serves as a diagnostic tool, auditing the validity of the process–product relationships used to establish the process control scheme.

1.6.4 Transitioning to Process Control

The transition from product inspection to process control is not done quickly or simply. The changeover requires planning and study. One model for the transition is the following six step process (to be used only after the work force has been trained in the fundamentals of SPC AND team problem-solving):

1. Flow chart the process
2. Characterize the process
3. Explore process–product correlations
4. Validate the measurement system
5. Implement SPC
6. Continuously improve the process

Flowchart the Process During product inspection, the focus was on the process output, the product. As we install SPC, we must shift attention to the process and the relationship between the product and the process. Flowcharting (flowcharting techniques are discussed in Section 2.2) breaks the process down into connected subprocesses and provides us with a pictorial view of the process–product relationships.

Characterize the Process Remember, in Section 1.3, we discussed how each process (and subprocess) has an input, value-added, and an output (Figure 1.23), and how the value-added in each process comes from the material, machines, methods, environment, and people (Figure 1.24). We characterize the process by identifying all of the value-added components of each subprocess. Using the flowchart from step 1, add all of the process parameters and variables (temperatures, pressures, flow rates, raw material streams) to the flowchart. Turn the flowchart into both a mass balance and a throughput rate model.

Explore Process–Product Correlations With product control, it was relatively easy to determine what to test. We tested those final product performance properties mutually agreed as meaningful with the customer. When moving to process control, we have to understand how the process affects the product, or the cause and effect relationships between the

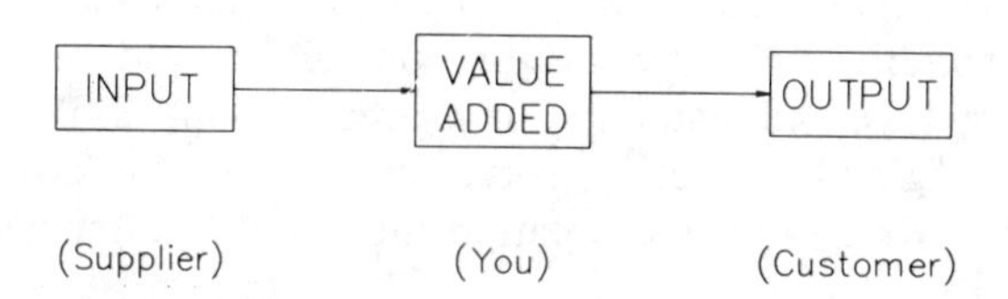

FIGURE 1.23. The process sequence.

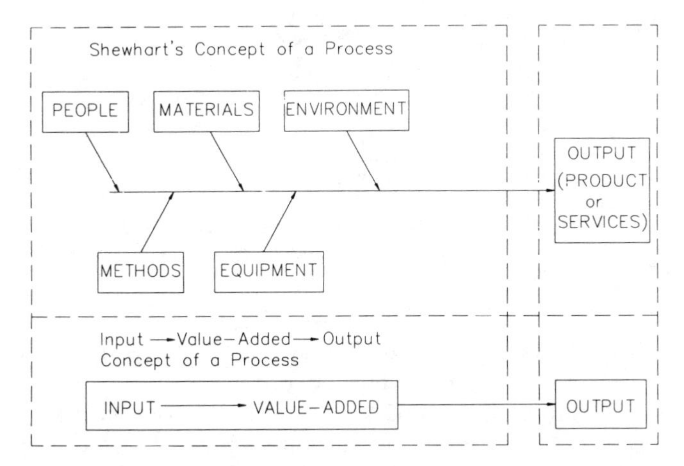

FIGURE 1.24. Tying two concepts of a process together.

process variables (the causes) and the product (the effect). We also need to identify those controlling process variables that create significant product variation. Once we understand the process–product correlation (or cause and effect), we can develop a model of process control that yields a controlled process output (or product). By controlling the process (of making the product), we can control our process to make the product right the first time, reducing the need for product control (final inspection).

This may well be the toughest step in the transition. We may think we know our processes, but we find out how much more we have to learn about them when we try to identify the process–product correlations. Usually, this study is not a one time event. Finding the correlations becomes a dynamic process in its own right; as we learn more about the process, we better understand the process variables that control product variability. Advanced statistical techniques, such as screening experiments (covered in Chapter 5), are often used in sequential iterations of a process–product correlation study.

Let's take the process–product correlation one step further. Earlier, we said it was relatively easy to determine what to test when we used a product control approach. It is easy only because the customer (sometimes with supplier input) took an educated guess to specify product performance properties meaningful to his/her (the customer's) process. But, are those product properties really meaningful? How many billions of dollars have been spent trying to make a product to the "guesstimated" specification, when the specification targets didn't correlate to the needs of the next process anyhow? How much "better" (e.g., more cost effective, with less inherent variation, and yielding a higher performance product) could

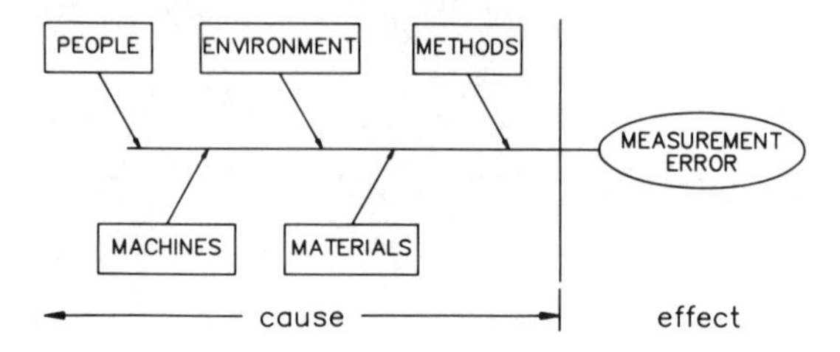

FIGURE 1.25. The measurement system as a process.

two sequential processes operate if they were studied as one extended process? Use the concept of establishing process–product correlations not only "inside" your internal process but extend it backward to your supplier's process and forward to your customer's process to "strip the filters" (eliminate the misunderstanding of needs) out of the entire extended process.

Validate the Measurement System Whenever and whatever we test (measure), the test result is an observed value that may, or may not, bear close resemblance to the true value. Measuring is a process just like our oil pipeline example is a process; purchasing is also a process. The measurement process (Figure 1.25) has inputs, value-added, and an output. The output is a test value. We put lots of stock in test values, but sometimes they don't tell us what we think they do.

Although we expect the measurement system to tell us how the process we are testing is doing, sometimes we get fooled by the test. The measurement system output is a combination of the measure of variation in the tested process plus the variation in the measurement process. If the variation in the measurement process is big, say one-half of, or even equal to, the variation in the process under test, we really don't know what the tested process is doing. Our test results are clouded by the variation in the measurement system. We have to improve the measurement system before we can move on. Chapter 3 discusses measurement system error and techniques to determine where the error is coming from.

Implement SPC Once a process–product correlation is set and the measurement system is validated, a statistically based process control model can be developed. The key process parameters may be variables or attributes; SPC can be used with either. Chapter 4 provides a solid foundation in SPC–control chart techniques.

Control charts for variable data (e.g., $\overline{X}$ charts, R charts), for attribute data (e.g., p charts, c charts), as well as special control charting methods (e.g., short run SPC, batch operations, group control charts, precontrol) are covered.

We haven't discussed SPC and team problem-solving training except to say we presume training in these fundamentals has been provided. Full involvement of the work force is necessary for SPC to work. The entire work force needs to be trained in basic statistical concepts and team problem-solving. (Not everyone needs to be trained in statistical techniques, but everyone involved in the process should receive training in the same team problem-solving methods and basic statistical concepts.) The training helps first to establish a common language for improvement within the organization and then provides the skills (the statistical toolbox) to start the continuous improvement cycle.

Continuously Improve the Process Step 6—improve the process. Many people ask, "but doesn't step 5, SPC, do that?" No, SPC alone doesn't improve the process (other than giving us a warning when special causes visit). SPC is a way of looking into the future by using data from the past. With SPC, the future should be the same as the past, no worse, but also no better *unless* we do something to change the process.

SPC will help us easily find when special causes of variation come into the process. Our process will be out of control, and with SPC we will have an early warning signal to act to find the special cause, possibly even before our process makes bad product! In that sense, SPC improves upon the past. It will help us reduce the size of the hidden plant (or the hidden office) by reducing the quantity and scope of defects. But, without incorporation of a continuous improvement cycle, we can't "dismantle" the hidden plant.

We talked about the continual improvement cycle in Section 1.3. The Shewhart–Deming PDCA cycle of plan, do, check, act is a straightforward improvement model. The tools of improvement are the basic team problem-solving techniques discussed in Chapter 2 coupled with the statistical improvement methods discussed in Chapter 5. Together, these tools will work on the chronic common causes of variation within the process and support the spiraling cycle of continuous improvement. But don't jump to the statistical process improvement (SPI) tools before implementing SPC. The SPI toolbox should be opened only after the process is stabilized with SPC.

1.7 SPC—WHY AND HOW

Statistical process control (SPC) is a statistically based approach for monitoring, controlling, evaluating, and analyzing a process. The target of SPC is to first achieve stable, predictable process performance. Whether the stable performance yields a process output (product) that has an acceptable level of variation (e.g., meets specification) is not the issue, at

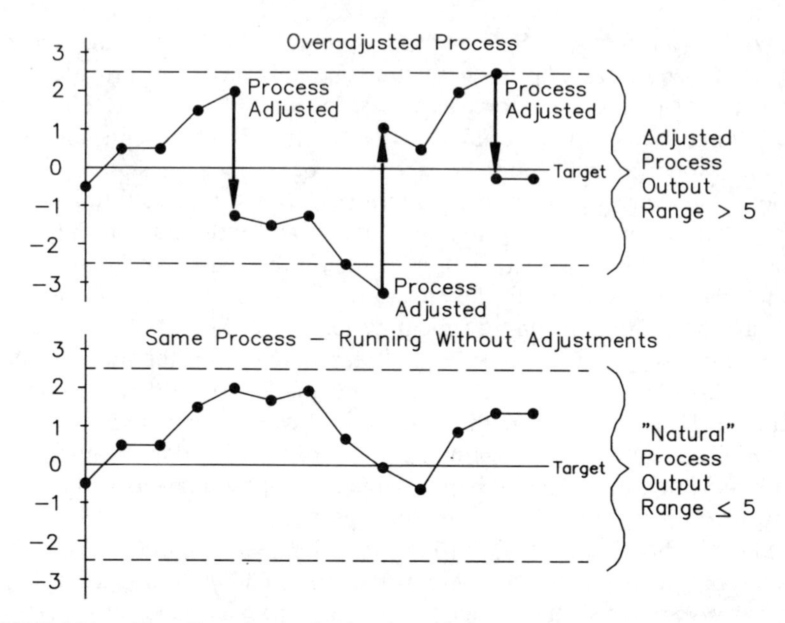

FIGURE 1.26. The effect of overadjusting the process.

least not initially. Achieving process stability, or getting the process under control, precedes process improvement.

1.7.1 Special Causes, Common Causes

With an SPC approach, special causes of variation that are acute or not part of the process design, can be flagged. When we have a special cause at work on the process, we will know to take action. We would adjust or shut down the process, then find the source of the special cause. Common causes of variation that are chronic or inherent in the process design do not call for action or process adjustment. SPC tells us when to leave the process alone, preventing unnecessary process adjustments. Overadjusting a process will tend to increase the variability of a process, not decrease it (see Figure 1.26).

1.7.2 Back to the Future

SPC takes the emotion out of process control. It separates opinion from fact by providing a clear set of processing guidelines by measuring the process and its output against realistic and meaningful measures of past performance. SPC really provides a journey "back to the future"; SPC control limits are predictors of the future based on the experiences of the past.

1.7.3 SPC Creates Operator Ownership

Some people think SPC is keeping control charts on the process performance—and little else. Charting is only (part of) the visible part of SPC. (The balance of the visible part is the return to the bottom line through a cycle of reduced defects, higher productivity, lower costs, and increased marketplace competitiveness.) The real power of SPC comes from responding to the control charts, taking action on the information the charts provide; and who better to use the information than the person operating the process.

SPC gives ownership of the process to the operator. SPC gives the operator the tools, in the form of valid information, on the process performance, plus guidelines on when to act and what to do to perform quality work. Without SPC, the operator is in a quandary. Should he/she adjust the process? Will the supervisor agree? (Definitely, if all turned out right; maybe not, if the adjustment made things worse.) Separation of opinion from fact is good for an operator–supervisor relationship and good for business.

1.7.4 Using the Statistical Toolbox

SPC gives the operator tools for quality, but he/she needs instructions on how to use the tools first. Operator training in basic statistical concepts plus team problem-solving training is needed before SPC is rolled out to the work force. Remember, in Section 1.1 we talked about boundaries of freedom? Opening the statistical toolbox to the operator, with the guidelines on what control charts are, how to fill them out, interpret them, and apply their messages begins to set the boundaries. Training in team problem-solving, establishing a problem-solving language and approach the whole organization speaks and uses completes setting the (initial) boundaries. Then the freedoms begin to unfold.

For the operator, freedom to control the process is fueled by the knowledge of what the process is doing, what it is saying, what it needs. The operator becomes comfortable in taking ownership, because with SPC he/she finally has the tools to do the job right, not guess at it.

1.7.5 Freedom to Focus on Improvement

For the engineer, the technical staff, SPC provides freedom to shift focus from control to improvement. With SPC, the operator runs the process. He/she has the tools to keep it running on target. SPC helps the operator fight off change and maintain the status quo. The engineer is free to investigate, to experiment, and to find that good type of change that breeds improvement.

1.8 IMPROVING THE STATUS QUO WITH STATISTICAL PROCESS IMPROVEMENT

In discussions of quality, we have a tendency to use "control" and "improvement" interchangeably, but the implied value of the two are fundamentally different. By nature, control mechanisms prevent change. But improvement is change—specifically, change for the better. Tools of control, SPC (covered in Chapter 4), are designed to prevent change, not cause it. SPC maintains the status quo. The statistical tools of improvement that accompany the tools of control are experimental designs. Experimental designs, or statistical process improvement (SPI) tools (covered in Chapter 5) are the antithesis of control.

1.8.1 SPC First, Then SPI

Before we use SPI to improve our process, we should first assure that the process is stable. Without process stability, the experimental results may be confounded with special causes of variation. SPI techniques are based on our ability to determine if one response is statistically different from another. If our process is unstable and full of special causes of variation, we may have difficulty in determining if differences of experimental responses are due to the experiment or due to the instability of the process.

Application of SPC will give us the process stability we need to effectively use SPI. SPC provides a solid foundation for experimentation. A stable process is one in control and predictable without special causes of variation. It sounds so good, why should we introduce change and try to improve? We strive for improvement because we live in a TQM environment. We look for never-ending improvement.

SPC establishes a stable base, a status quo. But, regardless of how good the status quo performance seems to be, over time our competitors will improve. If we stand still and don't improve, to our customers, it will look like we are going backward, changing for the worse. SPI will help us reduce common causes of variation and can become our lifeline to future survival.

1.9 THE *KAIZEN* MINE

Dr. Deming has been quoted as saying, "Statistics are essential—but relatively unimportant." Using the statistical toolbox (SPC for stability, SPI for improvement) is essential. Once in place, the power of the statistical tools is not just the statistics, but, more importantly, the boundaries of freedom they establish. With boundaries of freedom in a TQM environment, employee involvement drives the continuous improvement cycle faster and faster.

1.9.1 Singles Count Too

The Japanese term "*kaizen*" has found its way into our vocabulary. *Kaizen* means improvement, continuous improvement, everyday, by everyone. *Kaizen* embraces the "little" ideas, the small, at first seemingly insignificant improvement. The western world is enamored with the big, new innovation, the quick fix, the home run. But think about it. Not many baseball games are won by home runs alone. Most are won by consistent, steady players hitting singles. In baseball, there are lots and lots of singles for every home run. Throw away the singles, and a ball team would be hard pressed to win many games.

1.9.2 Revolution WITH Evolution

The United States is the home of innovation, the technology breakthroughs that create revolutionary improvement in product performance or productivity (see Figure 1.27). But Japan is the home of *kaizen*, the slow but steady evolutionary change that compounds improvement on top of improvement on top of improvement (Figure 1.28). It may not look like it day to day, but the evolutionary way can be just as powerful an improvement model as the revolutionary approach. How about putting them together? (See Figure 1.29.) The rate of improvement would be fantastic. Remember, the productivity improvement rate comparisons discussed in Section 1.5? Couple the evolutionary *kaizen* model with the big hit revolutionary innovations and the productivity growth, the rate of improvement, becomes substantially higher than either method alone.

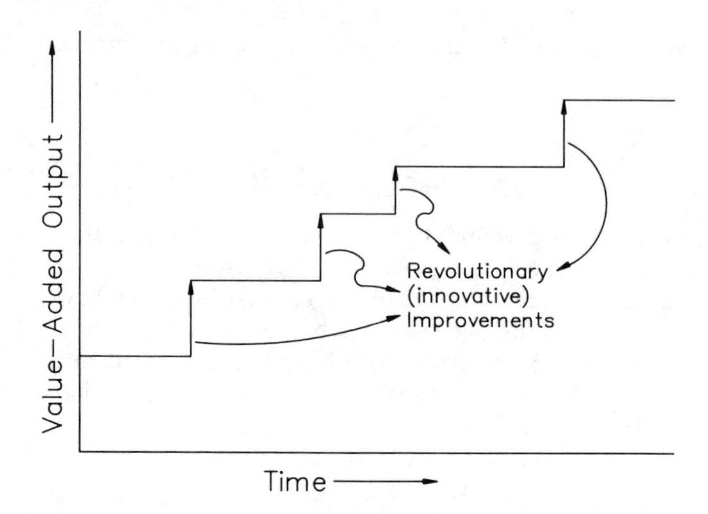

FIGURE 1.27. Revolutionary improvement alone.

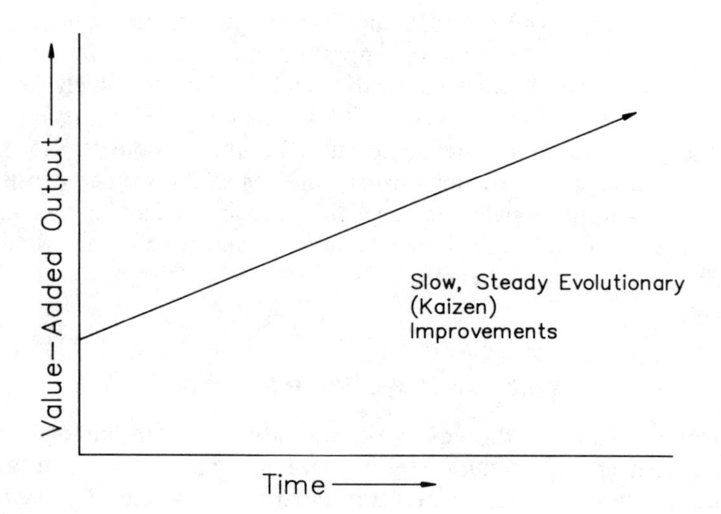

FIGURE 1.28. Evolutionary improvement alone.

1.9.3 Opening the *Kaizen* Mine

Innovation works with involvement from few people. The technical community and the marketing arm of an organization match breakthrough change with marketplace needs (that are sometimes still unperceived), and top management provides the implementation capital. *Kaizen*, on the

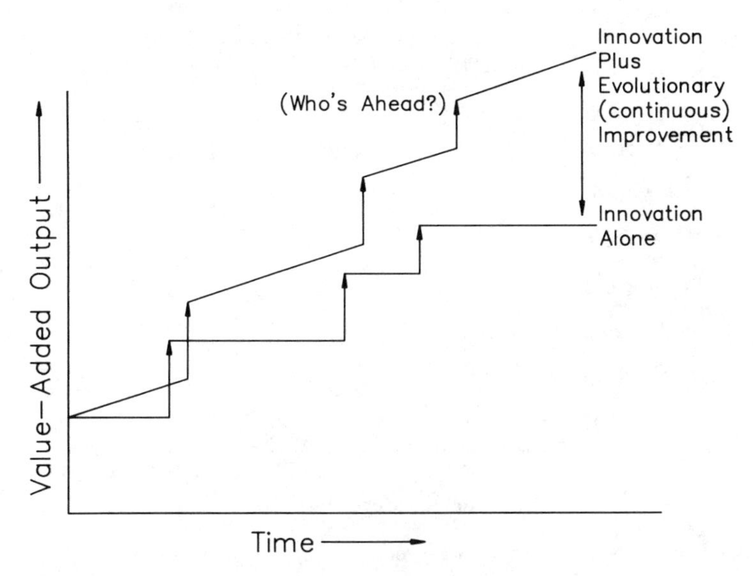

FIGURE 1.29. Revolutionary improvement coupled with evolutionary improvement.

other hand, involves everybody. The differences are marked by:

- Method (revolutionary and abrupt vs. evolutionary and gradual)
- Effect (dramatic vs. almost hidden from view)
- Pace [big step(s) vs. many little ones]
- Requirements (big investments vs. small, up-front costs)
- Time frame (one time vs. continuous)
- Orientation (technology vs. people)
- Involvement (few vs. everybody)

Probably the last three tell the biggest part of the story. *Kaizen* continuously involves people, all people, everybody in the improvement cycle. Let's look at the *kaizen* mine shown on Figure 1.30, largely untapped in the United States.

In an innovative driven society, the big ideas, the home runs, get the attention. Figure 1.30 has the "separation line" of innovation versus *kaizen* drawn at the $1000 per idea level. It's an arbitrary demarcation. In some organizations, the line is higher; in others, a little lower. But, regardless of whether it's at $10,000, $1000, or even $100, there are many ideas below the line.

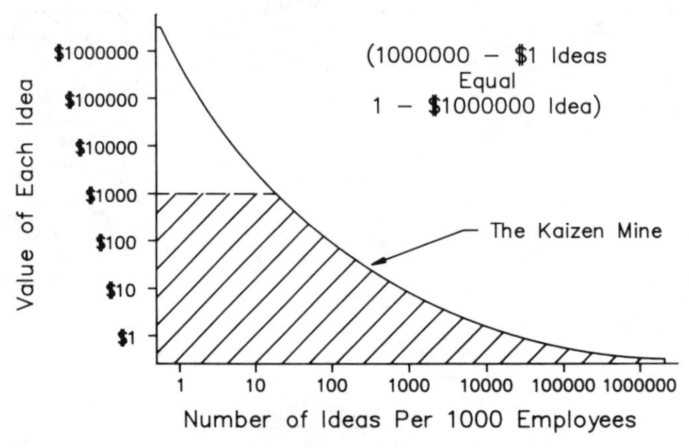

FIGURE 1.30. The *kaizen* mine, largely untapped in the United States.

1.9.4 Empowerment—Boundaries of Freedom in Action

You may say that it's unproductive for management to work on all of those ideas. Isn't sponsoring so many ideas in direct conflict with the Pareto principle that tells us to focus on the vital few? But are these ideas really part of the trivial many? They may be, to managers and engineers, but the *kaizen* idea most assuredly is a vital idea to the person who came up with it. It may make his or her job easier, safer, and more efficient. To the employee, it is probably the hottest project going. But to the manager or engineer, its significance can be almost lost. The importance or ranking of an idea depends on your perspective. We don't want to dilute the efforts of the engineer, but we also don't want to turn off the creative improvement machine of the work force. How can we answer both sets of priorities and reap both the innovation and *kaizen* rewards? With boundaries of freedom.

In a TQM environment, the continuous improvement toolbox is extended to and used by all employees. All members in the organization learn the new common language of team problem-solving and how to use basic statistics to separate opinion from fact to drive the continuous improvement cycle. The boundaries include the team-based continuous improvement procedures, the language, and the techniques that all are trained in, that all use. Of course, not all people become proficient in all statistical methods, but all do need to learn and use the team methods —that's part of the "entry fee" to stay a part of the TQM organization.

The boundaries create freedom. The freedom is the empowerment to work the improvement cycle within the procedural guidelines. With the boundaries, and the corresponding freedom, all employees are empowered to work the improvement cycle, to implement their ideas. The engineers are freed to spend more time on innovation. The work force works the *kaizen* mine, calling on engineering for support when needed. Innovation and *kaizen*—they aren't mutually exclusive activities. The engineers can't do both, there's not enough time, but the TQM organization, with empowered employees, can.

1.10 THE ENGINEER AS A CHANGE AGENT

Most organization's structures are built to withstand and fend off change. Their purpose is to maintain the status quo and keep the business running smoothly. Consequently the odds are that the momentum of the status quo will defeat most any type of change. It doesn't matter whether it's change for the bad or change for the good, the organization is designed to fight it. It's not that "the organization," or anyone in it, doesn't want to change for the good; it's just that the basic structure works to keep all change out.

The engineer plays the role of change agent for many organizations. The engineer investigates, experiments to find beneficial change, the change called improvement, and then incorporates that improvement into the process. But, the improvement can't just be brought to the process; the changes will bounce off of the status quo. The improvement must be incorporated in such a way as to change the process itself.

"Holding the gains" is an exercise in frustration. Checking, cajoling, and confronting to make sure someone remembered to "do the new thing" works only while the improvement is hooked to the life support of the change agent. Changing the process so that the improvement becomes part of it stops the frustration by taking the change off life support and into the mainstream. Changes (improvements) are worked through the PDCA cycle. Every new performance plateau becomes a new starting point for improvement. The engineer brings change into the organization by challenging the present and, eventually, by resetting (moving out) the boundaries of freedom.

The boundaries are a function of how full and how well used the continuous improvement toolbox becomes. Using SPI, the engineer supplements SPC to reset the limits of the process. With SPI, the engineer can work to reduce variability, to improve productivity, to enhance (product) performance, to increase the level of involvement of others in the continuous improvement cycle. The toolbox can be expanded, adding tools of improvement such as just-in-time (JIT) manufacturing methods and cellular manufacturing, quality function development (QFD), and total productive maintenance (TPM).

The statistical–continuous improvement toolbox never gets totally filled. But the filling does need to go slow to be effective. Everyone needs to be trained in the purpose and the use of each tool. The training should be in a "just-in-time" mode—people should receive the training just in time to apply it on the job—and the application needs to be integrated into the everyday work practices, incorporated into the continuous improvement processes before moving on to new tools. Addition and integration of more tools resets the boundaries, simplifies the structure, gives birth to more freedom, and fuels more improvement. The engineer can best serve the role of change agent by expanding employee involvement. The more people using the tools of improvement, the greater the multiplier for improvement. By helping people become the primary source of value-added in a process, the engineer, as the agent of change, multiplies his/her own outputs.

2

The Basic Tools

2.1 UNDERSTANDING VARIATION

Everything varies. Whether it be in industry or everyday life, no two things are exactly alike. Even things such as twins or baseballs, which appear on the surface to be identical, are never the same. Some aspects will differ from the other twin or other baseballs. In the case of "identical" twins, this difference may be physical such as a slight difference in build. It may be more subtle such as a difference in tastes or in drive for achievement. Baseballs may look the same. Looking from afar, all baseballs in a bucket appear to be the same. Yet, if examined closely, there would be minute differences in weight or diameter or roundness or some other property.

The same is true in business and industry. There are differences in every product and service. We call these differences variation. Variation arises from differences in the processes that create our products and services. The goal of every business must be to understand, to control, and to improve (reduce) the differences in our processes in order to reduce the variation in our products and services. Achieving this goal is the greatest difficulty facing all industries.

Figure 2.1 shows Dr. Shewhart's concept of a process. The process is made up of the methods, materials, manpower, and equipment that produce the product or service, the environment the preceding four elements operate in, and the measurement system that translates the actual process output so that it can be understood. Each of these factors varies the magnitude that each varies can be similar or vastly different. In a highly automated refinery, much more of the variation in the process can be attributed to the crude oil (materials) than to the operators (manpower).

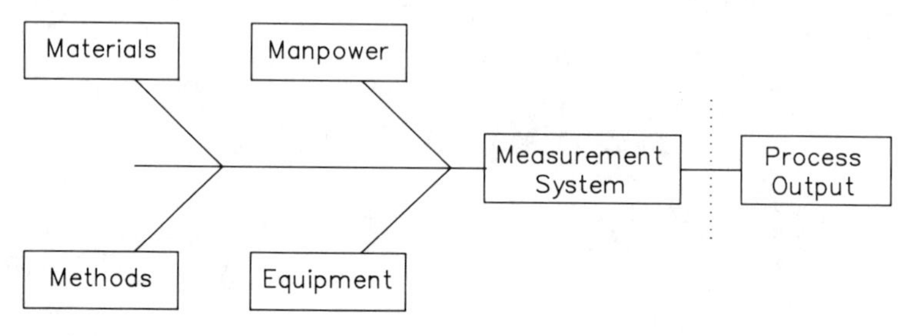

FIGURE 2.1. Shewhart's concept of a process.

In a manual assembly line, more of the variation can arise from variation in the methods than from variations in equipment.

2.1.1 Sum of Variation

The total variation of a process output (product or service) is the sum of the variation of the components that make up the process:

$$V_{\text{output}} = V_{\text{methods}} + V_{\text{materials}} + V_{\text{equipment}} + V_{\text{environment}} + V_{\text{people}}$$

where V is variation (or variance, in mathematical terms).

Normally, the variation of the product/service cannot be determined directly, because the actual process output is filtered by the measurement system. By measuring the process components, the measured variation can be calculated. This measured variation is the sum of the true variation of the product (or service) plus the variation of the measurement system:

$$V_{\text{measured output}} = V_{\text{actual product output}} + V_{\text{measurement system}}$$

When this book (and many others) use the word "variation", it refers to the measured variation of the product, not the true product variation. It is critical to understand this relationship, because in almost all cases, product specifications are based upon the measured variation. If the variation of the product is much greater than the variation of the measurement system, ($V_{\text{product}} \gg V_{\text{measurement system}}$), the measured variation will be close to the true product variation.

In most industries, however, the measurement of a product is a complex, but often overlooked, process in itself. It may involve sampling, sample or test specimen preparation, conditioning, and then testing. For example, measuring the tensile strength of a plastic or rubber compound involves sampling the product, molding test specimens, conditioning the specimens at the proper temperature and humidity, fixturing them, and then finally, testing them. The variation in the measurement system may be much greater than the variation of the product. If this is the case, how can we know the actual product variation? It can be calculated from

$$V_{\text{actual product output}} = V_{\text{measured output}} - V_{\text{measurement system}}$$

This value is needed to know whether to concentrate continuous improvement efforts on variation in the (main) process or in the measurement process. Calculating the measured variation ($V_{\text{measured output}}$) will be covered in Section 2.8; the variation in the measurement system ($V_{\text{measurement system}}$) will be covered in Chapter 3.

2.1.2 Common Causes

Variation in a process may either be random or nonrandom. Random variation is the result of chance events. These events are routine and

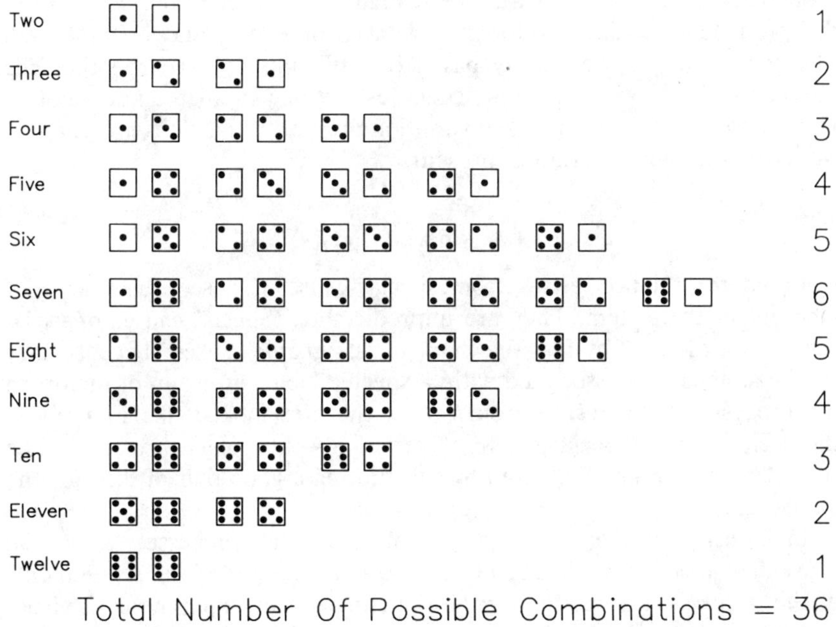

Two 1
Three 2
Four 3
Five 4
Six 5
Seven 6
Eight 5
Nine 4
Ten 3
Eleven 2
Twelve 1
Total Number Of Possible Combinations = 36

FIGURE 2.2. Dice combinations. (*Reprinted from SPC In Action, copyright 1990, Resource Engineering, with the permission of the publisher, Quality Resources, White Plains, New York.*)

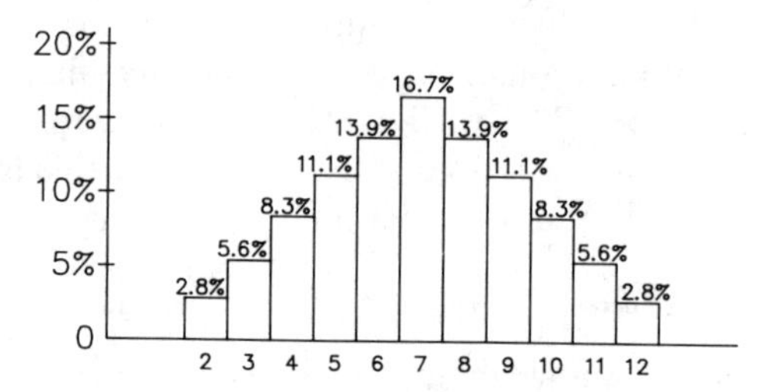

FIGURE 2.3. The chance for each roll.

based upon probability. Dr. Deming refers to these chance events as common causes of variation. They are the causes of variation inherent in the process or the ones that are always present.

A pair of dice can be used to demonstrate random variation. The possible combinations of the dice are shown in Figure 2.2. When we roll the dice, we have the highest probability of rolling a 7 and the lowest probability of rolling either a 2 or 12. Figure 2.3 shows in bar graph form the predicted percentages for the chances of each value being rolled. These percentages are strictly based on probability. If we roll the dice often enough, the values should occur close to their predicted percentages. If this happens, we have random variation present; if this does not happen, we have nonrandom variation present.

2.1.3 Special Causes

Nonrandom variation arises from special causes. These causes are not inherent in the system. They are unpredictable. (Special causes of variation were described by Dr. Shewhart as "assignable causes" because they can be assigned, or associated with, a specific factor or group of factors in the process.) If the results from rolling the dice looked like Figure 2.4, there clearly is a special cause of variation—the dice are loaded. The special cause can be eliminated by substituting a good pair of dice for the loaded pair.

In industry, special causes can also arise in all processes. A special cause may arise from unpredictable variation from any of the six elements of the process—the methods, materials, manpower, equipment, environment, and measurement system. Two operators following different operating procedures would be a special cause of variation. The operators would be making products that may vary greatly from each other. Likewise, a problem with increasing molecular weight from batch to batch of a

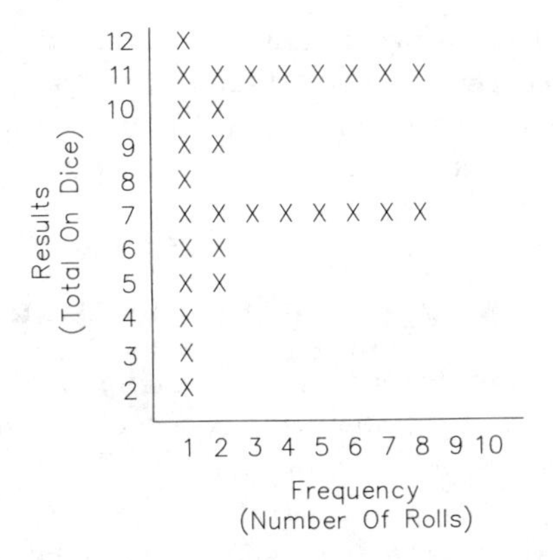

FIGURE 2.4. A special cause of variation: loaded dice. (*Reprinted from SPC In Action, copyright 1990, Resource Engineering, with the permission of the publisher, Quality Resources, White Plains, New York.*)

polymer could be traced to a special cause of variation. This special cause could be scale buildup in the cooling jacket of the polymerization reactor. Over a period of time, it would get increasingly more difficult to quench the reaction. The polymerization would continue longer each batch, thereby creating the problem. The special cause could be eliminated by instituting a descaling operation for the cooling jacket at regular intervals.

2.1.4 Measurement System Variation

It is important to identify and eliminate all special causes of variation in the process. However, it may be most important to identify and eliminate all special causes of variation resulting from the measurement system. Special causes of variation from the measurement system will send false signals about the process and product. It will appear that the process changed when, in actuality, it did not. A great deal of effort may be wasted on resetting the process, reworking (or scrapping) product that "appears" to be defective, or worse, on a large problem-solving effort on the process. Reacting to a false signal from the measurement system may even result in defective product as we adjust to false signals from the measurement system.

To achieve statistical process control, the initial goal is to eliminate special causes of variation. To do that, we need to have a process that operates with only common causes of variation present. The balance of this chapter discusses concepts and tools to understand and begin to limit variation. Measuring variation will be covered in Section 2.8. Sections 2.2

to 2.7 lead up to measuring variation, covering techniques needed to understand·the process, to collect data, and to organize that data. These basic tools and techniques will serve as the foundation for determining the variation in our processes, for bringing the variation into control, and for reducing the variation.

2.2 CLARITY WITH FLOWCHARTS

Before we examine a process, we must first understand it. The flowchart is a visual representation of the steps in a process. This simple tool aids the understanding of the process by providing information on the key process steps or equipment, critical product features at each step, and the process measurement–control scheme. It is the starting point for troubleshooting and process improvement activities.

In the chemical process industry (CPI), the process and instrumentation diagram (P&ID), supplied as part of the start-up package, can serve as the basis for a flowchart. Creating a flowchart should never be an office exercise. In many cases, the P&ID reflects the process design and not the actual installation of the process. Even if the P&ID is updated at start-up, it will not reflect the constant change that goes on in most operations. To build an accurate flowchart, we must both examine the process in the facility and discuss our findings with the local work force.

Figure 2.5 shows a flowchart for a papermaking operation. This flow chart reflects the author's preference for format and symbols. Standardization of symbols for flowcharts has largely been unsuccessful. Attempting to standardize symbology, Veen authored an article published in *Quality* magazine in 1971. However, many companies have established their own symbology. We recommend companies use the same conventions as they've established for P&ID drawings, adding appropriate symbols needed for decision-making points and product parameters used in the flowcharting process. For example, Ford Motor Company uses an inverted delta (∇) on their drawings to indicate a key product-control characteristic.

When creating a flowchart, don't skimp on paper or white space. Allow enough space for all of the information you will be adding to the chart. After drawing in each process step or piece of equipment, draw or list process measurements or control loops for each. Again, be sure to check the installed process to verify the control scheme. There is no need to draw in conveyors or other material handling equipment unless they affect the product. For example, a pneumatic conveying system may have a humidity control system associated with it, especially if it conveys a hydroscopic (water-absorbing) material, and it would be included because it represents a process control step. However, a roller conveyor used to

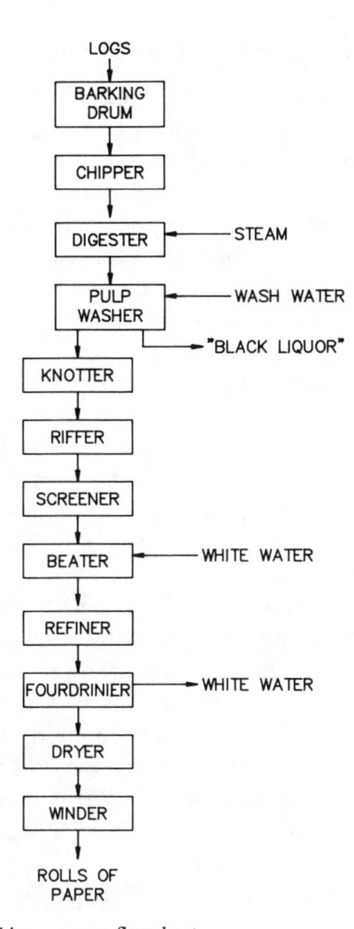

FIGURE 2.5. Paper-making process flowchart.

move full drums of resin from the filling station to a palletizing unit need not be included because it does not directly affect the process output (product) characteristics.

For simplicity, the key product characteristics can be listed alongside each process step. Product characteristics would include measured in-process critical cosmetic requirements, particle size distributions, and dimensional tolerances, among others. The measurement method for each characteristic should also be described. If the flowchart does not have enough spaces to list all of the information, use reference numbers or letters on the actual flowchart and list the information on an attachment.

2.2.1 Flowchart the Office

Flowcharts should be drawn for every process done in an organization. The word "process" historically brings to mind a manufacturing process. But, in a TQM environment, this definition is too narrow. A process is a series of operations (input, value-added, output) that results in the production of an output. The process output may not always be a physical product. Administrative procedures are processes too. The output is not a physical product, but a service. For example, the output of a purchase order procedure (or process) would be a completed requisition with an order placed with the selected supplier. Hiring is a process where there are a series of steps starting at the time the need is established until the right person joins the organization. Taking a measurement on a product is a process in itself and, as will be discussed in Chapter 3, deserves major study and analysis. A flowchart of the steps in making the measurement can ensure that it is done properly each time andcan serve as a reference for both troubleshooting measurement system problems and improving the measurement system.

Flowcharts for "office processes" are similar to flowcharts for manufacturing processes in that they visually show the process flow. One big difference, however, with office processes is that we usually can't "see" them. Manufacturing processes are usually bolted down to the floor; we can visually trace the process flow from input through the value-adding steps to the output by looking at the equipment in place. The flow of most office processes isn't visual, it's mental. We can use flowcharting techniques to help put these mental process flows into pictorial form.

2.2.2 Top-Down Flowcharts

For office processes, we'll use a different flowcharting method than we use for manufacturing processes. We will use a two-phase flowchart approach called the "top-down" flowchart. First, the top-down flowchart shows the major steps of the process. Second, the flow within each of those major steps is charted. Major steps in a top-down flowchart are represented by a double rectangle.

Although there are many different symbols used in flowcharting language, we will only be using a few of the most common ones for top-down flowcharts. The following symbols shown in Figure 2.6 are the only ones needed in order to flowchart your office process.

Arrow — An arrow shows the *direction of the flow* in the process. The arrow points to the next step in the process.

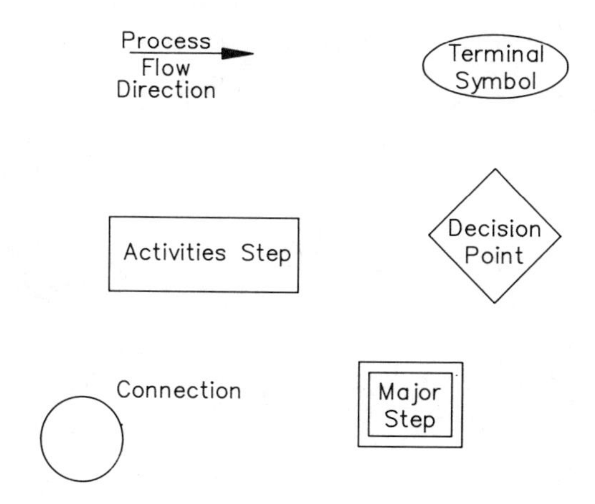

FIGURE 2.6. Basic flow chart symbols.

Oval — The oval (or terminal symbol in flowcharting language) signifies *the beginning or end* of the process. Inside the oval, an instruction, "Start," or "Stop," is written. There is normally only one start step and for our flowcharts only one stop step.

Rectangle — The rectangle is the symbol for an *activity*. A brief description of the activity should be written in the rectangle. The rectangle should not be used if the activity involves a decision to be made. An activity rectangle may have one or more incoming arrows, but will have only one outgoing arrow.

Diamond — This shows a *decision-making point*. The decision question should be written briefly inside the diamond. A decision diamond will have one incoming arrow and two or more outgoing arrows. These outgoing arrows show the different paths depending upon the answer to the question.

Many companies try to use only questions that can be answered "yes" or "no" so that there are not too many paths out of a decision diamond. This simplifies the flowchart. For example, Figure 2.7 shows how a complex decision diamond could be broken down into three "yes" or "no" questions.

Circle — A circle is a symbol for *a connection*. A connection is used when the flow chart is complex and takes

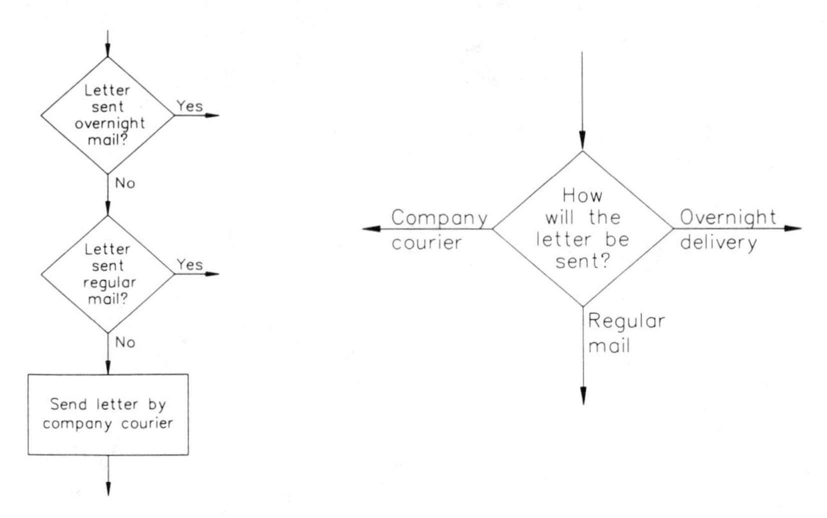

FIGURE 2.7. Alternate approaches to decision steps with multiple outcomes. (*Reprinted from First Class Service, copyright 1991, Resource Engineering, with the permission of the publisher, Quality Resources, White Plains, New York.*)

several pages. The circle at the edge or bottom of the page will be assigned a letter A, B, C, and so on. Another circle with the same letter will restart the process on another page. The process flow will continue from there, as shown in Figure 2.8.

Double rectangle — Inside this double rectangle, a brief description of the *major steps* should be written. The double rectangle will have one incoming arrow and one outgoing arrow.

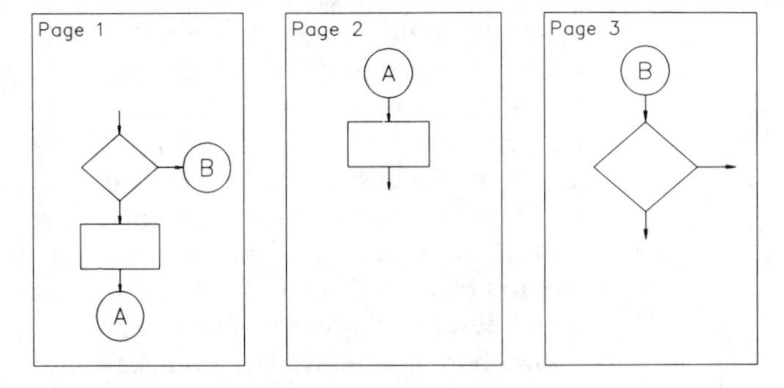

FIGURE 2.8. Using connection symbols. (*Reprinted from First Class Service, copyright 1991, Resource Engineering, with the permission of the publisher, Quality Resources, White Plains, New York.*)

Now we'll look at the steps to construct a top-down flowchart.

As simple as it may seem to construct a flowchart, there can be a lot of arranging and rearranging of major steps, activities, and decision points. It is rare to start out with a clean sheet of paper and end up with the final flowchart on that same paper. It is recommended each of your major steps, activities, or decision points be listed on self-stick notes. Tape one or two pieces of flipchart paper to the wall and post the self-stick notes onto them. This way activity can be rearranged easily if need be.

Identify the Major Steps in the Process Be careful to include only the major steps that, when put together, make up the process. In order to qualify as a major step, several activities and possibly some decision points may be required before moving on to the next major step.

Identify the Substeps for Each Major Step Obviously, if the major steps were all that was flowcharted on the process, it would not be very useful for understanding the process. In addition, it would be impossible to improve the process from a flowchart of only major steps. To understand the process, the substeps within the major steps must also be identified and flowcharted. (It may help the flowchart organization to list the substeps under the corresponding major steps.)

Identify the Flowchart Symbols For Each Step Once all the steps are listed out and agreed upon, assign one of the flowchart symbols for each step. This sounds easy, but take some time to think about it. Many steps we think are activity steps, may, in fact be decision steps. Because we know that activities have only one possible outcome (one arrow leading out), any step that has two possible outcomes is a decision step.

If the flowchart has been constructed using the flipchart paper and self-stick notes, simply draw the appropriate shape (double rectangle, rectangle, or diamond) around the self-stick notes.

Don't forget to draw in the direction arrows. The direction arrows coming out of a decision diamond need to be labeled (e.g., "yes," "no").

Complete the Flowchart When the draft is complete, transfer the work to a regular size sheet of paper so that everyone can have copies of it.

Test the Flowchart Look over the flowchart and have others familiar with the process check it too. Then, put it to the acid test. Work through the process by using the flowchart exactly as drawn. One may even want to have someone unfamiliar with the process use it. Can they follow and understand the flowchart? Was a step forgotten? Did two steps get put in reverse order? Don't worry, this often occurs on the first draft; now's the time to correct it.

Periodically Review and Revise The more complete the flowchart is, the more helpful it will be. As the flowchart is used, change it to reflect steps left out or update it to show any changes made to the process. It

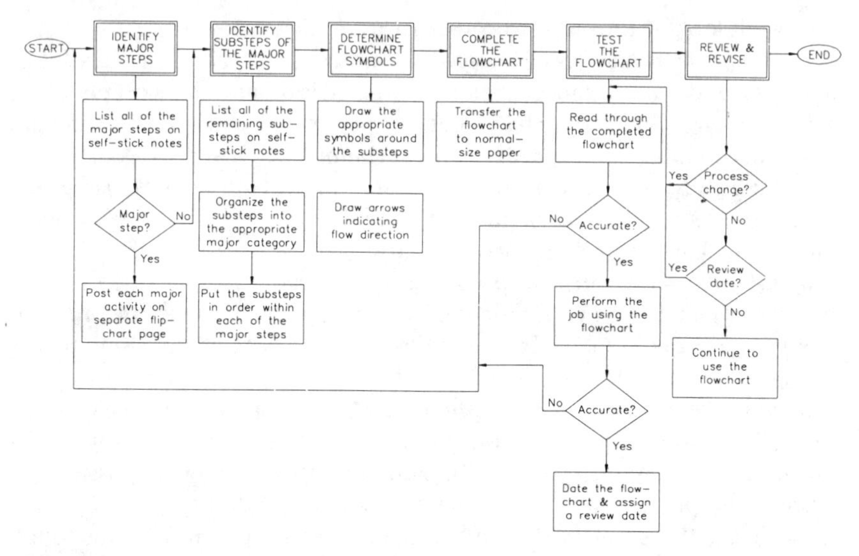

FIGURE 2.9. Flowchart of the top-down flowcharting process. (*Reprinted from First Class Service, copyright 1991, Resource Engineering, with the permission of the publisher, Quality Resources, White Plains, New York.*)

might be helpful to show a review date on the flowchart as a reminder to update it periodically.

Figure 2.9 flowcharts the top-down flowcharting process.

2.3 SAMPLING

A sample is a small selection of product taken from a process or from a larger group of the product. The measurement data from the sample can be used to predict the performance of the process or to estimate the properties of the larger group. This large group of product or all of the product coming out of a process is known as the population. A small sample is used to predict the performance of a larger population. Decisions on the process or population can be made based on the measurement data from the sample.

2.3.1 100% Inspection

There is an alternative to sampling in order to understand the properties of the population. The alternative is 100% inspection. This can be done either manually or with automatic test equipment. In many industries, such as the chemical process industry, 100% inspection cannot be done. There is no method to 100% inspect surfactant exiting a process for all of its critical properties. Many inspections consume (destroy) the product. Even if it could be done, 100% inspection often adds cost, not value, to the product. 100% manual inspection is only about 80% effective. Not only can defective product be missed, but good product can be sorted out inadvertently. An example of the ineffectiveness of 100% inspection we often given to disbelievers is an F test.

In this F test, the "inspector" is requested to read the following sentence, once and only once, and count the number of Fs it contains:

FEDERAL FUSES ARE THE RESULT OF YEARS OF SCIENTIFIC STUDY COMBINED WITH THE FIRST HAND EXPERIENCE OF FIFTY YEARS.

How many Fs did you find? The distribution (Figure 2.10) of the results of many inspections show that relatively few "inspectors" found the correct number of the Fs. One engineer even got 24 because, in his mind, each E contained an F. All of the "F inspectors" had years of training on what an F looks like and still didn't inspect the product correctly. (By the way, there are nine Fs.) Imagine the difficulty of production operators or inspectors who don't have all of that training, who are looking at multiple aspects of the product, who are often rushed in their job, and whose working environment is usually not adapted for inspection work. Estimating manual inspection as 80% effective may be generous.

Automatic measurement equipment for 100% inspection should give adequate guarantees of good product (or really "good" sorting precision) provided the inspection system does not go out of calibration. Some companies use automatic inspection in place of SPC techniques for quality control. Many of the *poka-yoke* (mistake-proofing) techniques employed

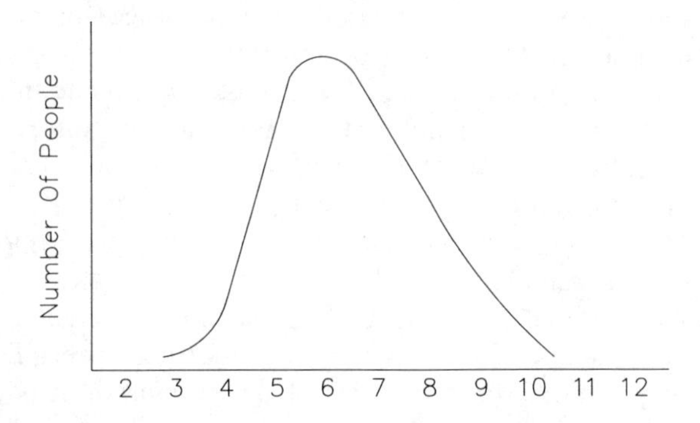

FIGURE 2.10. Results of F-inspection test. (*Reprinted from SPC In Action, copyright 1990, Resource Engineering, with the permission of the publisher, Quality Resources, White Plains, New York.*)

by Japanese companies are actually automatic 100% inspection devices. The drawbacks of these devices are that they are often costly, they must be rigorously maintained and calibrated, and they cannot be used for QC tests that consume (destroy) product. When destructive testing is being used, taking samples from the lot or the process is the only adequate sampling solution.

2.3.2 Random Sampling

In order to get the best idea about the lot of material or about the process, samples should be taken at random. When each part, or unit of material in the lot of material has been thoroughly mixed, such as in a continuously stirred tank reactor (CSTR), then any sample taken would be considered as being random. More likely, the lot has not been thoroughly mixed, and it is impractical to do so. A random-number generator can be used to ensure random samples are taken if every item in the lot is numbered. The Appendix contains a random-digits table and describes its use. If each item or lot of product is not numbered, the best way to randomly sample is to sample in a nonbiased manner. Some suggestions are:

1. Don't select only items that appear to be good or avoid those that appear bad.
2. Don't take samples from just one container if the lot is in multiple containers.
3. Don't just take samples off the top of the pile, tank, or container.

Avoiding bias in sampling is part training in proper techniques and part having the proper tools on hand. For example, sampling a rail car of resin could be biased because particle size segregation may have occurred during transport. The resin in the rail car is no longer homogeneous, so randomness cannot be obtained with a sample from the top or a sample port. In this case, an automatic sampler installed in the pneumatic conveying system could take small amounts at programmed intervals to get a more representative sample of the contents of the rail car as it is unloaded. Even better would be to have the supplier install an automatic sampler in the loading system. The sample could be segregated in the shipment and the sample tested before the tail car was unloaded. For large lots or batches of material, it is much more efficient to sample in process or in transport, than to sample a pile or container. This will also produce a sample more representative of the population it is taken from.

The automatic sampler would be taking a composite sample. A composite sample is basically a series of small samples taken periodically from a lot or batch and combined for testing. These small samples make a representative sample of the entire lot. Composite samples can also be taken manually. One technique is to simply take a small scoopful out of each barrel or drum of material. The scoopfuls are placed in a sample bag or container. Once the lot has completed processing, the contents of the sample bag or container are mixed thoroughly and a portion taken for testing. A sample "thief" also takes a manual, composite sample of a drum of solids. The "thief" is a wand with a series or chambers that fill when the thief is pushed into the drum. The chambers fill with particles from different levels of the drum, thereby creating a composite sample.

Stratified sampling may be used when there are multiple product streams making up a population. In stratified sampling, the number or amount of product taken from each stream is proportional to the percent that stream makes of the overall production. These subsamples from each stream must still be taken randomly. Stratified sampling is not the same as random sampling though. Theoretically, sample-to-sample variation will be less for stratified sampling because the sample will predict less variation in the population than there is in actuality. However, in an industrial environment, this difference is minor compared to other factors that may cause bias.

Selected sampling is a technique frequently used in process industries. In contrast to random samples, selected samples are pulled at a selected time or time intervals, at selected quantities or quantity intervals, or from specific parts from the product (e.g., sampling strictly from the edge of a continuous web to minimize damage to the web and therefore costs). Industries use selected sampling because of its ease. In deciding on a sampling plan, consideration must be given to the training and discipline it takes to obtain a sample that gives an unbiased view of the population. In a production run, it takes less discipline to sample every hour than to randomly sample at random time intervals. This sampling technique would seem to have more bias than random sampling. In actuality, it may produce less bias than an undisciplined random sampling plan that would be likely to occur in a manufacturing setting. Selected sampling is used with the variable control chart described in Section 4.6.

2.4 DATA COLLECTION FORMATS

Data are defined as information on the product or service and the process it came from. Collection of data allows for decisions to be made. All sound management, technical, manufacturing, sales, and other business decisions are based on data. Without data there is no way to separate perceptions from the true facts: All decisions would be based on "gut reactions", and we all know how often these are incorrect. One would frequently take incorrect actions if data to analyze were not available.

In order to make decisions, and make correct ones, data must not only be available, but must also be correct. To be correct, it must be timely, accurate, and tell the whole story. Untimely data may lead to decisions and actions that are no longer appropriate at the time of implementation. Inaccurate and incomplete data increase the probability of incorrect decisions and actions.

Timeliness, as a function of data collection, depends on the purpose for which the data are being collected. Obviously, there are vastly different time constraints for data collected for on-line process control and for data collected on employee absenteeism. Both are important for overall business purposes, but there will be little affect on the business if the data on employee absenteeism takes a few days to reach the Human Resources Department for collection. However, there could be major impacts on the business if data critical for process control were delayed in reaching the instrumentation or the operators. Imagine the potential problems if the temperature data on a catalytic cracker (used to make gasoline) in a refinery were only collected once an hour. Timely data collection in this case would include continuous monitoring, transmission, and control.

In order for data to be accurate, they must be collected the same way each time. For example, on a three-shift operation, all three operators must collect data on the process and product using the same method. They must also collect data with the same, or with correlated, measurement devices. These measurement devices should undergo checks and calibration periodically to ensure their accuracy. (Refer to Chapter 3 on the measurement system.)

Once collected, data must be complete and tell the whole story. Sometimes, it is easier to collect some data rather than others. If there are two pressure gauges on an autoclave and one is hard to reach, the operator may only read the easier gauge and "estimate" the reading on the other. These partial data may give an incomplete picture and lead to incorrect decisions on the operation of the autoclave. Telling the whole story also depends on the clarity of purpose for collecting the data. The purpose for collection determines the plan for collections. The clearer the purpose, the easier to establish a plan for collecting the necessary data.

Data should only be collected if they are going to be used for decisions or actions. In many cases, data are collected simply because an engineer or manager thinks it might be useful in the future. This is not a legitimate purpose. If it is known that the data are not used, there will not be the same level of dedication to collecting it completely, accurately, and in a timely manner. This waste of effort may color the collector's perception of all data collection, which may negatively impact data collection that does have a clear, important purpose for the operation.

There are three types of data—variables, attribute, and subjective. All three are important; however, subjective data should only be used when no variable or attribute data are available or applicable.

Variable data are also known as measurement data. It is data that can be measured on a continuous scale. Examples of variable data include dimensions, temperature, pressure, viscosity, and time. Because they are measured on a continuous scale, variable data are expressed in whole numbers and fractions. For example, a temperature could be measured at 17.8°C. In contrast, attribute data are counted, so they are measured in positive integers. Examples of attribute data include results from "go–no go" gauges, number of defects in a part, number of parts defective in a lot, and number of accidents. Attribute data may also be expressed as percentages. If 25 parts out of 200 were deemed defective, the lot might be said to contain 12.5% defectives.

Subjective data are not "hard" data like variable and attribute data; they are "soft" because they are based on how one feels about something. Subjective data cannot be measured like variable data, nor counted like attribute data. Judging in Olympic figure skating or gymnastics are examples of subjective data. Here, judges score the athlete from 0.0 to 10.0 based on their feelings on the performance. Tallying this subjective data determines the medalists. Similarly, in an industrial environment, operators, supervisors, and engineers may have many different ideas for improving the process. The group may all vote on which ideas they feel are best. Their voting is subjective. The ideas with the most number of votes are tried first. This is an example of decisions and actions based upon subjective data.

Operators may also be asked to judge how well batches of material processed. In a rubber-compounding facility, operators may be asked to rate how each batch processes on a two-roll mill. These data may be useless in a three-shift operation because the first shift operator's view of a batch running well may be different from that of the second or third shift operator. The data can be made more meaningful by giving the operators some guidelines and training to standardize what they are to look for. This is done in the Olympic events for the judges. The skaters must perform to certain standards that are known to the judges: They can't stumble or fall; they must hold certain lines while spinning; they must cover a certain amount of the rink.

Likewise, standards can be set for a process like rubber compounding to guide the operators' judgment: Does the sheet of rubber bond to the correct roll of the mill? Does the sheet of rubber hold together while it's stripped off? Depending on the operation, there may be many standards to guide the operators. Once standards are established, all of the operators must be trained to uniformly apply them. This would make it more likely that the third shift operator was judging similarly to the second and first shift operators and make the subjective data more meaningful.

No matter what type of data are collected, they must be organized into a form that makes it easy to understand and easy to use. Data collection sheets serve this purpose. Whenever possible, a data collection sheet should be designed specifically for a given process. (This may not be possible if the data collection sheet was mandated to meet regulatory agency requirements, for instance.) Using operators' inputs in designing the data collection sheet will make it easier for them to collect the data and make it less likely that they will make mistakes entering the data. The sheet should not be too congested or it will not be easy to understand or use.

There are many types of data collection sheets. The most common are process data collection forms, checklists, defective-item check sheets, defective-type check sheets, tally sheets, concentration diagrams, and control charts. Control charts will be discussed in detail in Chapter 4. The others will be briefly discussed here.

2.4.1 Process Data Collection Form

Figure 2.11 shows a process data collection form for a crystalline sugar-substitute granulating operation. This form is a communication tool from operator to operator and to the area engineers. It provides a picture of the setup for all of the equipment and how it is operating. It can prevent problems from occurring. For example, if the pressure drop (ΔP) across the bag house-type dust collector was creeping upward on first and second shifts, the third shift operator knows that the bags are blinding and that an eye needs to be kept on the dust collector. It also records the changes made to the processing parameters. The form captures the history of the process that the engineers can use to correlate with product performance. This may be especially useful in investigating a customer complaint a month after the product was made. Without the data, it will be much more difficult for the engineers to discover the cause of the complaint and to mistake-proof the operation.

2.4.2 Checklists

Figure 2.12 shows a checklist for ensuring the contents of a spill box are present. The checklist provides the site environmental coordinator with

GRINDING PROCESS DATA—COLLECTION SHEET

Lot Number: ________ Date: __/__/__

Lot Count: ________ Checked By: ________

Time										
Granulating										
Feed Rate										
Gran. Speed										
Gran. Amps										
Pneumatic Sys.										
Damper Setting										
Pressure—P1										
Pressure—P2										
Baghouse ΔP										
Blower Amps										
Rot. Valve Sp.										
Screening										
Top Screen										
Mid. Screen										
Iris Setting										
Bot. Screen										
Wt. Angle										
Packaging										
Vibr. Setting										
Wt. Setting										
Dust Collection										
Baghouse ΔP										
Rot. Valve Sp.										
Initials										

FIGURE 2.11. Granulating data collection form. (*Reprinted from SPC In Action, copyright 1990, Resource Engineering, with the permission of the publisher, Quality Resources, White Plains, New York.*)

data on whether items in the spill box require restocking. Checklists are the simplest and easiest-to-use data collection sheets. Often, they serve as memory joggers. Maintenance personnel use checklists when performing preventive maintenance to ensure they don't overlook a step. Interviewers can use checklists to determine if a candidate meets all requirement for a job. Design engineers use them to ensure they've considered critical

SPILL BOX CONTENTS CHECKLIST		
Item	Quantity	Present?
Respirators	2 ea.	
Boots – large	2 pr.	
Boots – small	2 pr.	
Aprons	2 ea.	
Faceshields	2 ea.	
Plastic Shovel	1 ea.	
Speedi-dry	20 lbs.	
Pigs	8 ea.	
Plastic bags – large	8 ea.	
Plastic bags – small	10 ea.	
Twist ties	18 ea.	
Pails	5 ea.	
Lids	5 ea.	
Checked by		
Date		

FIGURE 2.12. Spill box checklist. (*Reprinted from SPC In Action, copyright 1990, Resource Engineering, with the permission of the publisher, Quality Resources, White Plains, New York.*)

CEMENT LABORATORY ANALYSIS
DATA COLLECTION SHEET

GRADE: ________

BATCH	DATE	%CaO	%SiO_2	%Al_2O_3	Fe_2O_3	MgO	SO_3
	MIN VALUE	61.0%	18.0%	4.0%	1.5%	1.0%	1.0%
	MAX VALUE	67.0%	23.5%	7.5%	6.0%	5.5%	2.5%

NOTE: CIRCLE ANY VALUES OUT OF SPEC.

FIGURE 2.13. Product data sheet.

aspects of the project. Checklists are an easy way to mistake-proof an operation or a process.

2.4.3 Product Data Sheet

A simple form is needed to collect inspection data on the product, whether it is in the process or in the quality or engineering departments. The form should make it easy to record the data and to understand it. This aids in calculating results on the product properties. Figure 2.13 shows a typical product data sheet. These can be combined on the same sheet of paper with a checklist on product attributes. This would ensure that all critical properties are inspected.

2.4.4 Histogram or Tally Sheet

If the data being collected are going to be presented in the form of a histogram (see Figure 2.14), collect the data in that form to eliminate duplication of effort. This data collection sheet is also known as a tally sheet because one tallies the entries. (Refer to Section 2.7 for information on histograms.)

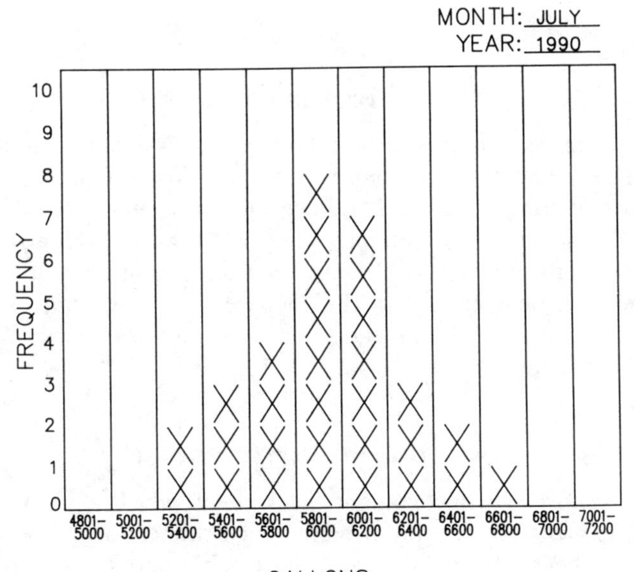

FIGURE 2.14. Histogram.

RIBBON ROLLER DEFECTIVE-TYPE
CHECK SHEET

Lot Number: ________ Date: __/__/__
Lot Count: ________ Checked By: ________

TYPE	CHECK	SUBTOTAL
Voids		
Tears in Foam		
Width Small		
Width Over		
Foam Diameter Small		
Foam Diameter Over		
Foam Location		
Foreign Matter		
Insert ID Small		
Insert Damaged		
Misassembled		
Other		
	Grand total	

Remarks: ________

FIGURE 2.15. Defect-type check sheet. (*Reprinted from SPC In Action, copyright 1990, Resource Engineering, with the permission of the publisher, Quality Resources, White Plains, New York.*)

2.4.5 Defect-Type Check Sheet

In order to provide better information for analysis and elimination of defects, a defect-type check sheet will often be used. It is not enough to know that parts are defective; it is necessary to know the types of defects present. A product may experience 90% yields two days in a row yet the defect type might be totally different each day. An investigator could be looking for a single cause for the low yields rather than two causes. Knowing the defect types can lead to quicker solutions. Figure 2.15 shows a defect-type check sheet.

2.4.6 Defective-Item Check Sheet

Figure 2.16 shows a defective-item check sheet, which is used to identify which components in the multicomponent products are defective. Like defect-type check sheets, this aids in troubleshooting the process to eliminate defects. The components most often defective are identified on this data collection form so that the investigation can be focused.

DEFECTIVE–ITEM CHECK SHEET

Model No.________ Lot No.________

Description________

Serial Number →									
↓ Defective Item									
Left Front Burner									
Left Rear Burner									
Right Front Burner									
Right Rear Burner									
Burner Panel									
Front Oven Panel									
Oven Light									
Top Oven Coil									
Bottom Oven Coil									
Pan Tray									
60–minute Timer									
Oven Timer									
Oven Controls									
Lf Fr Brnr Cntrl									
Lf R Brnr Cntrl									
Rt Fr Brnr Cntrl									
Rt R Brnr Cntrl									
Control Panel									
Chrome Burner Liners									
Checked by									
Date									

FIGURE 2.16. Defective-item check sheet. (*Reprinted from SPC In Action, copyright 1990, Resource Engineering, with the permission of the publisher, Quality Resources, White Plains, New York.*)

2.4.7 Summary

Data can be collected in a variety of types and in a variety of manners. Because data are the key to making informed decisions and serving as the basis for action, it is critical that the data be correct, complete, and in a form that is easily understood. Most bad decisions can be traced back to data that were incorrect, incomplete, or difficult to understand. Without meaningful, planned data collection, most techniques described in this book will be useless.

2.5 THE POWER OF CONCENTRATION DIAGRAMS

When dealing with defects in a product, one tends to think that the defects are randomly distributed in the product. This is often untrue. In many cases, the defects are concentrated in specific areas. Documenting their locations on a *concentration diagram* will, as the name implies, show where the defects are concentrated on the part. This will allow investiga-

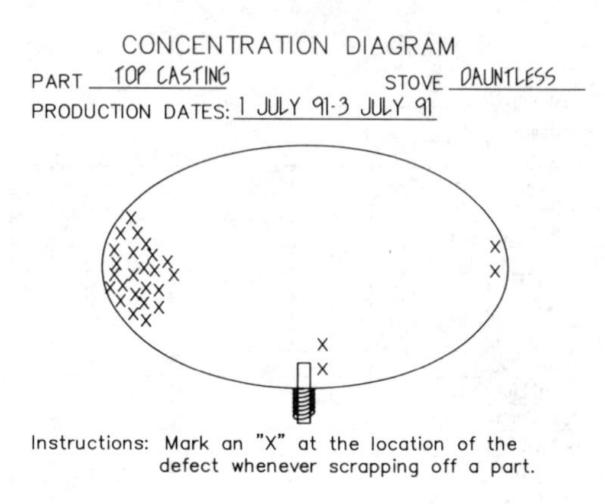

FIGURE 2.17. Stove defect concentration diagram.

tion into the causes of the defects to be more focused and to lead to their quick elimination.

Figure 2.17 shows a concentration diagram from foundry operation at Vermont Castings, Inc. Vermont Castings is the leader in quality and market share for wood-burning stoves. On some parts that were difficult to cast, Vermont Castings was forced to sort parts at the foundry to maintain quality. Within one month of implementing concentration diagrams for the part shown, corrective measures, which reduced scrap castings by 60% and avoided thousands of dollars per year in scrap costs, were installed.

2.5.1 Constructing a Concentration Diagram

1. Draw the part showing all necessary views. (It may be possible to simply use a copy of the print!) The drawing may be a two- or three-dimensional view, depending on the part complexity. A flat part, such as a laminate, will have only one or two views—a front view or front and back. Other parts can have up to six views—top, front, right side, left side, bottom, and back.
2. Create an information section. This should include the part name or number, the date, and a section for comments that may aid the investigation. Other information often desirable to collect includes operator identification, lot number, number of pieces produced, and number of pieces defective. The latter two would be collected after completing the diagram.
3. Collect defect data. Wherever a defect occurs, an X is marked on the diagram.

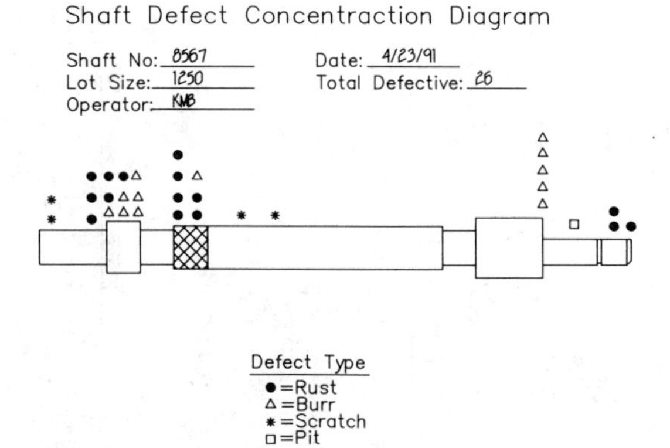

FIGURE 2.18. Use of symbols with concentration diagrams. (*Reprinted from SPC In Action, copyright 1990, Resource Engineering, with the permission of the publisher, Quality Resources, White Plains, New York.*)

2.5.2 Modified Concentration Diagrams

Concentration diagrams can be modified to provide more valuable data on the product defects. Symbols can replace the X on the diagram to denote defect types. An example is shown in Figure 2.18. Numbers or letters can replace the X to show the sequence of defects (Figure 2.19). Numbers or letters can also be used for recording the shift on which the defect

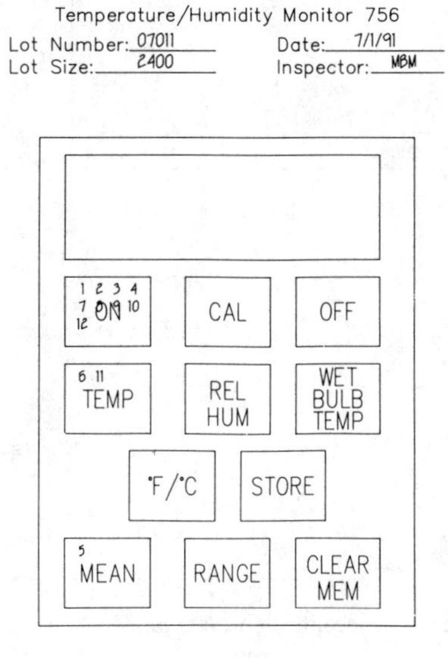

FIGURE 2.19. Use of sequencing on a concentration diagram.

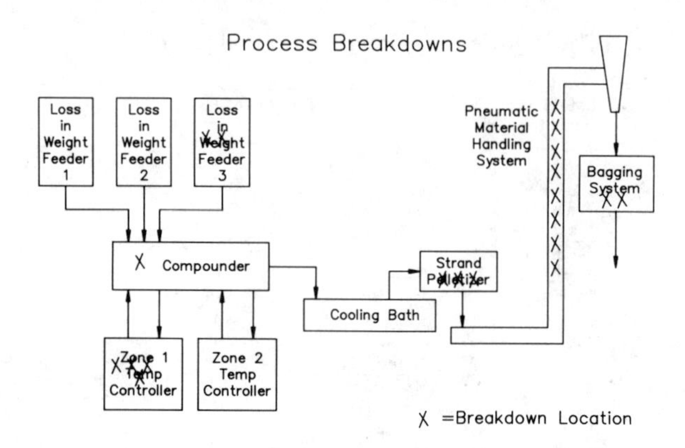

FIGURE 2.20. Equipment downtime concentration diagram.

DATE OF ORDER	LOT NUMBER	CUST. ORDER NO.
		X

CUST. CONTROL	F.O.B.	SALESMEN	TERMS
	XX		XXXXX

INVOICE NUMBER	INVOICE DATE	DATE SHIPPED

SHIP VIA:	SHIP METHOD:
XX	X

SHIP TO
Company Name
Street Address XXXX
City, State Zipcode

SOLD TO
Company Name
Street Address X
City, State Zipcode

ORDER QUANTITY	DESCRIPTION	UNIT PRICE	QUANTITY SHIPPED	BALANCE DUE	INVOICE AMOUNT
XX	XXXX	X	XX		XXXX
			XX		XXXX

FIGURE 2.21. Invoice errors concentration diagram.

occurred, the day of the week, and the operator, among others. These additional data help narrow the focus of the investigation into defect causes so that quicker solutions can be put in place.

2.5.3 Applications

This section has focused solely on part defects but concentration diagrams are also efficient tools for locating process "defects" such as equipment downtime or accidents. Figure 2.20 shows a concentration diagram for equipment downtime. A team chartered to improve process uptime would use this diagram to target the *material handling system* to work on first.

Concentration diagrams are also useful for manufacturing support groups and the service industry to locate paperwork "defects". A copy of the actual paperwork is used with an X marked in the space where the error occurred. Figure 2.21 shows a concentration diagram for invoice errors. Insurance companies also use this technique to assist in redesigning their myriad of forms. Eliminating paperwork defects will result both in less waste in correcting them and in improved service to their customers.

The concentration diagram is a basic, visual improvement technique. Because it is basic, many companies often overlook it in favor of more advanced SPC techniques to control defects. This is a mistake. Concentration diagrams are easy to construct and use. They should be employed first, and then followed by action on the product or process if they indicate a concentration of defects. Once the improvements have been made, then put the SPC techniques to work to control the defect level.

2.6 DATA GROUPING: CAUSE AND EFFECT DIAGRAMS

Data will not always be in the form of numbers. At times, they will be in the form of information on the process. Included in this information are the factors or parameters within the process that affect the process output. Like numerical data, these factors need to be organized to be put to use.

Factors could be simply organized into a list. However, a list may not be easily understood. The contribution a factor makes toward the process output or how one factor relates to another cannot be seen when factors are lumped together in a list. Organizing factors by listing them in groups would be an improvement, but this may not be enough either. How one factor relates to another still cannot be seen. The cause and effect diagram (CEDiagram) was developed not only to organize factors into groups, but also to further break them down into subgroups. The CEDiagram is a visual data grouping tool.

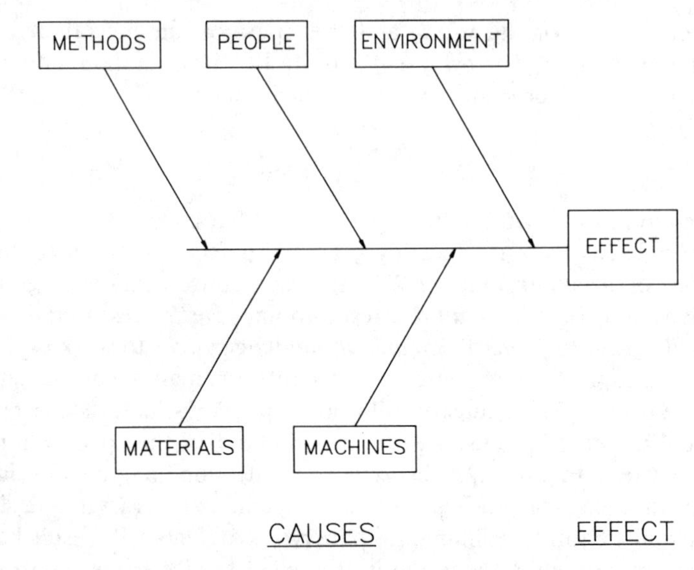

FIGURE 2.22. Cause and effect diagram.

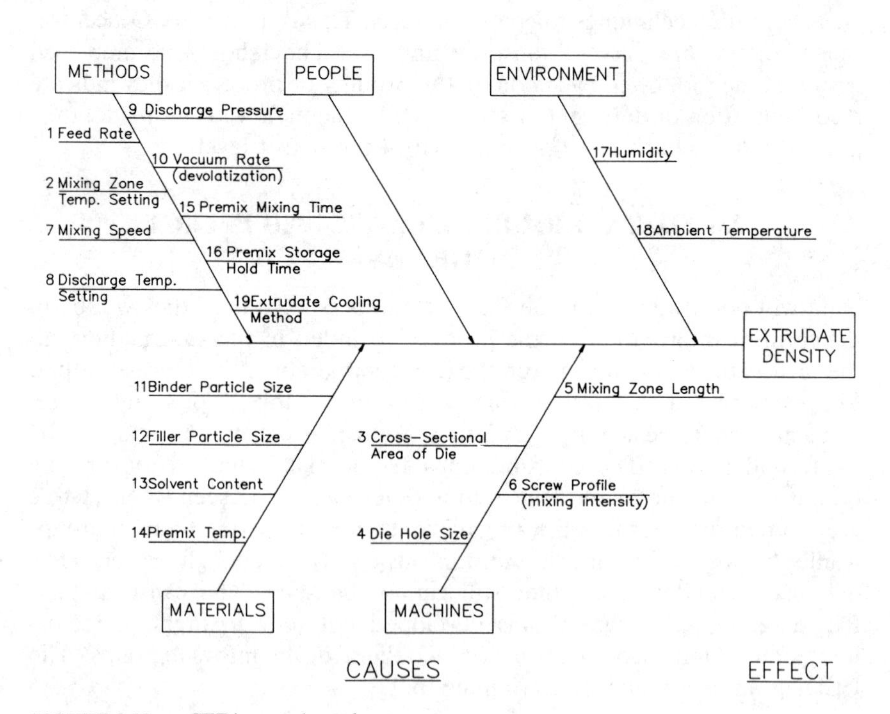

FIGURE 2.23. CEDiagram branches.

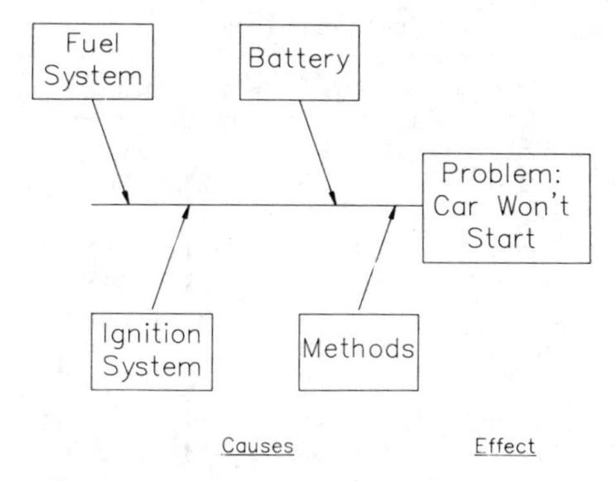

FIGURE 2.24. CEDiagram showing nonstandard categories. (*Reprinted from SPC In Action, copyright 1990, Resource Engineering, with the permission of the publisher, Quality Resources, White Plains, New York.*)

Dr. Kaoru Ishikawa developed the CEDiagram. Ishikawa's CEDiagram uses Shewhart's concept of a process (Figure 2.1) as the backbone for organizing the factors affecting a process output. The CEDiagram (Figure 2.22) lists the process output at the right. The factors affecting the process output are grouped on the backbone. Generally, the factors are grouped into one of five groups: manpower, materials, methods, equipment, and environment.

As factors are categorized into these five groups, it may be seen that some factors are really subsets of other factors. These factors are listed as branches off of the main factor (Figure 2.23). This branching helps identify factors that may interact; that is, factors that are dependent on one another.

The five groupings may not be appropriate for all CEDiagrams. The CEDiagram for a given process may not use all five of the groups; the factors may fit into only three or four. For some processes, it may be more appropriate to develop groups specific to the process (Figure 2.24). This will make it easier to group the factors affecting the output for that process and will make the CEDiagram easier to understand.

CEDiagrams are also used to group ideas from problem-solving brainstorming. Grouping ideas makes it easier to identify those most likely to solve the process problem. The technique for grouping brainstorming ideas using a CEDiagram are covered in Section 5.2.3, which also discusses the method for constructing CEDiagrams.

2.7 DATA ORGANIZATION

Once the process has been sampled and data collected, the data must be organized in a form that can be easily understood. If data are easy to

Table 2.1 Conventional Data Table

Daily Steam Usage for May (in pounds)		
1. 10,080	11. 11,240	21. 9,650
2. 17,140	12. 15,520	22. 14,780
3. 12,660	13. 9,210	23. 7,530
4. 8,630	14. 13,940	24. 11,110
5. 6,800	15. 11,990	25. 9,090
6. 14,090	16. 17,490	26. 16,830
7. 10,720	17. 18,220	27. 10,200
8. 8,140	18. 13,720	28. 6,790
9. 13,310	19. 10,300	29. 12,880
10. 13,400	20. 11,980	30. 15,650

understand, it will be easy to use that data to direct our actions on the process (although, the data may also tell us not to take action on the process).

Many engineers simply list their data in a table such as Table 2.1. Although all of the data are present, it is difficult to understand. Questions like "What's the most common value?" or "How much spread is there in the data?" can be answered, but only with a lot of hunting. If the data were presented pictorially (graphically) so that they could be seen, the questions posed above could be quickly answered.

This section presents three techniques for visually organizing data: histograms, Pareto diagrams, and scatter diagrams. Each of these is used for different applications. Histograms are used to show how often a particular data value or event occurs. They describe how the process output varies. Pareto diagrams show frequency of events or severity of impact of an event on a priority basis. This enables action to be taken on the most frequent or most severe events first. Scatter diagrams show if there is a relationship between two variables. They can tell if one variable is dependent on another. More details on each are described in the three following subsections.

2.7.1 Histograms

Histograms are a visual tool for presenting attribute and variable data. They organize the data to describe the process performance. Additionally, a histogram of variable data shows the amount and the pattern of the variation from the process. Histograms do not show how the process varies over time. They offer a snapshot in time of the process performance. Histograms are useful for understanding a process, for deciding on the level of need for improvement, and for improving the process.

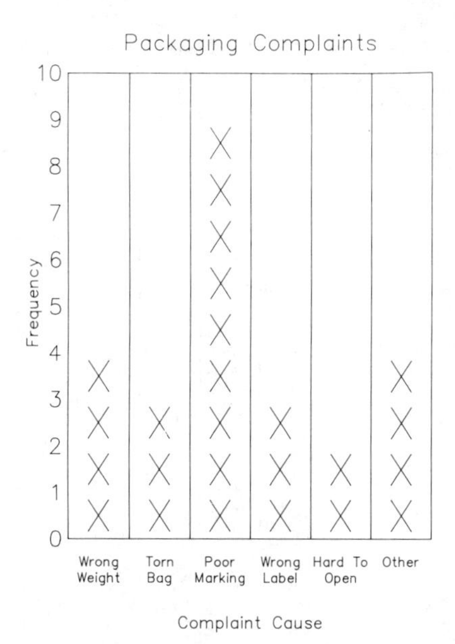

FIGURE 2.25. Packaging complaints histogram.

Basic histograms for attribute data (Figure 2.25) are simple to construct. The attribute types are listed on the *x* axis; the frequency of occurrence is the *y* axis label. For each occurrence of an attribute type, an X is drawn on the graph. In the data collected for Fig. 2.25, complaints from customers occurred eight times for incorrect bag weights. Care should be taken that each X is the same size. Otherwise, mistakes could be made in interpreting which attribute type occurred most frequently.

Modifications to histograms for attribute data can provide more information and aid in analysis. The X could be replaced with a 1, 2, or 3 to represent on which shift the attribute occurred (Figure 2.26). It could also be replaced with the product number (Figure 2.27) or date (Figure 2.28) or sequential numbers (Figure 2.29), each of which provides additional data on the process to help identify the causes of the attributes. The attributes that occur most frequently are the first ones to be targeted for improvement.

Histograms for variable data (Figure 2.30) are similar to those for attribute data, except that the events are replaced with measures. The measures are listed on the *x* axis and the frequency of occurrences on the *y* axis. The number of columns of measures depends on the number of data points. Ishikawa, in the *Guide to Quality Control* (1984), suggests the convention shown in Table 2.2. If there are more individual values than the number of columns, then the data must be grouped together for

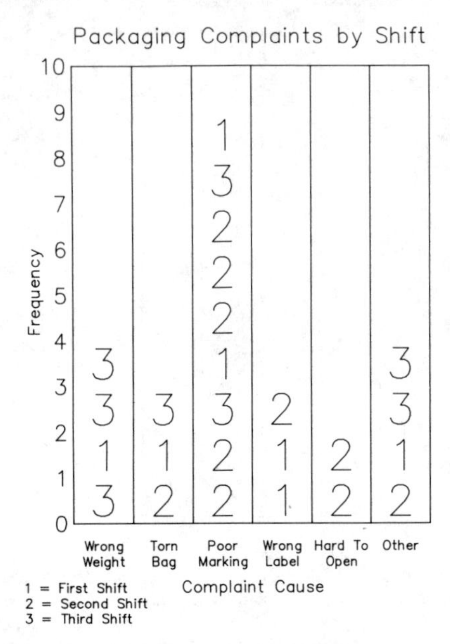

FIGURE 2.26. Packaging complaints showing shifts.

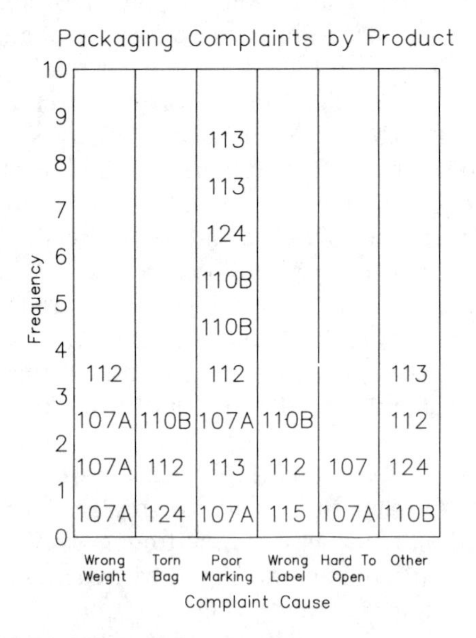

FIGURE 2.27. Packaging complaints using product number.

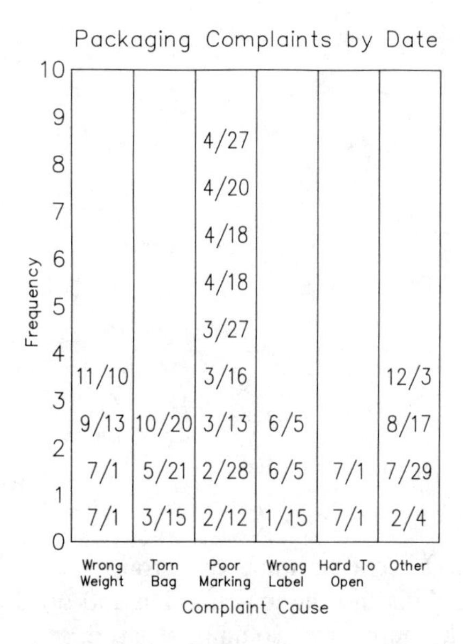

FIGURE 2.28. Packaging complaints using date.

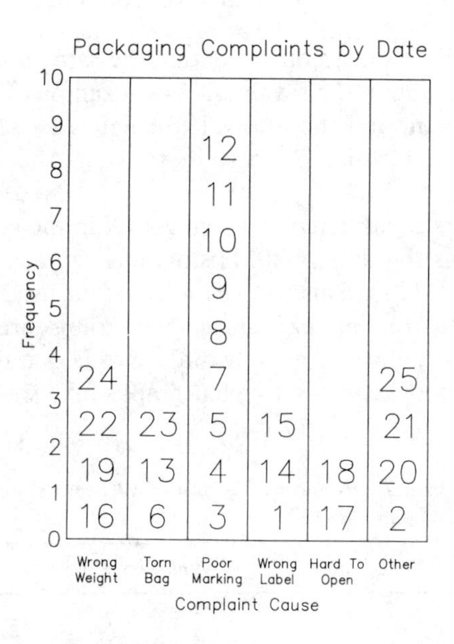

FIGURE 2.29. Packaging complaints in sequence.

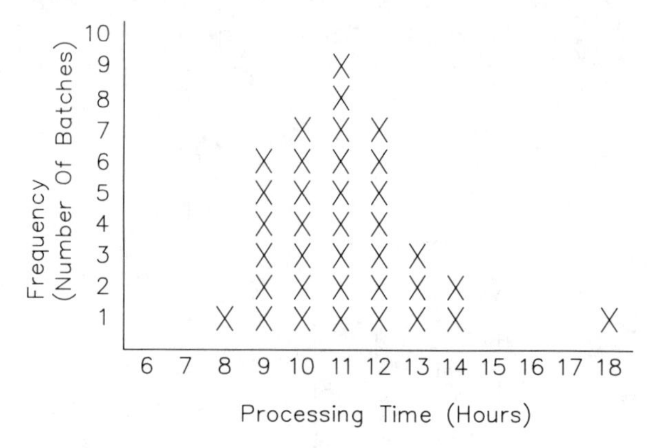

FIGURE 2.30. Histogram with variable data.

plotting (Figure 2.31). To determine the group size or interval, find the highest value in the data and the lowest value and divide the difference of these numbers by the number of columns desired:

$$\text{Interval} = \frac{\text{highest value} - \text{lowest value}}{\text{number of columns}}$$

The value should be rounded off to contain the same number of significant digits as the individual data values. For example, if the parts were measured to 0.001 in. and the interval calculated to be 0.0043 in., round the size down to 0.004 in. Be certain to keep the interval the same throughout the histogram.

The histogram is constructed by placing an X in the column whose value or interval contains the data point. Histograms can also be constructed as bar graphs (Figure 2.32). This form is most often used in report writing. The basics of constructing bar graph histograms are identical to the techniques described above. The only difference is that the Xs are replaced with bars. Some people constructing bar graphs take short cuts that should

Table 2.2 Number of Columns for Constructing Histograms

Data Points	Number of Columns
< 50	5–7
50–100	6–10
100–250	7–12
> 250	10–20

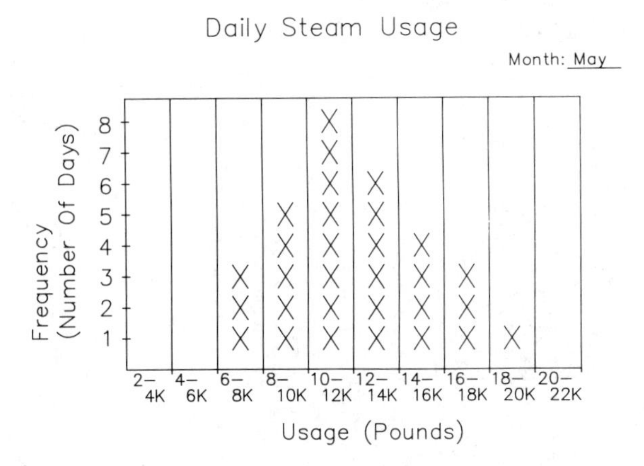

FIGURE 2.31. Grouping data into cells.

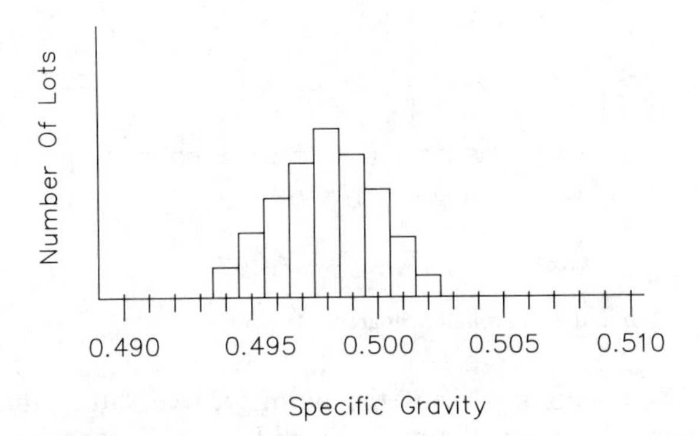

FIGURE 2.32. Histogram in bar graph form.

be avoided because they could lead to misinterpreting the data:

- Keep the interval size the same throughout each histogram.
- Include intervals even if no data fall within them.
- Keep the frequency scale continuous. Do not put breaks in.

Histograms are constructed to show the pattern of variation or the distribution of the process output. Remember in school how much ado was made about the bell curve? The bell curve showed the distribution of grades for students (Figure 2.33). It shows that a few students earned As, a few got Fs but most of the students earned Bs, Cs, and Ds. If a curve is drawn around the data on the histogram for specific gravity (Figure 2.34), it would look very similar to the bell curve for students' grades. In

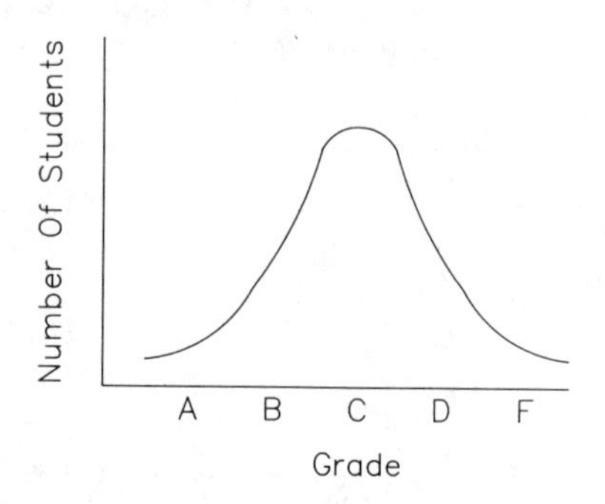

FIGURE 2.33. Bell curve of students' grades.

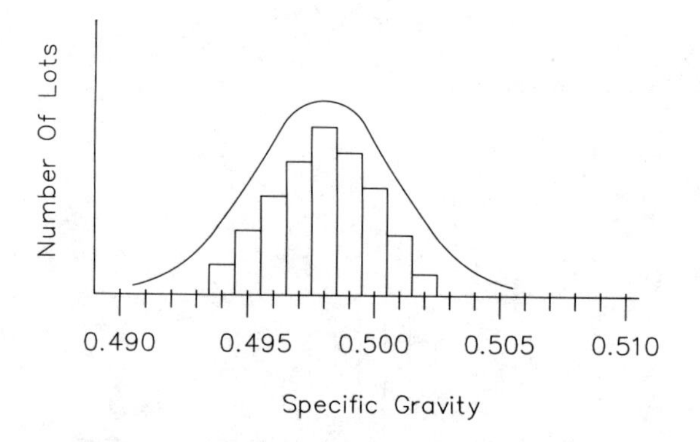

FIGURE 2.34. Curve drawn around histogram.

statistics, this curve is known as the normal curve rather than the bell curve. The normal curve is the basis for all basic SPC techniques.

The normal curve shows that we have a "normal" pattern of variation or rather, a normal distribution. The normal curve is always shaped the "same": It is a continuous curve, symmetrical around the center. That means there is an equal chance of a new data point falling above the center of the curve as there is of it falling below the center. There's also a greater chance the new data point will fall near the center of the curve. If the specific gravity was sampled again, there is a greater chance of the composition falling between 0.496 to 0.500 than there is of it falling below 0.493.

Many distributions of data for the process industries are not normal. Skewed distributions (Figure 2.35) can be caused if there is an upper or lower limit for the data. One example would be if the histogram were plotted for length of downtime for a process. Downtime has a lower limit because it can never be below zero. But if there were a serious problem, it could be down to a long period of time.

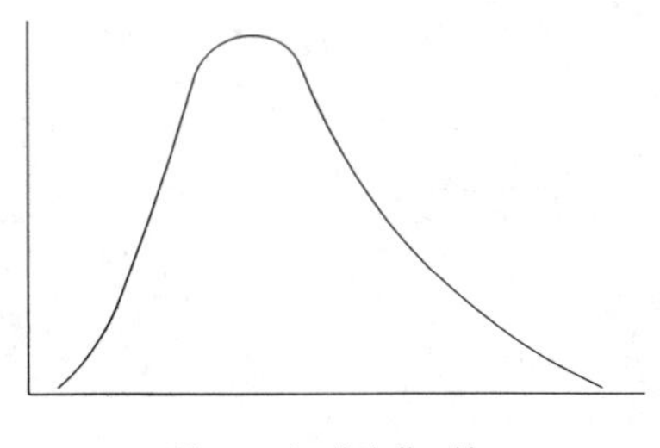

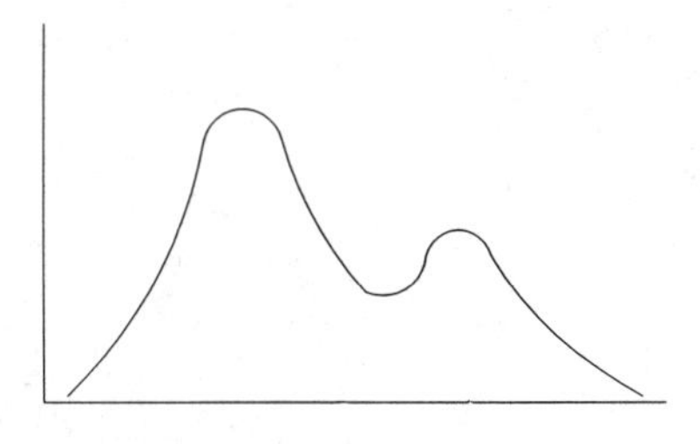

FIGURE 2.35. Skewed and bimodal distribution.

Another pattern of variation seen is the bimodal distribution (Figure 2.35). This may be caused by a shift in the process output during the time period the data were collected to the histogram. It could also be caused by measuring the product from two different processes making the same thing.

Because the normal curve is the basis for most standard SPC techniques, we'll have a problem using those tools if the histogram shows the pattern of variation to be nonnormal. Processes showing nonnormal distributions must be investigated for the causes of the variation. For some nonnormal patterns, such as the bimodal distribution, problems in the process or in the sampling techniques will be found as the cause. However, skewed distributions may not have a special cause for the nonnormal variation. There are some techniques discussed in Chapter 4 that will allow SPC techniques to be used even if the distribution is skewed.

2.7.2 The Pareto Principle

In the late 19th century, an Italian, Velfredo Pareto, studied the distribution of wealth and found that this distribution was not equal across the populace. He found that the majority of wealth was concentrated in relatively few hands. Dr. J. M. Juran recognized that this unequal distribution applied to many fields. For example, most of our home expenses are from only a few of our expense items (rent or mortgage payment, taxes, food, and auto costs). In most businesses, a few of the customers account for a majority of the sales dollars. Most of the problems with a process come from a few sources, and only a few sources of variation create most of the total variation. Dr. Juran termed these few sources the "vital few" and all of the other sources the "trivial many". He called this phenomenon the Pareto principle.

The application of the Pareto principle is easy. If we can organize our data to identify the major sources of variation or our major problem areas, we can concentrate our efforts on those vital few first and leave the trivial

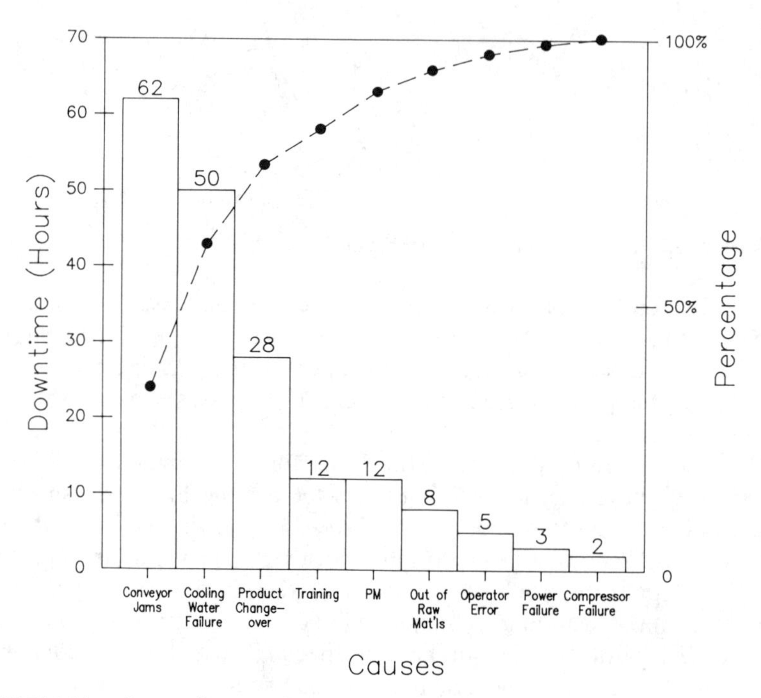

FIGURE 2.36. Pareto diagram of process downtime.

many for later. This ensures we get maximum returns on our efforts. Working on processes is just like working with customers: A $30 million company should be spending much more time on one or two $5 million accounts than on each of its one-hundred $10,000 accounts. Manufacturing and engineering should be expending more effort on the vital few sources of variation, not the trivial many.

To help separate the vital few from the trivial many, the Pareto principle can be applied graphically in a bar chart called a Pareto diagram. The Pareto diagram shows components of a total population ranked in descending order of importance. Figure 2.36 shows a Pareto diagram for downtime of a process. The causes of the downtime are graphed with the cause that created the most downtime (conveyor belt jams) first, and the cause of the second next largest amount of downtime (cooling water failures) second, and so on. We can look at the Pareto diagram and see clearly the vital few causes to be concentrated on. Between them, the conveyor belt jams and the cooling water failures account for 61% of the process downtime. Resources should be concentrated on them.

The Pareto diagram helps to separate fact from opinion or emotion. Power failures may seem to be a major problem to a manufacturing engineer because of the high activity level needed to bring a process back on-line. Without a Pareto diagram to show where the effort should be applied, a great deal of time and money may be spent to install a backup generator. But in our example, that investment would only reduce the downtime by less than 3%.

To construct a Pareto diagram:

1. Define the problem, cost, feature, or attribute to be studied. In Figure 2.36, it was process downtime. (It could have been yield losses by product, data entry errors per department, or accidents by plant area, among many things.)
2. List the categories that contribute to the problem. For process downtime, these categories are power failures, product changeover, conveyor jams, cooling water failure, preventive maintenance, out of raw materials, training, compressor failure, and operator error.
3. Make a table of the categories and their contribution to the problem. List them in descending order from highest to lowest.

Category	Hours of Downtime
Conveyor jams	62
Cooling water failure	50
Product changeover	28
Training	12
Preventive maintenance	12
Out of raw materials	8
Operator error	5
Power failure	4
Compressor failure	2
Total	183

4. Calculate the frequency of each category as a percent of the total.

Conveyor jams	62/183 × 100% =	34%
Cooling water failure	50/183 × 100% =	27%
Product changeover	28/183 × 100% =	15%
Training	12/183 × 100% =	7%
Preventive maintenance	12/183 × 100% =	7%
Out of raw materials	8/183 × 100% =	4%
Operator error	5/183 × 100% =	3%
Power failure	4/183 × 100% =	2%
Compressor failure	2/183 × 100% =	1%
		100%

5. Calculate the cumulative frequency percent. For each category, this is

the summation of its frequency percentage and those above it.

Conveyor jams	34% = 34%
Cooling water failure	27% + 34% = 61%
Product changeover	15% + 61% = 76%
Training	7% + 76% = 83%
Preventive maintenance	7% + 83% = 90%
Out of raw materials	4% + 90% = 94%
Operator error	3% + 94% = 97%
Power failure	2% + 97% = 99%
Compressor failure	1% + 99% = 100%

6. Construct a combination bar-line chart with the data above. The categories are listed on the x axis. The hours of downtime is the left y-axis label and cumulative frequency is the right y-axis label. The downtime hours are presented in bar graph form and the cumulative frequency data in line graph form.

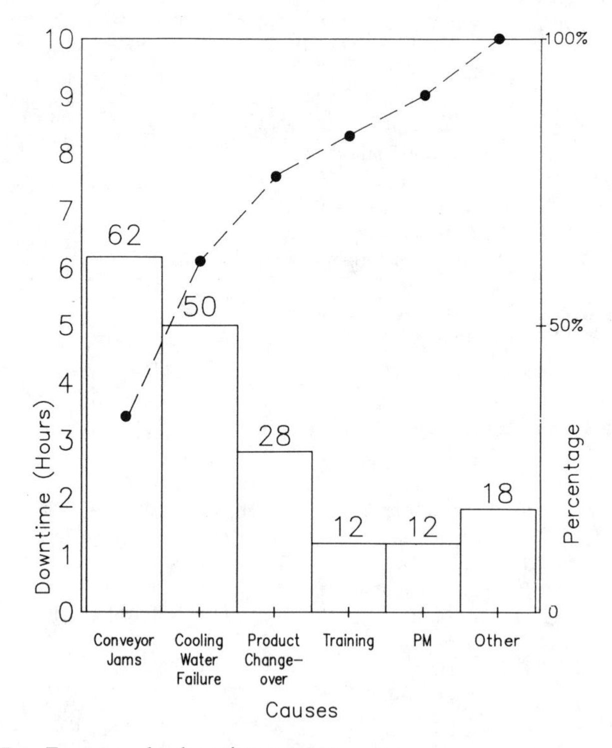

FIGURE 2.37. Frequency by downtime category.

In this example, there are nine categories, but five of them contribute to 90% of the downtime. These are the "vital few". The other four account for only 10% of the total downtime. These will be worked on after we have reduced the downtime due to the vital few.

In many cases, there will be more than nine categories. If this occurs, the "trivial many" will often be lumped together into an "other" category. Figure 2.37 shows the process downtime organized this way.

2.7.3 Scatter Diagrams

A scatter diagram organizes data from two variables to show if there is a relationship between them. It is actually just a simple *x-y* graph. A scatter diagram normally looks at the relationship between an independent variable and a dependent variable. The independent axis would be plotted on the *x* axis and the dependent variable on the *y* axis. There may also be cases where a scatter diagram is plotted for two independent variables, in order to look at their relationship, or between two dependent variables. Some texts refer to independent variables as causes and dependent variables as effects.

Scatter diagrams are useful for:

- Showing how a dependent variable responds to change in an independent variable; for example, how a blower output (cubic feet per minute) changes with a change in the speed of the motor.
- Determining if one variable is dependent on another.
- Determining if two dependent variables respond similarly to a change in an independent variable. (This would be useful to know if one of the dependent variables is very difficult to measure or to reduce the number of measurements being taken in a process.)
- Predicting the response of a dependent variable to a setting of the independent variable that has not been measured before.

Figure 2.38 shows a scatter diagram for viscosity versus agitator speed for a solution. From this diagram, it can be seen that there is a relationship between the viscosity and the agitator speed: As the speed increases, the viscosity increases.

To construct a scatter diagram:

1. *Develop a data collection form.* Set up the form to collect the data so that it's easy to plot. For example, organize the form with columns to record the independent variable data (x) on the left and the dependent variable (y) on the right.

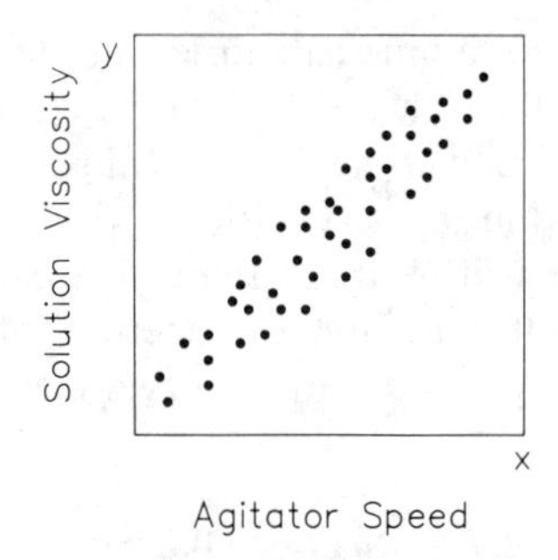

FIGURE 2.38. Scatter diagram of solution viscosity versus agitator speed.

2. *Collect the data.* For best results, collect 30–100 sets of data. The data should run through the operating range of the independent variable. If it is not known whether one variable is independent from another, take data over the entire operating ranges for both variables. Because variation in the process is expected, take multiple readings at each value selected from the operating range. These readings should be made randomly.
3. *Prepare the graph.* Draw the x and y axes on a piece of graph paper and label them. Determine the spread of the data for each variable. The scales of the axes should be set up to cover the entire spread of the data.

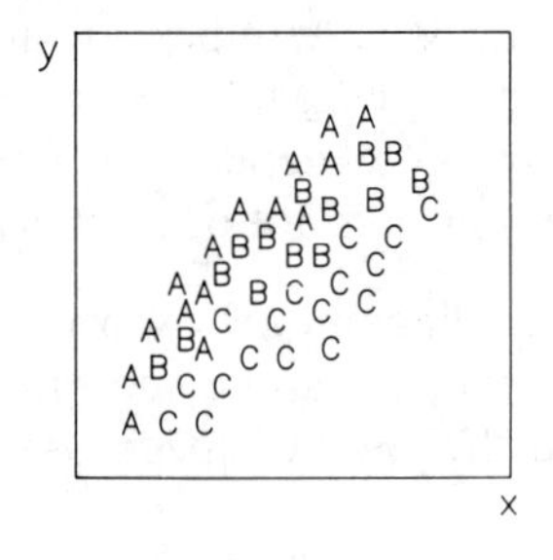

FIGURE 2.39. Scatter diagram indicating supplier of material.

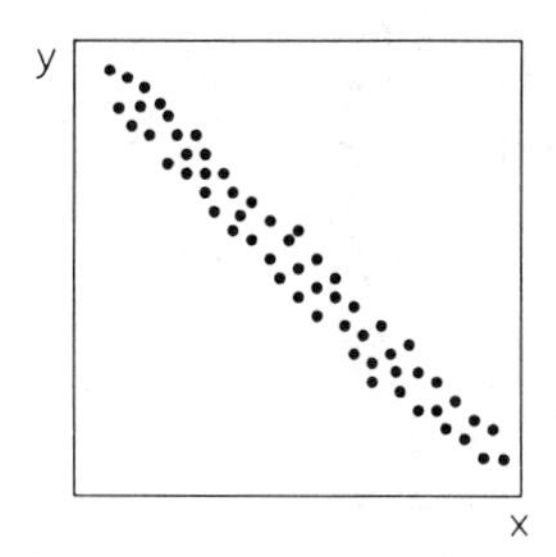

FIGURE 2.40. Scatter diagram with a linear correlation.

4. *Plot the points.* Use a dot to mark the point unless there is the desire to segregate the data. For instance, if there are two different suppliers of a key raw material, data collected using material from supplier 1 can be marked by a circle and from supplier 2 by another symbol. Figure 2.39 shows a scatter diagram using this approach.

If two variables show a relationship, they are said to be correlated. A correlation exists between them. This correlation may be linear or nonlinear. Figure 2.40 shows a linear correlation, which is also a negative correlation because the dependent variable decreases as the independent variable increases. Figure 2.38 showed a positive, linear correlation. If there is no correlation between the two variables, the data points should be scattered on the graph (Figure 2.41). Figure 2.42 shows a nonlinear correlation between the two variables.

The scatter diagram visually shows the relationship between two variables. This relationship can also be described mathematically. Section 5.4.7 on correlation covers the mathematics more thoroughly. Some relationships are clear enough that a line describing the relationship can be drawn on the scatter diagram (Figure 2.43); others may not be so clear

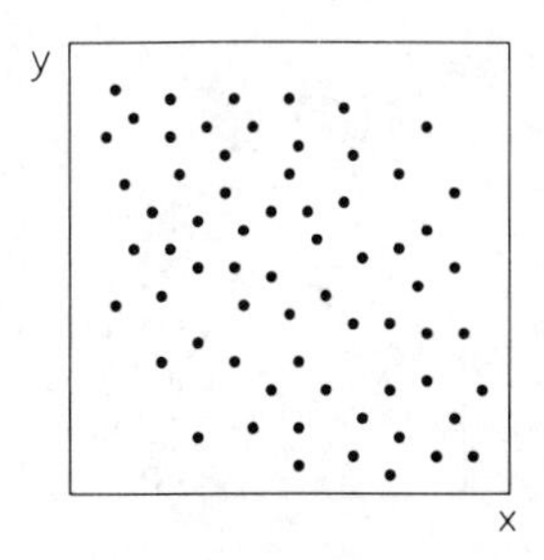

FIGURE 2.41. Scatter diagram with no correlation.

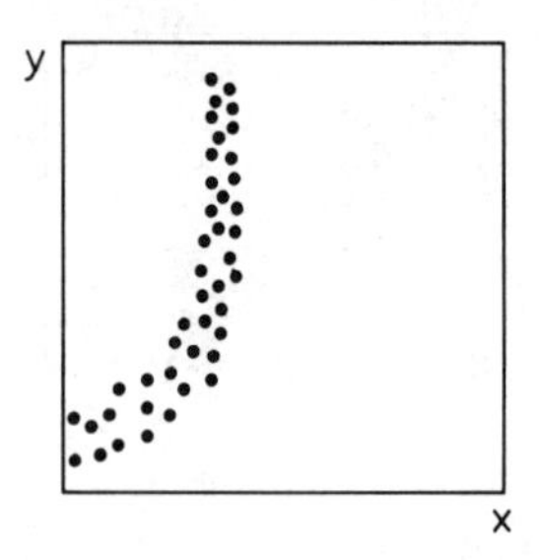

FIGURE 2.42. Scatter diagram with nonlinear correlation.

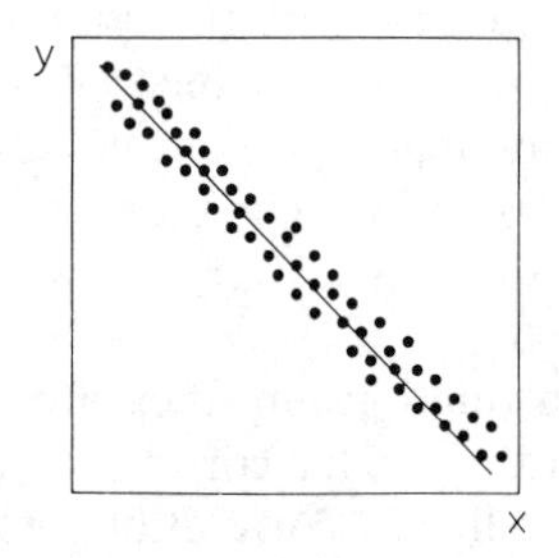

FIGURE 2.43. Scatter diagram showing the line defining the relationship.

(Figure 2.44). If it is not clear, we can use mathematics to try to describe the relationship rather than a scatter diagram. A calculator or computer can be used to define a linear or nonlinear equation that fits the data. It will also describe how well the equation describes the correlation. Mathematics can show whether or not a relationship exists if the simple and fast pictorial method of scatter diagrams cannot.

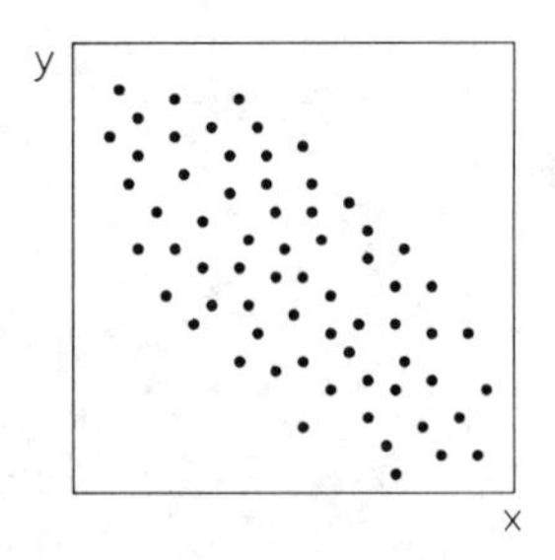

FIGURE 2.44. With limited correlation, the relationship is difficult to define exactly.

2.8 DATA ANALYSIS—MEASURING VARIATION

A histogram visually shows the distribution of data taken from a population or a sample. Although this information is useful, it is often necessary to mathematically describe the distribution. The math can describe both the location of the process and the amount of variation in the process. This is analogous to having a map of an area to look at: The map shows the distribution of the area. However, it does not provide enough information to describe where that area is located on the globe or how much of the globe is taken up by the area. Both the location of the area and its spread on the globe can be described mathematically using latitude and longitude coordinates. In statistics, we use values of the mean, median, or mode to describe the distribution location and the range or standard deviation to describe the spread of the distribution.

2.8.1 The Mean

Normally, the mean is used to describe the location of the distribution of a population or a sample. The mean is the arithmetic average of the data. It is calculated by summing the data values and dividing this total by the number of data values:

$$\text{Mean} = \frac{\Sigma X_i}{n}$$

where X_i = data value for data points 1 to n
n = number of data points

The mean of the set of numbers {11.4, 12.8, 11.9, 10.6, 11.3} is 11.58. The sum of the numbers is 57.9 (11.4 + 12.8 + 11.9 + 10.6 + 11.3) and there are five data points. Dividing 5 into 57.9 gives 11.58. Notice that the values in the set of numbers have three significant digits whereas the mean has four. Normal rounding convention is that the mean has one more significant digit than the individual values of the data. The mean of the set of numbers {120, 130, 120, 140, 110} is 124. Each value in the set of numbers has two significant digits so here, the mean has three.

The Greek symbol μ represents the mean value of a population. The value for μ is rarely known because of the difficulty in measuring all parts of a population. In most process industries, it is impossible to measure all of the output from a process to calculate μ. Imagine trying to capture μ

for the particle size of a powdered detergent; there is no practical way to run sieve analyses on all of the product. As described in Section 2.3 on sampling, samples are measured to predict the measurements of populations. The sample mean, $\bar{x}$ (x-bar) is used to estimate the population mean μ. As the sample size increases, $\bar{x}$ becomes a better estimate of μ.

2.8.2 The Median

The median is used infrequently in describing the location of a distribution. The median, represented by $\tilde{x}$, is the middle value in a set of numbers when that set is arranged from lowest to highest values. The median of the set of numbers {125, 100, 90, 150, 200} is the middle value, 125. If there is an even number of values in a set, the median is the average of the two middle values. The median for the set of numbers {95, 105, 102, 55, 72, 85} is 90. When the set is arranged from lowest to highest, {55, 72, 85, 95, 102, 105}, the two middle values are 85 and 95. Their average is 90.

The median is not as good a measure of location as the mean. However, some companies use it because there are no calculations involved, and they accept that it is not quite as good. The median is a favorite tool of politicians. They often use the median to describe personal income in an area. They use a statement such as the median annual income for a family of four in Chicago is $28,000. This means that half the people make more than $28,000 and half the people make less. (The mean annual income would be different than $28,000, and would be affected by the number of millionaires in Chicago.)

2.8.3 The Mode

The mode is simply the value that appears most frequently. There is no mode in any of the sets of numbers we used before in this section because no value appears more than once in each set. The mode for the set of numbers {6.5, 7.0, 5.5, 8.5, 7.0, 6.0, 6.0, 6.5, 7.0, 7.0, 5.5} is 7.0 because that value occurs most often. It occurs four times whereas no other number appears more than twice.

The mode should only be used to describe the location of a severely skewed distribution. It cannot be used with small sample sizes because of the likelihood that each value in the sample would only occur once. The mode is the same as the mean and the median for a normal distribution. For other distributions, all three can differ. For example, the mode of the set of numbers above is 7.0, the median is 6.5, and the mean is 6.59.

2.8.4 The Range

The range is a simple measure of variation, and it is denoted by the symbol R. Given a set of numbers:

$$R = \text{highest value in the set} - \text{lowest value}$$

For the set of numbers {0.007 in., 0.004 in., 0.005 in., 0.002 in., 0.004 in.}, the range is 0.005 in. because 0.007 in. is the highest value and 0.002 in. is the lowest.

2.8.5 Standard Deviation

The range describes only the absolute spread of the data. It tells nothing about how much the data values vary from the mean. The standard deviation is a more useful measure of variance because it is based on the difference of the data values from their mean. The symbol σ denotes the standard deviation for a population and s denotes the standard deviation of a sample:

$$\sigma = \sqrt{\frac{\Sigma(x - \mu)^2}{N}}$$

where x = data values
μ = population mean
N = number of pieces in the population

$$s = \sqrt{(\Sigma(x - \bar{x})^2)/(n - 1)}$$

where $\bar{x}$ = sample mean
n = sample size

Like the population mean μ, the population standard deviation σ can rarely be determined. The sample standard deviation s is used to estimate σ. Having the term $(n - 1)$ rather than N in the denominator of the calculation for s compensates for the bias of using $\bar{x}$ rather than μ in the calculation. The larger the sample size, the better s estimates σ. As n approaches N, $\bar{x}$ will approach μ and s will approach σ.

The standard deviation for the following set of measures taken from samples {0.297, 0.299, 0.296, 0.297, 0.296} is 0.0012. This is calculated in the following manner:

$$\bar{x} = \frac{\Sigma x}{n}$$

$$\bar{x} = \frac{0.297 + 0.299 + 0.296 + 0.297 + 0.296}{5}$$

$$\bar{x} = 0.2970$$

$x - \bar{x}$	$(x - \bar{x})^2$
0.297 − 0.2970 = 0	0
0.299 − 0.2970 = 0.002	4×10^{-6}
0.296 − 0.2970 = 0.001	1×10^{-6}
0.297 − 0.2970 = 0	0
0.297 − 0.2970 = 0.001	1×10^{-6}
Sum	6×10^{-6}

$$s = \sqrt{\frac{\Sigma(x - \bar{x})^2}{n - 1}}$$

$$s = \sqrt{\frac{6 \times 10^6}{5 - 1}}$$

$$s = 0.0012$$

Like the mean, the standard deviation is also normally calculated to one significant digit greater than the data values. Although the preceding example shows the standard deviation calculated by hand, this is rarely done today. Texas Instruments, Hewlett-Packard, and others make low-cost, hand-held calculators that can calculate the mean s and σ. They can be calculated by simply entering the data points and then pressing the correct keys. The small investment in a scientific calculator will have an almost immediate payback in time savings when calculating standard deviations.

The standard deviation is most useful when the distribution of the process output is normal. The combination of the mean μ and the standard deviation σ totally describes mathematically a normally distributed population. We use $\bar{x}$ to estimate μ and s to estimate σ. The $\bar{x}$ describes the location and the value of s describes how much the output varies from $\bar{x}$. For a normally distributed process output, 99.7% of the output lies within a distance of $\pm 3s$ from the process mean $\bar{x}$. Within

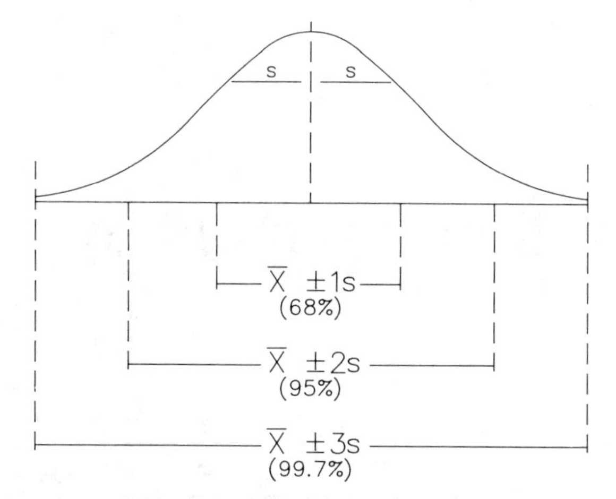

FIGURE 2.45. Normal curve distribution.

$\pm 2s$ of $\bar{x}$ lies 95.4% of the distribution and 68.3% lies within $\pm 1s$. Figure 2.45 shows this. This concept is the basis for process control charts, which will be discussed in Chapter 4. These control charts will graphically look at the location and the variation of the process output. Control charts are used to identify the appearance of special causes so that they can be eliminated.

Before we get into control charts and SPC techniques, we will look at the measurement system in Chapter 3. If we don't have a good measurement system, we can never be sure that our data are correct. Incorrect data send false signals on the location and variation of the process output. This may lead to the product being out of specification or to adjustments made to the process that are not only not necessary, but cause more variation.

3

The Measurement System

3.1 THE MEASUREMENT SYSTEM AS A PROCESS

When a process output, or product, is tested against a specification, the test itself is rarely thought about. Usually the focus is on the product test data. These test data are compared to the specification and conclusions are drawn. But, unless the test process (the measurement process) is studied and understood, there is no way to know if the product test result is a true representation of the product. The test may be flawed and may be giving an erroneous measure of the product.

The test may indicate the product is in specification when in fact the product is out of specification. In this case, the test gives a false sense of security by showing the product is within specification when it is not. Testing schemes cannot be blindly trusted. They must be used to test the products off of a process to show how the process is running. The test or measurement system must be validated to assure that it can be trusted to tell the truth about the product and process. If this doesn't occur, a flawed measurement system can lead to a costly wild goose chase as work is done to stabilize, control, and improve the process.

To understand and validate the measurement system, it must be recognized that it too is a process. Figure 3.1 shows Shewhart's concept of a process as it relates to the measurement system. Measurement variation or error can come from:

- The test equipment (machines)
- the test method (methods)
- the test sample (materials)
- the sample preparation or conditioning (environment)
- the tester (people)

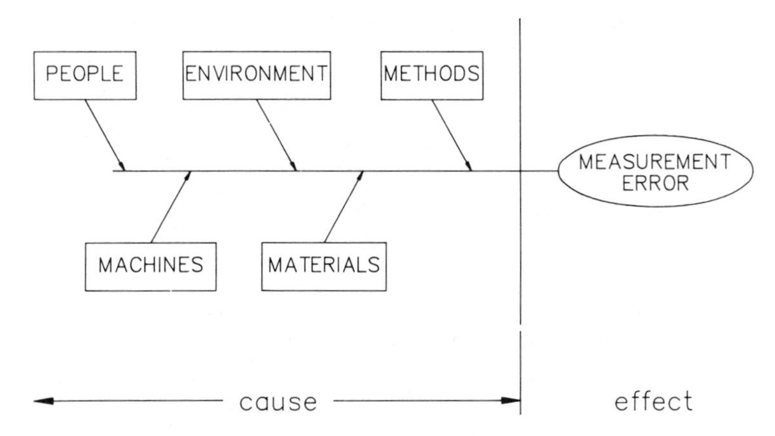

FIGURE 3.1. Shewhart's concept of a measuring system.

Measurement error can come from any one or all of these five elements of the measurement process.

If the measurement process is relatively simple (e.g., a linear measurement of a flat piece of wood), measurement error will usually be small. However, some measurement systems, such as a detailed quantitative chemical analysis or a test of the rheology of a plastic, are complicated processes. The measurement error may be so large that it clouds the view of the process and its output.

The error, or variation, in the measurement system can be greater than the total tolerance of a product specification! Take for example a product with a specification of 70 ± 5 units. If the variation (or error) in the measurement system used to test the product is 15 units, can it ever be known if the product is in specification? No, of course not. The measurement system can't distinguish between 60 and 75! Until this measurement system is studied as a process, and improved, manufacture of the product is risky. The manufacturer will be looking through a thick fog (the incapable measurement system) to aim the process at a target (see Figure 3.2); essentially shooting at the target, but never really know if it has been hit or not.

Some of the concepts discussed in this chapter on the measurement system cannot be fully understood until the reader has completed Chapter 4, Statistical Process Control (SPC). However, the "measurement system" has been placed before the SPC section to emphasize the need to study the measurement system as a process, to understand the measurement error or variation, and to relate it to the needs of the process before undertaking an effort to apply SPC to the process.

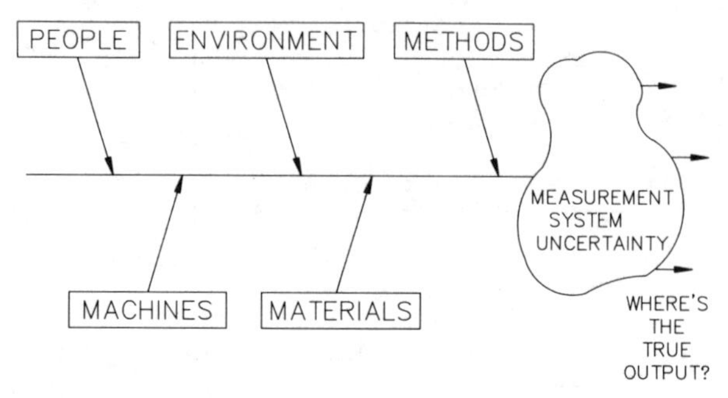

FIGURE 3.2. Incapable measurement system.

3.2 UNITS OF MEASURE—ARE THEY ADEQUATE?

Although the bulk of this chapter deals with the different types of measurement error and methods to measure them, there is one fundamental aspect of measurement systems that is often neglected: the adequacy of the units of measure. Many times, a measurement device that is chosen is not sensitive enough to give a true picture of product or process variation. Other times, a measuring device may be adequate enough but the sensitivity is reduced by the way the results are interpreted. Would a ruler be the tool of choice to measure the tight tolerances involved in machining engine parts? Would a grocery scale, used to measure the weight of vegetables in pounds, be suitable for weighing grams of raw material for pharmaceuticals. Of course not. The units are inappropriate and the sensitivity of the equipment is not adequate. These are obvious examples. Most occurrences are not as obvious and, therefore, slip by.

The real problems of using incorrect units and identifying that the units are incorrect can be shown in an example.

Example A company that produces baked goods uses a water content test as an in-process check to determine if the cookies are cooked. The machine automatically heats a cookie (placed in the unit by an operator) at a given temperature for a given time and records a percent weight loss, which is attributed to the water evaporating off. The machine, which the manufacturer claims is very accurate, gives results out to the second decimal place. The operators chart the results on a control chart (control charts will be discussed in Chapter 4) and if the chart has indicated a problem with the process, they make the necessary adjustments to the process. In order to make it easier for the operators (and because of limited space on the charts themselves), they are instructed to round up or

Table 3.1 Data for Cookie Weight Loss with Operators Weighing

Subgroup						$\bar{x}$	R
1	3.4	3.4	3.4	3.4	3.4	3.40	0.0
2	3.4	4.3	3.3	3.4	3.3	3.36	0.1
3	3.4	3.4	3.4	3.4	3.4	3.40	0.0
4	3.4	3.5	3.4	3.4	3.4	3.42	0.1
5	3.3	3.3	3.4	3.4	3.4	3.36	0.1
6	3.4	3.4	3.4	3.4	3.4	3.40	0.0
7	3.4	3.4	3.4	3.4	3.4	3.40	0.0
8	3.4	3.4	3.4	3.4	3.4	3.40	0.0
9	3.4	3.5	3.4	3.4	3.4	3.42	0.1
10	3.4	3.4	3.4	3.4	3.4	3.40	0.0
11	3.4	3.3	3.5	3.5	3.5	3.44	0.2
12	3.4	3.4	3.4	3.4	3.4	3.40	0.0
13	3.4	3.4	3.4	3.4	3.4	3.40	0.0
14	3.4	3.4	3.4	3.4	3.4	3.40	0.0
15	3.4	3.4	3.4	3.4	3.4	3.40	0.0
16	3.4	3.4	3.4	3.4	3.4	3.40	0.0
17	3.4	3.4	3.4	3.4	3.5	3.42	0.1
18	3.4	3.4	3.4	3.4	3.4	3.40	0.0
19	3.4	3.4	3.4	3.4	3.4	3.40	0.0
20	3.4	3.4	3.4	3.4	3.4	3.40	0.0
	$\bar{\bar{x}} = 3.401$		$s = 0.0177$		$\bar{R} = 0.035$		

down (depending on the last digit) and record only out to one decimal place.

The operators check five cookies every 30 min and record the average and range on an $\bar{x}$ and R control chart. The data for one day of running are shown in Table 3.1.

Figure 3.3 shows the $\bar{x}$ and R charts with the data plotted. The circled data points indicate problems with the process and the product that it is making. This shows that the process should be adjusted.

Shown in Table 3.2 are the actual data from the test equipment without the numbers being rounded off.

When these data are put into a control chart (Figure 3.4), it indicates that the process has no problems, so it does not require any adjustments.

Due to the fact that the values were rounded (the units were made smaller) by the operators, there were some instances during the day that would have prompted them to adjust the process. When the data from the measurement equipment are analyzed, it can be seen that the process requires no adjustments. The fact that the units were too small after rounding caused overadjustment of the process.

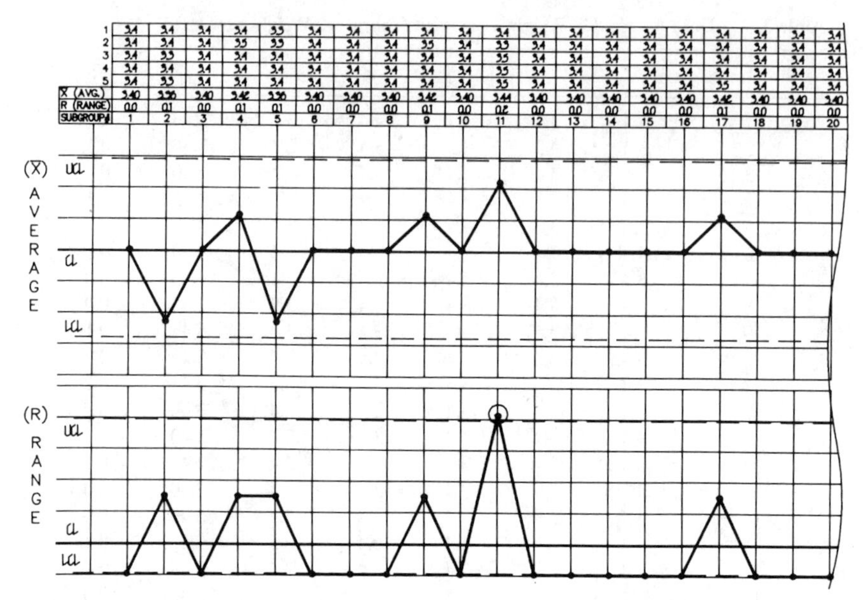

FIGURE 3.3. Operator data—charted.

Table 3.2 Data for Cookie Weight Loss with Automated Test Equipment

Subgroup						$\bar{x}$	R
1	3.45	3.45	3.37	3.38	3.40	3.410	0.08
2	3.42	3.37	3.34	3.40	3.32	3.370	0.10
3	3.44	3.42	3.43	3.35	3.44	3.416	0.09
4	3.42	3.47	3.37	3.42	3.38	3.412	0.10
5	3.33	3.32	3.44	3.45	3.41	3.390	0.13
6	3.40	3.43	3.37	3.34	3.35	3.378	0.09
7	3.42	3.42	3.43	3.40	3.33	3.400	0.10
8	3.42	3.42	3.39	3.41	3.42	3.412	0.03
9	3.37	3.47	3.42	3.37	3.35	3.396	0.12
10	3.37	3.37	3.42	3.43	3.41	3.400	0.06
11	3.39	3.33	3.47	3.48	3.49	3.432	0.16
12	3.43	3.37	3.45	3.37	3.38	3.400	0.08
13	3.36	3.42	3.40	3.39	3.37	3.388	0.06
14	3.40	3.45	3.42	3.39	3.37	3.406	0.08
15	3.37	3.42	3.42	3.45	3.43	3.422	0.08
16	3.38	3.45	3.41	3.37	3.41	3.404	0.08
17	3.38	3.43	3.43	3.45	3.46	3.430	0.08
18	3.37	3.45	3.44	3.37	3.40	3.406	0.08
19	3.42	3.44	3.40	3.38	3.43	3.414	0.06
20	3.40	3.45	3.43	3.44	3.38	3.420	0.07

$\bar{\bar{x}} = 3.4053$ $s = 0.0159$ $\bar{R} = 0.0865$

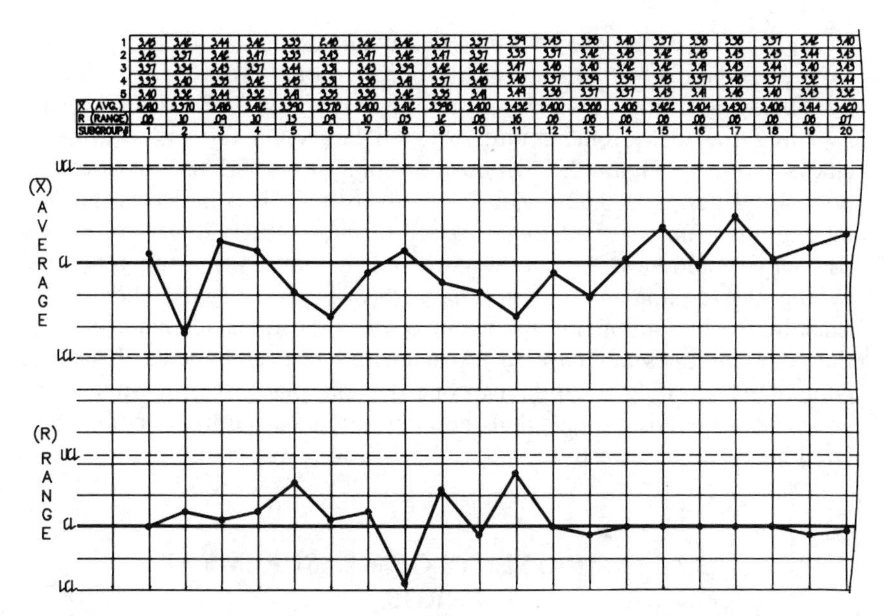

FIGURE 3.4. Test equipment data—charted.

How can a decision be made as to whether the units being used to measure a characteristic are incorrect or inadequate? Usually, this can be done simply by looking at the range (R) chart. One sign of possible problems in the magnitude of the measurement units is a situation where there are many range values equal to zero. Figure 3.3 shows that the data as rounded by the operators had many zero readings on the range chart. On the charts for the actual equipment readings, there were no range values of zero. Having many range values equal to zero indicates that the measurement system is unable to identify differences from sample to sample.

Another method for determining whether or not the units of the measurement system are inadequate is described by Wheeler and Lyday in *Evaluating the Measurement Process* (1984). Again, the range charts are used and not the $\bar{x}$ charts. To use their technique, count the number of possible values within the limits on the range chart. If there are only five values possible, then the units are borderline. If there are only four values, the units are too large and the control limits will be distorted. (Wheeler and Lyday also say that if your subgroup size is $n = 2$, then having only four possible values is the borderline case and only three values is a sign that the units are too large.)

This can again be shown from the cookie example. For the operator (rounded) data shown in Figure 3.3, there are only two possible values, 0 and 0.1, lying within the limits on the range chart (0, 0.194). This is a sure sign that the measurement units are too large. On the test equipment charts shown in Figure 3.4, within the limits (0 and 0.1705) there are 17 possible values: 0.01, 0.02, 0.03, . . . , 0.15, 0.16, 0.17. This indicates that the measurement units are adequate. Based on these observations, the cookie company should have the operators record the data as they are generated by the test equipment and not have the operators round the results. Luckily, the measurement equipment in place provides adequate units so the company does not have to invest in new test equipment. If the test equipment had only given data to one decimal place, then it would have given the same false signals that the operators incurred by rounding. New equipment would have to be ordered if this were the case.

3.3 COMPONENTS OF MEASUREMENT ERROR

A measurement system, like any other system, should be treated as a process with many discrete components. Each individual component has its own level of variation. The sum of the variation of the components will give the variation of the process. A measurement process can be described using Shewhart's concept of a process to better understand the sources of measurement error (Figure 3.5).

The traditional process as outlined by Shewhart is shown in Figure 3.5. The process output must be translated through the measurement system before the actual output is known. Therefore, it is important to know the variation due to the measurement system. The measurement system can also be looked at as a process as in Figure 3.6.

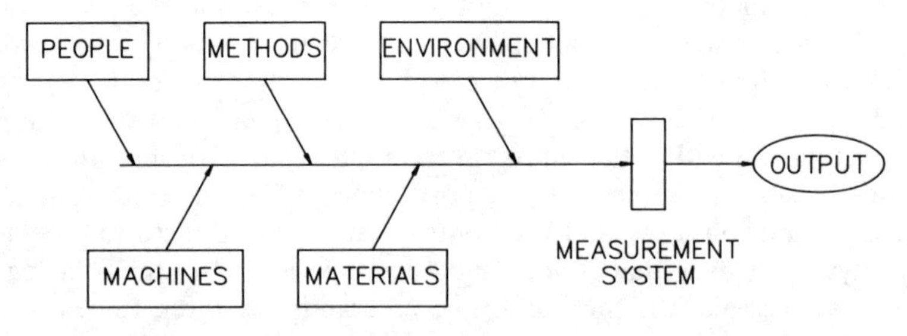

FIGURE 3.5. Shewhart's concept of a process.

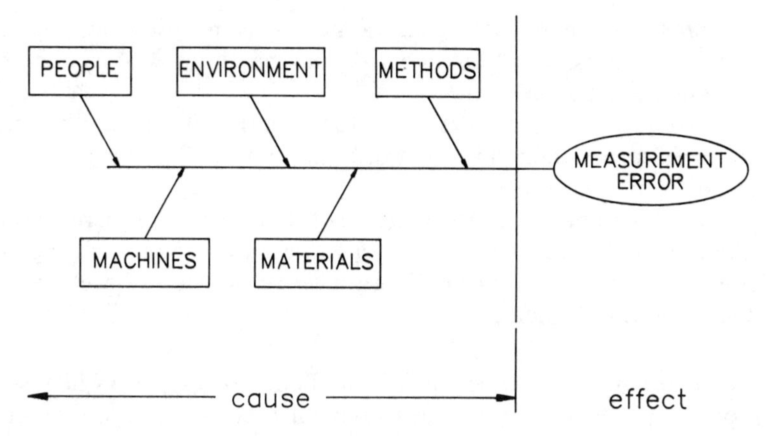

FIGURE 3.6. Shewhart's concept of a measuring system.

Each broad category can be broken down further into distinct causes. For example:

Manpower can represent	differences between operators differences between shifts differences between labs
Machines can represent	differences between gauges calibration frequency
Methods can represent	method for reading a gauge step-by-step test method differences
Materials can represent	within-part variation testing medium
Environment can represent	ambient conditions such as relative humidity, temperature, and cleanliness

Example An acetone drum filling operation uses a mechanical platform scale to periodically measure the validity of the drum-filling volumetric metering system. (Acetone is a low viscosity, clear, volatile solvent.) This measurement process could have errors that show that the volumetric metering system is not filling the drum to a proper weight and needs adjustment. These errors could come from many sources:

The *measuring equipment* (the scale) may not be calibrated or may have dirt built-up under the platform so that the scale shows a lower weight than the drum actually contains.

The *test method* may not include a step to compensate for the tare weight of the empty container: Again the drum will contain less acetone than it should.

One *tester* may look at the dial indicator from the left side and a second tester from the right side—both getting different readings from the same drum of acetone.

The *sample* drum of acetone may not have been properly sealed, allowing some acetone to evaporate. Because acetone is a volatile solvent, the sample (drum) filling process output may have less acetone than it did when filled.

This simple example of an acetone drum-filling process shows how some or all of the components in the measurement process may cause an error that signals that the overall process (drum filling) needs adjustment.

The first step then, is to identify all of the potential components of measurement error (using brainstorming) and organize them using a cause-and-effect diagram. Once these components have been identified, then the actual contribution to the error from each component can be studied. The sources contributing to the measurement error can further be broken down into five general categories: accuracy, repeatability, reproducibility, stability, and linearity.

Accuracy Measurement accuracy refers to the *difference* between the *observed average* of measurements and the *true value*. If the measurement

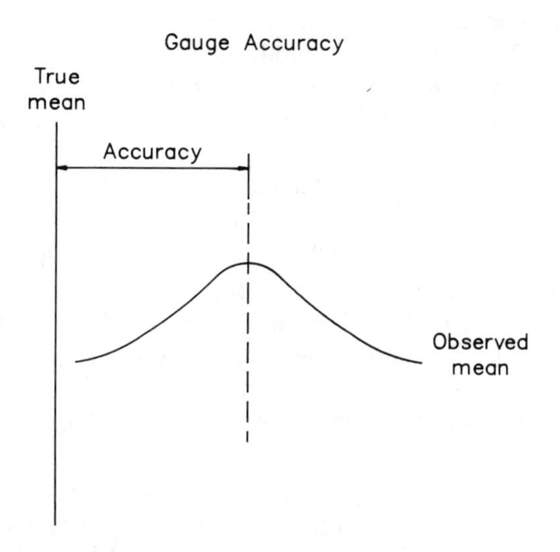

FIGURE 3.7. Gauge accuracy.

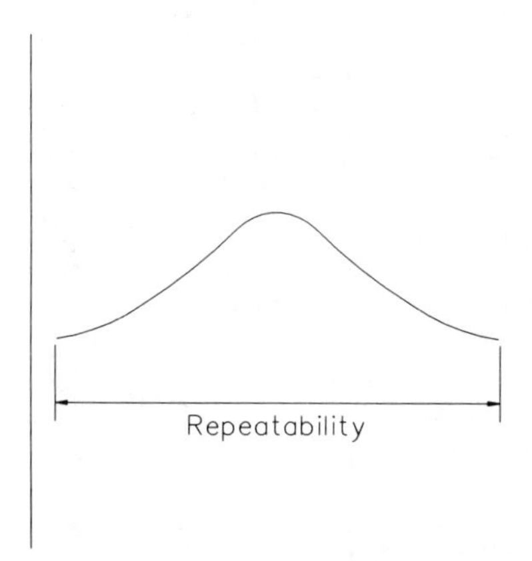

FIGURE 3.8. Gauge repeatability.

process is accurate, it is aimed correctly; if it is not accurate, the aim is off target and the measurement process is biased (see Figure 3.7). Accuracy is also referred to as bias or systematic error.

Repeatability Repeatability is the variation due to the test equipment. It is the variation in measurement found when the *same* operator uses the

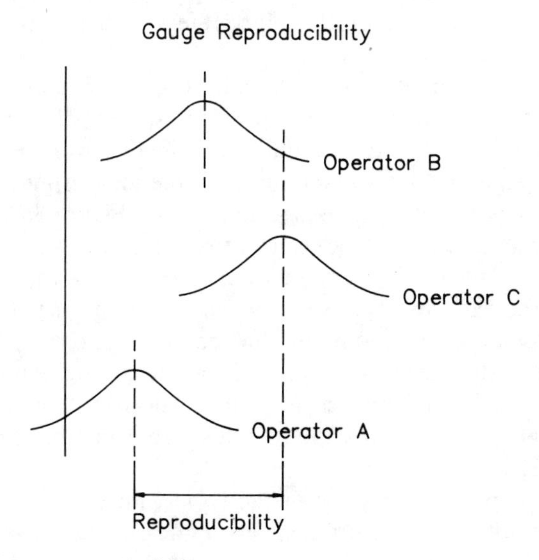

FIGURE 3.9. Gauge reproducibility.

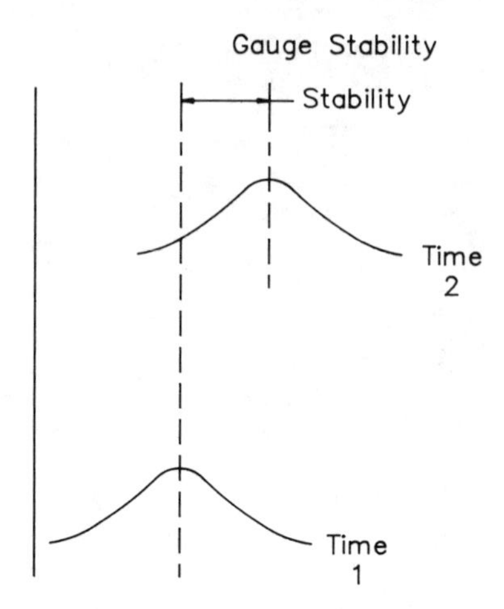

FIGURE 3.10. Gauge stability.

same measuring equipment to test the *same* process output parameter on the *same* sample (or a homogeneous sample). Repeatability is often referred to as precision (see Figure 3.8).

Reproducibility Reproducibility is the variation due to the test operator. It refers to the variation in the average of measurements made by *different* operators using the *same* measuring equipment measuring the *same* process output parameter on the *same* sample (or a homogeneous sample) (see Figure 3.9). Reproducibility refers to the consistency of the pattern of variation.

Stability Measurement process stability takes into account the influence of time on the measurement process, specifically the measuring equipment. Stability refers to the difference in the average of measurements taken at *different* times made on the *same* measuring equipment on the *same* samples (or homogeneous samples) (see Figure 3.10).

Linearity Measurement process linearity also focuses primarily on the capability of the measuring equipment. Linearity is the *difference in accuracy* over the expected operating range. (Often, measuring equipment that is accurate at the midpoint of the measurement range will not be accurate at the high and low extremes of the measuring equipment scale. Many manufacturers recommend using the measuring equipment over a band of 20 to 80% of the full scale to compensate for this type of error.) (See Figure 3.11.)

The relationship between some of the typical measurement error components and the type of measurement error they might cause is shown in Table 3.3.

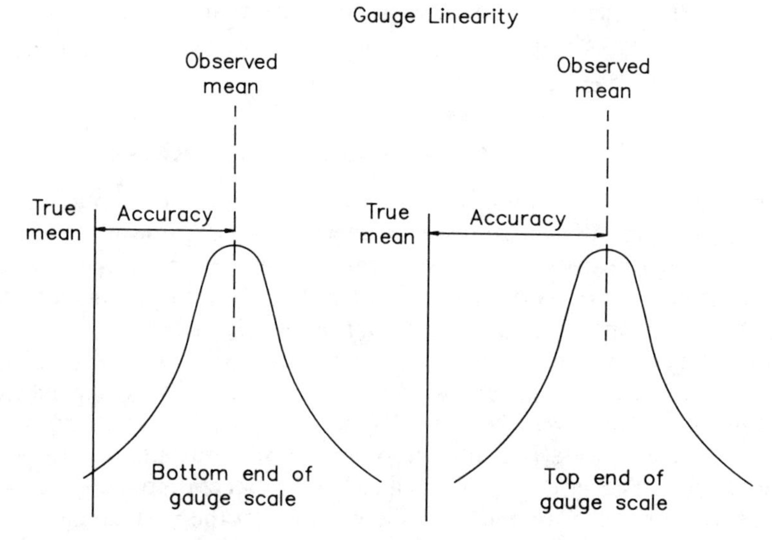

FIGURE 3.11. Gauge linearity.

Table 3.3 Relationship between Typical Measurement Error Components to Type of Measurement Error They Might Cause

If ______ is a problem, then ______ may be the cause.

Problem	Cause
Gauge accuracy	Worn gauge Improper calibration Improper use An error with the control
Gauge repeatability	Improper gauge maintenance Improper test method Excess gauge flexibility Inadequate gauge clamping
Gauge reproducibility	Inadequate operator training Inadequate units on the gauge
Gauge stability	Equipment warm-up Infrequent use/overuse Environmental conditions
Gauge linearity	Calibration not being done at lower and upper end of the scale Worn gauge An error in one of the controls

Once the components of measurement error are understood, gauge studies can be performed to determine the amount of variation in the measurement process.

3.4 LEVEL OF MEASUREMENT ERROR

Variation occurs in every process and in every process output. But, without studying the measurement system, it is not known whether the variation seen in the observed (or measured) process outputs is due to a variation in the process or due to variation from the measurement system. The measurement of the process output (e.g., weight, dimensions, flexural strength, and ash content) is itself a separate process. By looking at Shewhart's concept of a process with the measurement system included (see Figure 3.12), it can be seen that what is usually thought to be the output of the process is really the output as seen through the measurement system. The variability of the measured product output combines the true variation of the product with the variation of the measurement:

$$(\text{Measured output variation}) = (\text{actual product output variation}) + (\text{measurement variation})$$

or

$$\sigma^2_{\text{measured output}} = \sigma^2_{\text{actual product output}} + \sigma^2_{\text{measurement}}$$

Any view of how well the actual process is performing will only be as good as the measurement system.

There are a number of different methods for determining the level of measurement error that exists. They will be discussed in Sections 3.7 through 3.10. However, once the level of error present is calculated, it must be determined whether that level is acceptable or not. One method

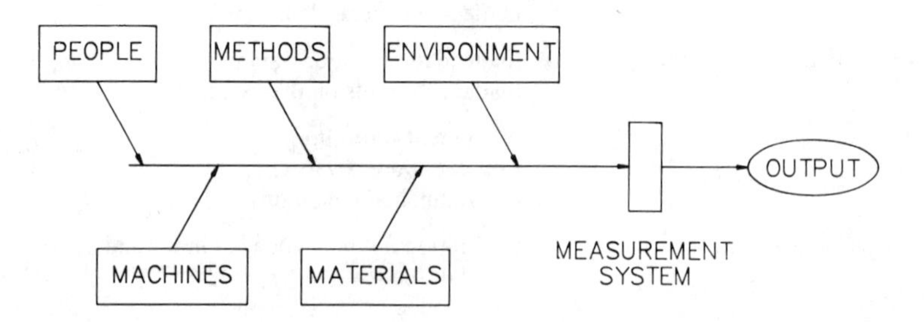

FIGURE 3.12. Shewhart's concept of a process.

uses the formula for the total variation and looks at what percentage of the total variation is due to the measurement system. This is known as the percentage of total variance (PTV) due to the measurement system:

$$\text{PTV} = \frac{(\text{measurement variation})}{(\text{measured output variation})} \times 100\%$$

If PTV < 10%, then the measurement error is considered acceptable. If the PTV is between 10 and 30%, the measurement error may be considered acceptable based on the cost of measurement system replacement. If the PTV > 30%, then the measurement system requires replacement or significant improvements.

There are other methods for calculating the measurement error besides the PTV. Regardless of which measure is used, it is critical to know the level of the error. Knowledge about the level of measurement error can be used to:

- Justify upgrading equipment or *not* upgrading equipment
- Lay the groundwork for a designed experiment to identify sources of excessive variation
- Most importantly of all, give a better understanding about the true product variability.

3.5 MEASUREMENT CONSISTENCY

A measurement process is consistent if repeated measurements of the same parameter display similar amounts of variation. When considering consistency of measurement, three of the five components of potential measurement error are grouped together: repeatability, reproducibility, and stability.

Consistency implies that there is little change in the measurement system variation over a period of time; hence, the inclusion of the stability components. An inconsistent measurement system is one that may need frequent recalibration because it may drift away from standard settings over time.

Repeatability and reproducibility are often linked, as in an R & R study. Both refer to the *dispersion* or range of variation within the measurement system.

Inconsistent measurement systems are problematic. There is a tendency to question the results and to lose faith in an inconsistent measurement system. Inconsistencies must be removed from the system by reducing the variation in the test method (reproducibility and repeatability) and establishing proper frequencies for calibration checks. If instability over time proves to be the major source of inconsistency, the measurement system needs to be studied and improved so it can hold calibration.

3.6 MEASUREMENT BIAS

Whereas measurement consistency refers to the dispersion or variation of a measurement process, bias refers to the location. An ideal measurement system will produce measurements identical to a known standard, or a measurement master. A biased measurement system will produce measurements some distance from the known "location".

Measurement bias covers two of the five components of potential measurement error: accuracy and linearity. If the measurement system is consistent (repeatable, reproducible, and stable), a biased measurement system sometimes can be accepted. To be acceptable, the amount of bias from the accuracy and the linearity errors must be known so that the measurement system can be adjusted by that amount. This essentially factors out the bias.

3.7 MEASURING TESTING VARIANCE

Gauge repeatability and reproducibility (GR & R) studies are simply studies of the variation in the measurement. Some texts refer to GR & R studies as gauge repeatability and reliability. Regardless of their name, these studies are designed to identify the variation of the measurement system. Knowing this allows the actual variation of the product to be determined.

3.7.1 Preparation for a GR & R Study

Good planning is the key to a good GR & R study. What will be measured? In Figure 3.3b, the components of the measurement system were shown as people, methods, materials, equipment, and environment. In preparing for a GR & R study, these components must be defined as well as possible to ensure a reliable study.

People — Select two or three typical users of the measurement equipment. The key is *typical*. Do not pick the best or worst operator and don't use engineers or inspectors unless they are the ones who normally conduct the measurement.

Methods — Define how the samples will be measured. For an existing measurement system, this should be just a matter of using a procedure that is already in place. For a new measurement system, determine the procedure before conducting the GR & R study. Make sure that all of the operators know the procedure. If they don't, one of the largest errors in the measurement system (and accordingly, in the measured product) may have been inadvertently uncovered. Let them know how the entire GR & R study will be conducted, not just how the measurements will be made.

Materials — Most GR & R studies use 10 samples from the process. Label them. If destructive testing is involved, more samples will be necessary. GR & R studies involving destructive testing will be discussed in Section 3.9. If samples from the process are not available, then standards traceable to National Institute of Science and Technology (NIST; formerly National Bureau of Standards, NBS) or other known samples can be used. Because the production of parts or materials may create some of the variation in the measurements, the use of standards may not completely define the errors in the measurement system. For example, a production part may not have a flat surface, which may create some of the measurement system error. It may be difficult to measure the part in exactly the same location each time.

Equipment — The test equipment should be capable of reading to one decimal place smaller than the tolerance. For example, if the tolerance is 1.02–1.09 g, the scale should be capable of reading down to 0.001 g. If the test equipment cannot measure down this finely, there may be false signals sent about the capability of the measurement system (refer back to Section 3.2).

Prior to starting the study, the test equipment should be calibrated.

Environment — Some test procedures must be conducted under defined ambient conditions, such as 21°C and 50% relative humidity, or the test specimens must be conditioned under certain ambient conditions. Environmental conditions may also affect the measurements in some test equipment. In either of these cases, the environment in which the measurement system study will be conducted should be defined prior to the study and adhered to strictly.

3.8 GR & R STUDIES—NONDESTRUCTIVE TESTS

The most traditional form of GR & R study is the type performed on nondestructive tests—those tests that do not damage the test piece. Nondestructive tests allow for measurement of the same sample many times, thus minimizing piece-to-piece variation and isolating the variation due to the reproducibility and reliability of the test. Nondestructive GR & R studies are easy to design, perform, and analyze. An example study follows.

Conducting the Repeatability and Reproducibility Study

1. Use a data collection form such as the one shown in Figure 3.13.
2. Designate the first operator as operator A, the second as operator B, and if there is a third, operator C.
3. Have operator A randomly select the sample parts and measure them. The measurements should be recorded in the appropriate slot in column 1 on the data collection form.
4. Repeat step 3 with operators B and C.
5. Repeat steps 3 and 4 once or twice more as desired. Each time the parts should be randomly selected. The greater the number of trials, the more accurate the results of the measurement system study will be.

Analyzing the Results

6. Calculate the range and the average for each of the 10 samples for each operator A, B, and C.
7. Calculate the average range $\overline{R}_A$ for operator A:

$$\overline{R}_A = \frac{\Sigma R_A}{n}$$

where R_A = the individual range values for each sample
n = number of samples (normally 10)

8. Repeat step 7 for operators B and C.

Measurement System Work Sheet

Test Characteristic: ___________

Gauge Type: ___________ Gauge Type: ___________

Operator	Sample Number	Test Value Trial 1	2	3	Repeatability (within samples) R_E	$\bar{X}$	Material (within operators) R_M	$\bar{\bar{X}}$	Reproducibility (between operators) R_O
A	1								
	2								
	3								
	4								
	5								
	6								
	7								
	8								
	9								
	10								
					$\bar{R}_A$=	$\bar{\bar{X}}_A$=			
B	1								
	2								
	3								
	4								
	5								
	6								
	7								
	8								
	9								
	10								
					$\bar{R}_B$=	$\bar{\bar{X}}_B$=			
C	1								
	2								
	3								
	4								
	5								
	6								
	7								
	8								
	9								
	10								
					$\bar{R}_C$=	$\bar{\bar{X}}_C$=			
					$\bar{\bar{R}}_E$ =		$\bar{R}_M$=		R_O=

FIGURE 3.13. Measurement system data collection form.

9. Determine that the measurement system was in control.
   a. Calculate the overall average range $\bar{\bar{R}}_{E'}$, from

$$\bar{\bar{R}}_E = \frac{\bar{R}_A + \bar{R}_B}{2} \quad \text{or} \quad \bar{\bar{R}}_E = \frac{\bar{R}_A + \bar{R}_B + \bar{R}_C}{3}$$

   b. Determine the upper control limit for the range (UCL_R):

$$UCL_R = D_4 \times \bar{\bar{R}}_E$$

where $D_4 = 3.27$ for two trials or
$D_4 = 2.58$ for three trials

   c. Check all of the individual range values for operators A, B, and C. If all are less than UCL_R, then the measurement system was in control. If any are above the UCL_R, then there is a special cause of variation present. Identify that cause and eliminate it. Then, repeat the measurements of the sample parts. Use these measurements to repeat the analysis steps 6 to 9.

10. Calculate the overall average $\bar{\bar{x}}_A$ for operator A using the sample averages for operator A found in step 6:

$$\bar{\bar{x}}_A = \frac{\Sigma \bar{x}_A}{n}$$

where $\bar{x}_A$ = sample averages for operator A's measurements
n = number of samples, normally 10

11. Calculate the range between operators R_o using the values of $\bar{\bar{x}}_A$, $\bar{\bar{x}}_B$, and $\bar{\bar{x}}_C$ from step 10:

$$R_o = \text{highest value of } \left\{\bar{\bar{x}}_A, \bar{\bar{x}}_B, \bar{\bar{x}}_C\right\} - \text{lowest value of } \left\{\bar{\bar{x}}_A, \bar{\bar{x}}_B, \bar{\bar{x}}_C\right\}$$

If only two operators has participated in the measurement system study, then the equation would be:

$$R_o = \text{highest value of } \left\{\bar{\bar{x}}_A, \bar{\bar{x}}_B\right\} - \text{lowest value of } \left\{\bar{\bar{x}}_A, \bar{\bar{x}}_B\right\}$$

12. Calculate the repeatability (the variation from the equipment) using $\bar{R}_E$ found in step 9:

$$s_E = \frac{\bar{\bar{R}}_E}{d_E}$$

where d_E = 1.13 for two trials or
d_E = 1.69 for three trials

Note that the values of d_E and d_o in this section depend on 10 samples being analyzed in the GR & R study. The values will change with the sample size. If it is necessary to use a sample size other than 10, then refer to Appendix A.6. These values (d_E and d_o) are all values of d_2. The subscripts have been changed here for clarity—to make a GR & R study easier to understand and to do.

Once the standard deviation for repeatability has been calculated, this value can be used to calculate the spread of the variation. Most companies use a spread covering 99% of the variation ($\pm 2.575s$) although some prefer to use $\pm 3s$ (99.7%):

$$\text{Repeatability} = 2 \times 2.575 \times s_E$$
$$= 5.15 \times s_E$$

13. Calculate the reproducibility (the variation from the operators) using R_o found in step 10:

$$s_o = \sqrt{\left(\frac{R_o}{d_o}\right)^2 - \frac{(s_E)^2}{nr}}$$

where d_o = 1.41 for two operators or
d_o = 1.91 for three operators
n = number of samples
r = number of trials

$$\text{Reproducibility} = 2 \times 2.575 \times s_o$$
$$= 5.15 \times s_o$$

14. Calculate the R & R, the total error from repeatability variation and reproducibility variation:

$$(\text{R \& R})^2 = (\text{repeatability})^2 + (\text{reproducibility})^2$$
$$(\text{R \& R}) = \sqrt{(\text{repeatability})^2 + (\text{reproducibility})^2}$$

15. Calculate the percent of the specification that the repeatability and reproducibility variation take up:

$$\%\ \text{R \& R} = \frac{(\text{R \& R})}{\text{total tolerance range}} \times 100\%$$

$$\%\ \text{repeatability} = \frac{\text{repeatability}}{\text{total tolerance range}} \times 100\%$$

$$\%\ \text{reproducibility} = \frac{\text{reproducibility}}{\text{total tolerance range}} \times 100\%$$

If the percent R & R is less than 30%, then the measurement system is adequate. If it is above 30%, the measurement system must be improved. The percent repeatability and percent reproducibility plus the data from the study will normally offer the clues necessary to show what area in the measurement system to concentrate on for improvement.

For example, if the percent reproducibility makes up the largest component of the percent R & R, then this would direct the improvement efforts to concentrate on the operators and their measurement techniques.

The data from the study may show that the variation from one operator is much greater than the others. The improvement effort would concentrate on what that operator does differently from the others. Correcting that operator's technique may be all of the improvement required.

Conversely, if one operator has significantly less variation than the others, the engineer conducting the measurement system study should also investigate why. The operator may have a knack that should be incorporated into the measurement procedure and taught to all of the other operators.

If all of the operators show similar measurement variation and the percent reproducibility is too high, then the whole measurement procedure is suspect. It will be necessary to use the statistical process improvement (SPI) techniques described in Chapter 5 of this book to improve the measurement procedure.

On the other hand, the percent repeatability may be much greater than the percent reproducibility. The piece of test equipment should then be looked at for improvement. The manufacturer or designer of the test equipment should be included in the improvement efforts. It may be that the SPI techniques in Chapter 5 can improve the repeatability, or it may be that piece of test equipment will never have the repeatability needed. In this case, another piece of test equipment must be designed or procured. If a new piece of test equipment is being purchased, have a GR & R study done prior to even committing a purchase order. This will ensure that the test equipment is adequate to do the job required.

If the percent R & R is above 30% and the percent reproducibility is nearly equal to the repeatability, then both the measurement procedures and the equipment must be investigated for improvement.

3.9 GR & R STUDIES—DESTRUCTIVE TESTS

In normal GR & R studies, the same sample is repeatedly tested. This is impossible when destructive testing is involved. This is a problem for much of the CPI because many of the measurements made consume or destroy the sample. To complete a GR & R study involving destructive testing, different samples must be used. This creates another source of variation in the measurement system—the variation between the samples. In order to properly evaluate the measurement system, this variation due to the samples must be quantified and minimized.

The GR & R study for destructive testing is set up exactly the same way as a normal GR & R study except for the sampling. When the study involves destructive testing, the sampling plan is critical to the success of the study. The sampling plan should be such that the samples taken are nearly homogeneous. In a nondestructive test GR & R, this means that 10 samples have to be nearly homogeneous. This is more difficult for destructive testing because if the study involves three operators and three trials, 90 samples must be nearly homogeneous.

This is most difficult for those operations where the samples must be prepared for testing. For example, the metals industry and plastics industry have to prepare test specimens for strength testing. The preparations of the test specimens are processes in themselves whether they are cutting the specimens out of sheet stock, or casting the specimens, or molding the specimens. These processes will vary, which in turn causes the specimens to vary.

Other CPI operations, such as those producing bulk powders or solutions, will not have the same degree of difficulty. In preparing for the GR & R study, a large enough quantity of the powder or solution can be taken from the process. This powder or solution can then be mixed well prior to pulling the individual samples for the GR & R study. These samples should be nearly homogeneous.

The GR & R study for destructive testing is conducted the same way as the standard GR & R study. The operators randomly pull samples and test them. The difference between the two lies in the analysis because the material variation must be calculated along with the repeatability and reproducibility.

The steps for analyzing a GR & R study for destructive testing are:

1. Calculate the range and the average for each of the 10 samples for each operator A, B, and C.
2. Calculate the average range for Operator A from

$$\bar{R}_{\mathrm{A}} = \frac{\Sigma R}{n}$$

   where R = individual range values for each sample measured by operator A
   n = number of samples (normally 10)
3. Repeat step 2 for operators B and C.
4. Determine that the measurement system was in control:
   a. Calculate the overall average range, $\bar{\bar{R}}_E$ from

$$\bar{\bar{R}}_E = \frac{\bar{R}_{\mathrm{A}} + \bar{R}_{\mathrm{B}}}{2} \quad \text{or} \quad \bar{\bar{R}}_E = \frac{\bar{R}_{\mathrm{A}} + \bar{R}_{\mathrm{B}} + \bar{R}_{\mathrm{C}}}{3}$$

b. Determine the upper control limit for the range (UCL_R):

$$UCL_R = D_4 \times \overline{\overline{R}}_E$$

where $D_4 = 3.27$ for two trials or
$D_4 = 2.58$ for three trials

c. Check all of the individual range values for operators A, B, and C. If all are less than UCL_R, then the measurement system was in control. If any are above the UCL_R, then there is a special cause of variation present. Identify that cause and eliminate it. Then, repeat the sampling and measurements of the samples, and start again.

5. Calculate the range (R_M) for the averages of the 10 samples for operator A:

$$R_{M,A} = \text{highest of } \{\bar{x}_1 \text{ to } \bar{x}_{10}\} - \text{lowest of } \{\bar{x}_1 \text{ to } \bar{x}_{10}\} \text{ for operator A}$$

6. Calculate the overall average $\overline{\overline{X}}$ for operator A:

$$\overline{\overline{X}}_A = \frac{\Sigma \overline{X}_A}{n}$$

where $\overline{\overline{X}}_A$ = grand average of the sample averages from operator A's measurements
n = number of samples, normally 10

7. Repeat steps 5 and 6 for operators B and C.
8. Calculate the average range ($\overline{R}_M$) for the material:

$$\overline{R}_M = \frac{R_{M,A} + R_{M,B} + R_{M,C}}{3}$$

[If only two operators were involved, then $\overline{R}_M = (R_{M,A} + R_{M,B})/2$.]

9. Calculate the range (R_o) of the overall averages of operators A, B, and C:

$$R_o = \text{highest of } \left\{\overline{\overline{X}}_A, \overline{\overline{X}}_B, \overline{\overline{X}}_C\right\} - \text{lowest of } \left\{\overline{\overline{X}}_A, \overline{\overline{X}}_B, \overline{\overline{X}}_C\right\}$$

If only two operators were involved, then:

$$R_o = \text{highest of } \left\{\overline{\overline{X}}_A, \overline{\overline{X}}_B\right\} - \text{lowest of } \left\{\overline{\overline{X}}_A, \overline{\overline{X}}_B\right\}$$

10. Calculate the repeatability using $\overline{\overline{R}}_E$ found in step 4:

$$s_E = \frac{\overline{\overline{R}}_E}{d_E}$$

where $d_E = 1.13$ for two trials or
$d_E = 1.69$ for three trials
Note that the values of d_E, d_M, and d_o in this section depend on 10 samples being analyzed. The values will change with the sample size. If it is necessary to use a sample size other than 10, then refer to Appendix A.6. These values (d_E, d_M, and d_o) are all values of d_2. The subscripts have been changed here for clarity—to make it a GR & R study easier to understand and to do.

$$\text{Repeatability} = 5.15 \times s_E$$

(This covers 99% of the variation. Some companies use $6 \times s_E$ to cover 99.7% of the variation.)

11. Calculate the material (sample) variation using $\overline{R}_M$ found in step 8:

$$s_m = \sqrt{\left(\frac{\overline{R}_M}{d_M}\right)^2 - \frac{(s_E)^2}{r}}$$

where $d_M = 3.13$ for two operators or
$d_M = 3.11$ for three operators
Material variation $= 5.15 \times s_m$

12. Calculate the reproducibility using R_o found in step 9:

$$s_o = \sqrt{\left(\frac{R_o}{d_o}\right)^2 - \frac{(s_m)^2}{n} - \frac{(s_E)^2}{nr}}$$

where $d_o = 1.41$ for two operators
$d_o = 1.91$ for three operators
n = number of samples
r = number of trials

13. Calculate (R & R) the total error from repeatability variation and reproducibility variation:

$$(\text{R \& R}) = \sqrt{(\text{repeatability})^2 + (\text{reproducibility})^2}$$

14. Calculate the percent of the specification that the repeatability and reproducibility take up:

$$\% \text{ R \& R} = \frac{(\text{R \& R})}{\text{total tolerance range}} \times 100\%$$

$$\% \text{ reproducibility} = \frac{\text{reproducibility}}{\text{total tolerance range}} \times 100\%$$

$$\% \text{ repeatability} = \frac{\text{repeatability}}{\text{total tolerance range}} \times 100\%$$

15. Calculate the percent of the tolerance taken up by the material variation:

$$\% \text{ material variation} = \frac{\text{material variation}}{\text{total tolerance range}} \times 100\ \%$$

For samples pulled from a solution or mixture of powders that appear to be homogeneous, the material variation should take up less than 10% of the specification. If it does not, then the sampling plan and thus the entire GR & R study are suspect.

As mentioned in the previous section, the percent R & R should be less than 30% of the specification. If it is above 30%, then the measurement system must be improved.

When the samples have to be processed to prepare them, the percent R & R and the percent material variation must be looked at together. The process of preparing the samples is part of the measurement system. In this case, the sum of the percent R & R and the percent material variation should be less than 30%.

If the sum is greater than 30%, then the measurement system should be examined. The investigation should start with whichever component makes up the greatest portion. If the percent material variation is greatest, then the investigation should start with the sampling plan and the process for preparing the sample. When the percent repeatability is the greatest, point the investigation toward the test equipment; if the percent reproducibility is the greatest, concentrate on the operator's techniques.

3.10 SIGNAL TO NOISE RATIO

The two preceding sections calculated the percent R & R based on the specification. Although it is important to know how much of the total tolerance the measurement system takes up, this may not be enough

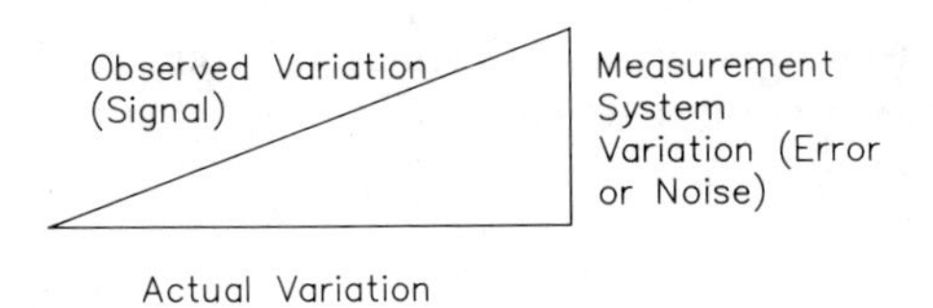

FIGURE 3.14. Total variance is the sum of the parts.

information to help direct our improvement efforts. The signal to noise ratio (S/N) is another measure of how good the measurement system is. This ratio is the inverse of the PTV discussed in Section 3–4. For the PTV, the lower the value, the better the measurement system. For the signal to noise ratio, the higher the value, the better the measurement system.

As stated in Section 3.4, the amount of variation observed is equal to the actual variation of the product plus the variation in the measurement system. This is shown pictorially in Figure 3.14.

The signal to noise ratio is the ratio of the observed product variation to the measurement system variation. The higher the ratio, the closer the variation observed in the product comes to reflecting the actual product variation.

The signal to noise ratio supplements the percent R & R to describe how well the measurement system performs (and in cases where there is no specification, such as in an R & D environment, it is the only method to describe how well the measurement system performs). The signal to noise ratio may show that it is a waste of money and effort to improve a measurement system even if it takes up 50% of the specification. Figure 3.15 shows a process that has a large signal to noise ratio. The observed product variation is very close to the actual product variation.

Even if the measurement system took up more than 30% of the specification in one case, it makes no sense to work on improving the measurement system. The effort should go into reducing the actual product variation. Once the variation has been reduced, then the measurement system can be worked on.

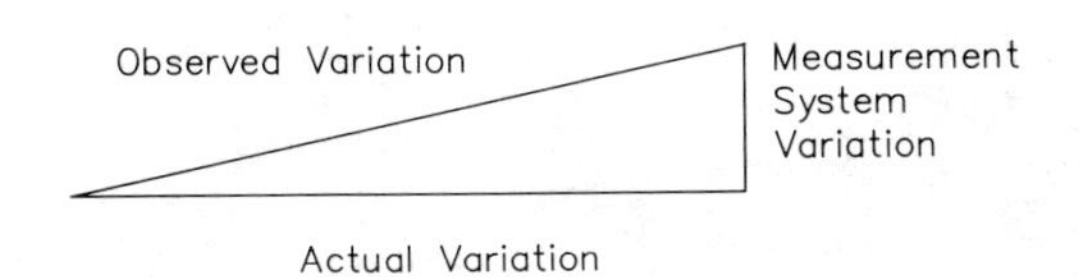

FIGURE 3.15. Large signal to noise ratio.

The signal to noise ratio is calculated from the process standard deviation and the standard deviations for repeatability (equipment) and reproducibility (operator):

$$\frac{\mathrm{S}}{\mathrm{N}} = \frac{s^2}{s_E^2 + s_o^2}$$

The value for the process standard deviation, s, can be calculated using historical data from $\bar{x}$ and R control charts:

$$s = \frac{\bar{R}}{d_2}$$

where $\bar{R}$ = average range from the control chart
d_2 = constant depending upon the sample size from the control chart

An alternative to this method would be to conduct a 40 piece capability study and calculate s from the individual measured values of the 40 pieces (refer to Section 4.5 on how to conduct a capability study).

4

Statistical Process Control

4.1 QUALITY IMPROVEMENT CYCLES

Conventional approaches to quality and productivity (Figure 4.1) don't really improve quality or productivity. They typically maintain yesterday's level of quality into tomorrow, or at best, show an ever so slight positive scope. Remember Section 1.5 where we talked about the small productivity gains the United States has had in recent years?

4.1.1 Putting Out the Fires

The conventional model of quality and productivity management has its basis in fire fighting. We're all so busy fighting fires every day, we never get to the root cause of the problem and extinguish the source of the fire.

A statistically based quality and productivity approach provides us with both a "fire break" and the tools to get to the source of "the fire", or causes of variation. Statistical process control (SPC) gives us the "fire break", providing a means to flag and attack special causes of variation when they first flare up, before they engulf our processes. Statistical process improvement (SPI) provides us with tools to work on the chronic common causes of variation, those causes inherent to the process. These are the smoldering embers, and, with SPI, we are able to put them out, one by one, for good. SPI gives us the means to bring beneficial change (improvement) with permanence.

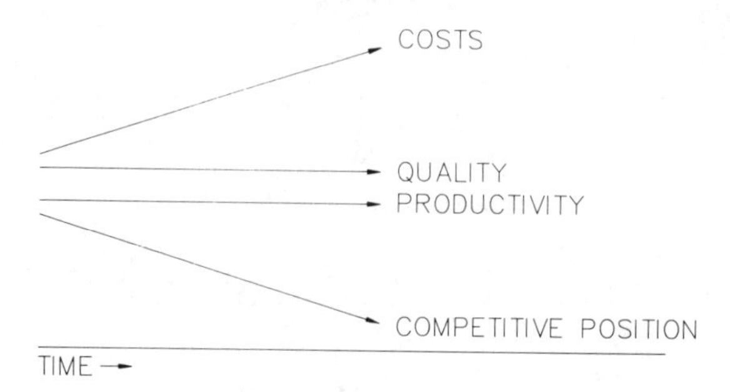

FIGURE 4.1. Conventional approaches to quality and productivity. (*Reprinted from SPC In Action, copyright 1990, Resource Engineering, with the permission of the publisher, Quality Resources, White Plains, New York.*)

4.1.2 The Tangible Benefits of SPC

After we embark on a statistically based approach to manage quality and productivity, we'll find sequential improvements in four tangible areas:

- *Quality Improves* (Figure 4.2). With the commitment to SPC and the understanding of the process that commitment brings, an almost immediate quality improvement is seen.
- *Productivity Rises* (Figure 4.3). Here's where quality drives productivity (Section 1.5). As quality improves, the hidden plant (or office) shrinks. The loss to the hidden plant is a direct productivity gain.
- *Costs Decrease* (Figure 4.4). With productivity increasing, total costs may stay the same, but unit costs drop. Shrinking the hidden plant drives the cost of scrap and rework right to the bottom line. Note that costs don't drop at first. The up-front costs of training are expensive. But they're really an investment (not a cost) with a huge return.

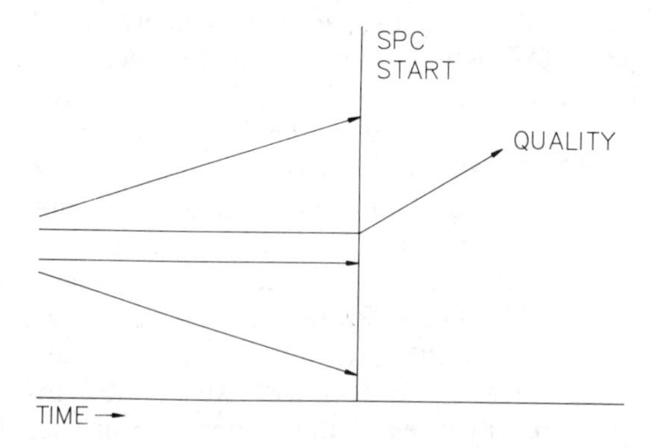

FIGURE 4.2. Benefits of SPC—quality improves.

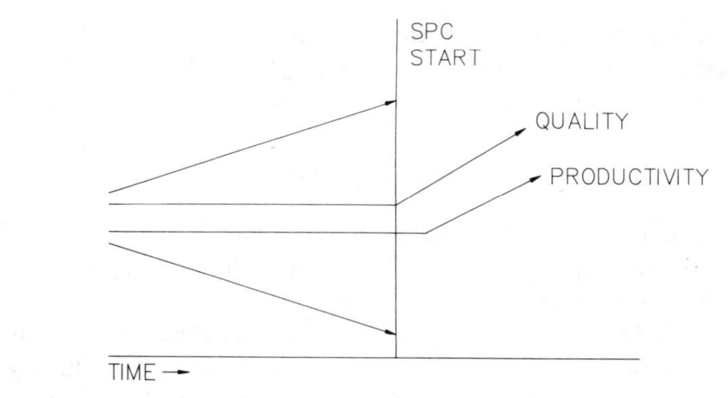

FIGURE 4.3. Productivity rises.

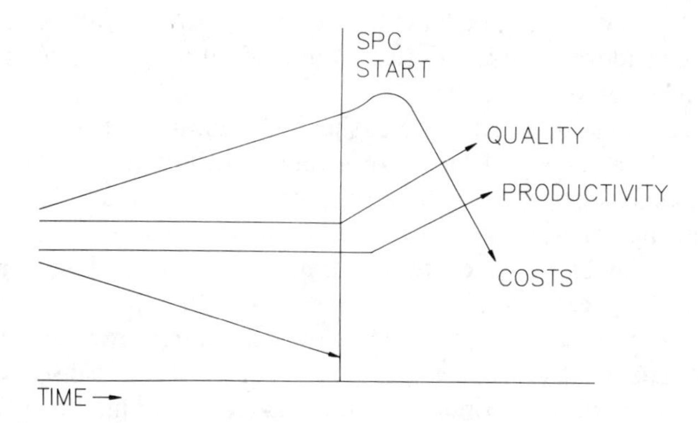

FIGURE 4.4. Costs decrease.

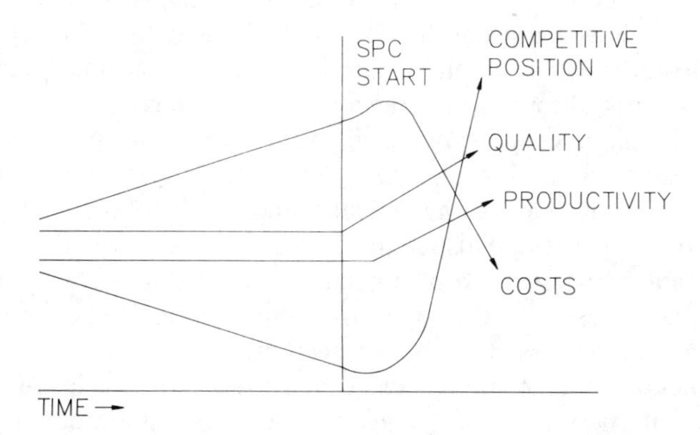

FIGURE 4.5. Competitive position soars.

- *Competitive Position Soars* (Figure 4.5). How can it not? With improving quality and productivity, with declining costs, you'll win the marketplace competition.

4.2 PROCESS STABILITY

The goal of SPC is to maintain a stable process, one where the pattern of variation for the process output is not changing. A stable process has only common causes of variation present, and no special cause variation. Many texts define a stable process as one that has a normally distributed output. This need not be the case. Often, the distribution of individual values from a process is not normal. Yet, collecting data on those processes periodically and then plotting the individual values on histograms may show the same distribution patterns time after time. This, obviously, indicates that the process is not changing due to special causes of variation. Such a process is considered stable despite its individual output values not being normally distributed.

Most processes that are stable show distributions that are normal, nearly normal, or skewed. Frequently, other distributions, such as bimodal, show that some type of special cause is present. For example, a bimodal distribution may be due to recurring shift in the process (e.g., operators with different techniques) or to the output from two similar pieces of equipment being collected in the same container. Even though it may appear that the process is not changing over time, the process is not stable because it has special causes present. These special causes could be eliminated by training the operators in the same technique or, in the latter case, by sampling and plotting the output from each machine separately.

A process that is stable and shows a normal or nearly normal distribution is a controlled process; it is in control. When the individual values of a process input show a normal distribution, it is clear that the process is in control. However, when the individual values are not normally distributed, the process is not always out of control; it may be in control. We can use a statistical technique called subgrouping to determine if the process output is normal. Subgrouping involves taking multiple samples from a process at a given time and treating the average and range of the subgroups as measures of the process. This technique will work to normalize the output only for stable processes. Applying subgrouping techniques to data from an unstable process still will show that the process lacks control. (Subgrouping will be discussed further in Section 4.3.)

Regardless of the technique used, if a process is stable and shows a normal distribution for the individual values of the output, or for the subgroup averages and ranges, the process is considered to be in control. If the process is changing (i.e., not stable) and its output cannot be normalized, then it is out of control. The special causes of variation that destabilize the process must be identified and corrected before SPC techniques can be used.

4.3 CONTROL CHART BACKGROUND

Control charts are the SPC tools we use to tell us when to adjust the process and when to leave it alone. As such, usually, they can only be applied to processes that are stable and whose output is normally distributed. Control charts are used for processes that are in control to let us know when the process has gone out of control. This allows immediate action to be taken to find and eliminate the special cause that forced the process out of control.

The key point is that the process must be stable and normally distributed to apply standard control charts. Unfortunately, in industry, the distribution of individual values of the process output is often not normal. Fortunately, Dr. Walter Shewhart developed the concept of subgrouping to overcome this problem with the distribution of individual values. Subgrouping is a technique of pulling two or more consecutive samples from a process and treating them together to monitor the process. Even though the distribution of the individual values may not be normally distributed, in most cases, if enough samples are subgrouped together, the values of the subgroup averages will show a normal distribution when plotted. For stable processes, the subgroup averages can be used to control the location of the process output. Similarly, the subgroup ranges can be used to control the amount of variation in the process.

In his work on subgrouping, Dr. Shewhart looked at two nonnormal distributions of individuals—rectangular and right triangular—and the normal distribution. Dr. Donald Wheeler later augmented Dr. Shewhart's work by also considering heavily skewed, heavily tailed distributions. Shewhart and Wheeler both compared plots of individual values with plots of subgroup averages for subgroup sizes of 2, 4, and 10. Histograms comparing these for the four distribution types are shown in Figure 4.6.

Note that for a subgroup size of 2, the distribution of the subgroup averages appear very similar to the distribution of the individuals. The distribution of the subgroup averages for a subgroup size of 4 become nearly normal for the rectangular and right triangular distributions. The heavily skewed, heavily tailed distribution continues to show a skew. By

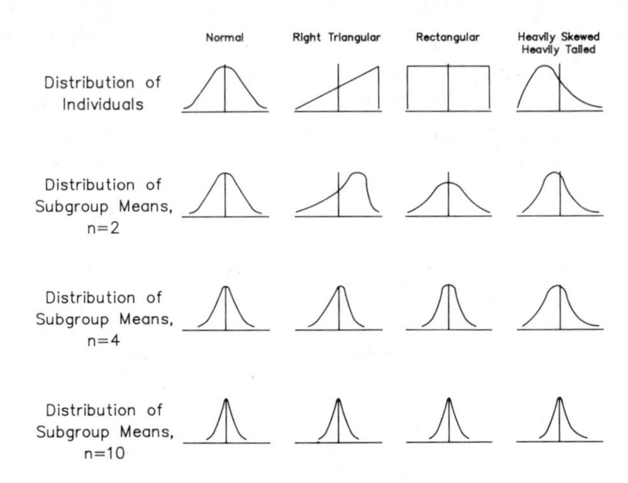

FIGURE 4.6. Individual values versus subgroups.

the time the subgroup size equals 10, the patterns for all four subgroup averages appear quite similar. (This is the central limit theorem in action). The heavily skewed, heavily tailed distribution still has a slightly larger tail than the others, but this tail accounts for only 1% of the distribution.

Control charts look for indications that the process is in control, that it shows a normal distribution. The 1% tail for the heavily skewed, heavily tailed process would fall outside the mean ± 3 standard deviations; it indicates this isn't a normal distribution. (Refer back to Section 2.8 for review.) If we tried (erroneously) to use standard control charts even with a subgroup size of 10 to monitor this process, we would still get a false alarm to adjust the process 1 time in 100. If we tried an even larger subgroup size, the heavily skewed distribution would look more like a normal distribution.

We used the histograms in Figure 4.6 to visually determine that distribution was normal or nearly normal. Most statisticians would not accept such a visual determination. There are mathematical parameters that can be used along with the mean and the standard deviation to determine normality. These parameters are skewness and kurtosis. Skewness compares the sizes of the two tails of a distribution. A distribution that is symmetric, like a normal distribution, will have a skewness of zero. Kurtosis compares the size of the two tails to the size of the rest of the distribution. This is an indirect measure of the "peakedness" of the curve. A normal distribution will have a kurtosis of 1.

The skewness and kurtosis can be calculated using integral calculus if the equation for the distribution curve is known:

$$\text{Skewness} = \int \frac{(x-\mu)^3}{\sigma^3} f(x)\,dx$$

$$\text{kurtosis} = \int \frac{(x-\mu)^4}{\sigma^4} f(x)\,dx$$

where $f(x)$ = equation for the curve
μ = population mean
σ = population standard deviation

In actuality, these parameters are rarely, if ever, calculated by hand. Statistical software packages are available that will calculate them for you.

4.4 IN CONTROL VERSUS IN SPECIFICATION

In the preceding sections, the concepts of stable and in-control processes were discussed. These concepts presumed that the process was not changing and was behaving in a normally distributed manner so that statistical control techniques could be applied. No mention was given to whether or not the process output would meet internal or external customer specifications. A process can be in control but not be able to meet the specification (Figure 4.7). It can also be out of control, but, for a given period of time, make product that is within the specification (Figure 4.8). Our goal is to have a process that is both in control *and* meets the specification (Figure 4.9).

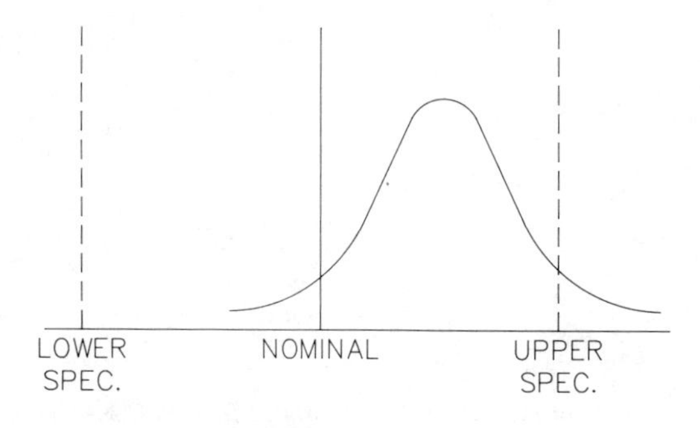

FIGURE 4.7. In control but not in spec.

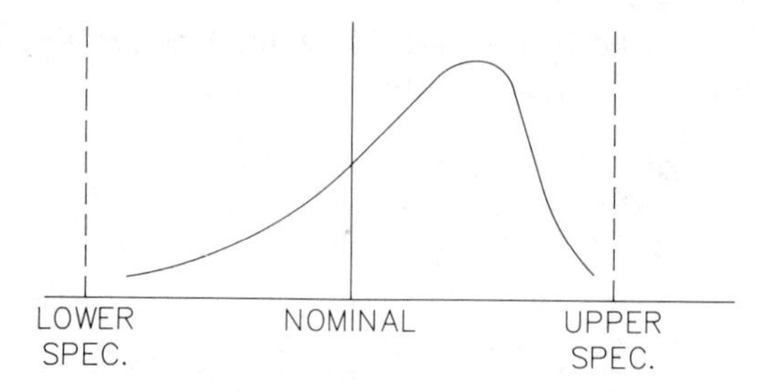

FIGURE 4.8. Out of control but in spec—for now.

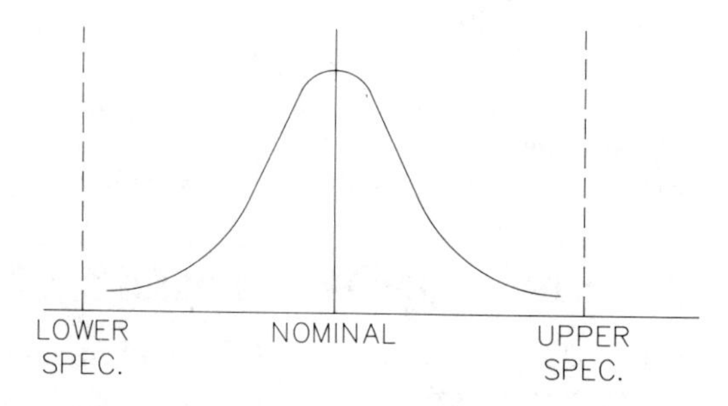

FIGURE 4.9. In control and in spec.

This goal is not always easy to reach. Look at the targets in Figures 4.10 to 4.13. The shots represent the process output of a sharpshooter. The specification is that all shots must hit the bull's eye. Target 1 (Figure 4.10) shows the process output scattered all over the target. This process is neither in control nor in specification. Targets 2 and 3 (Figures 4.11 and 4.12) show processes that are in control but cannot meet the specification.

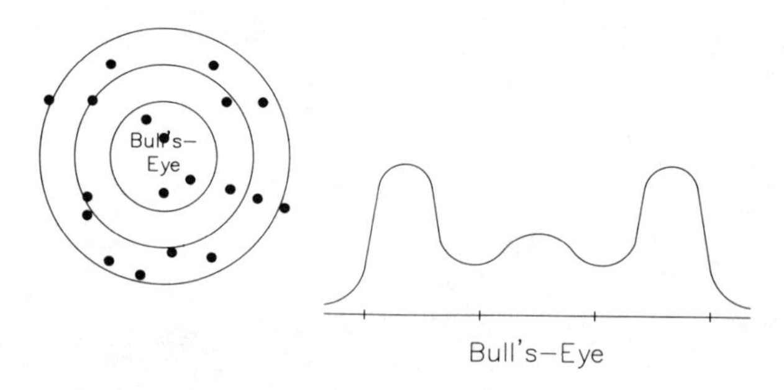

FIGURE 4.10. Process output scattered. (*Reprinted from SPC In Action, copyright 1990, Resource Engineering, with the permission of the publisher, Quality Resources, White Plains, New York.*)

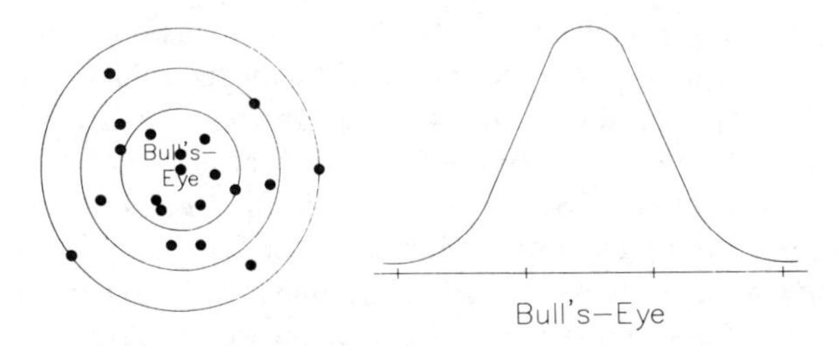

FIGURE 4.11. In control but not in spec target. (*Reprinted from SPC In Action, copyright 1990, Resource Engineering, with the permission of the publisher, Quality Resources, White Plains, New York.*)

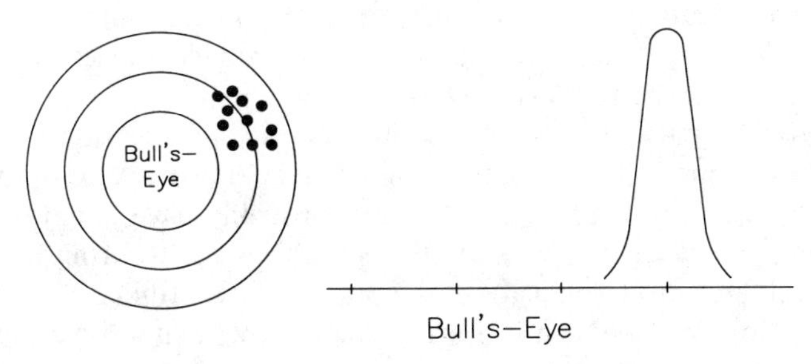

FIGURE 4.12. In control but not in spec but capable. (*Reprinted from SPC In Action, copyright 1990, Resource Engineering, with the permission of the publisher, Quality Resources, White Plains, New York.*)

Of these three shooters, only the shooter of target 3 can make an easy adjustment to meet specification. This shooter shows little variation in the shooting process but has not located the process correctly. The special cause in this case is that his/her rifle sights are incorrect. It is a simple process to resight the rifle and adjust the sights to bring the shooting process into specification (Figure 4.13).

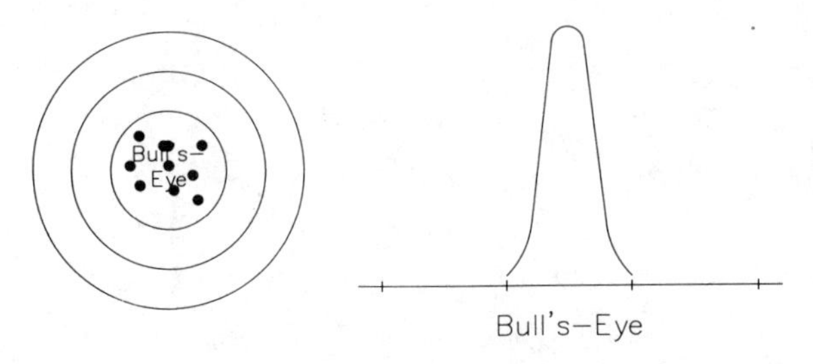

FIGURE 4.13. "Resighted" into spec. (*Reprinted from SPC In Action, copyright 1990, Resource Engineering, with the permission of the publisher, Quality Resources, White Plains, New York.*)

Shooter 2 has the process located correctly but has too much variation to meet specification. It will be more difficult to adjust this shooter's process to meet specification. This shooter has too much common cause variation, which is generally more difficult to attack than special cause variation. To correct this shooter's process, the problem-solving and design of experiments techniques described in Chapter 5 must be employed.

The improvement process for shooter 1 may be the most difficult. The shooting process for shooter 1 must be brought into control by identifying and then eliminating the special causes of variation. Then, the process must be reevaluated to see if it meets specification.

Eliminating the special causes may both bring the process into control and into specification, or it may just bring the process in control so that further improvement activities are needed to reduce the common cause variation to bring the process into specification.

Any process that is *not* both in control and in specification is a process that wastes money. The process will produce rework and scrap, have shipments of defective product, and most probably have situations where conforming product is rejected. If efforts aren't made to bring the process into control and into specification, this waste will continue. This drives up the costs of products, which in turn decreases the competitive position of the organization.

This is true both for our own processes and for our suppliers' processes. If a company receives a raw material that shows a distribution that is truncated to meet specification (Figure 4.14), the supplier must be sorting the process output. That supplier may have a process that is in control, but has too much variation to meet the specification. You, the customer may not think you are paying for the sorting operation, but in the long run, the supplier will pass on the cost of the sorting operation to you, the customer.

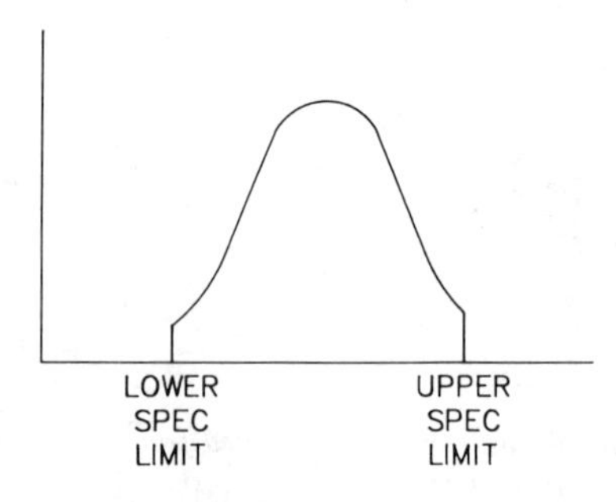

FIGURE 4.14. Process output sorted.

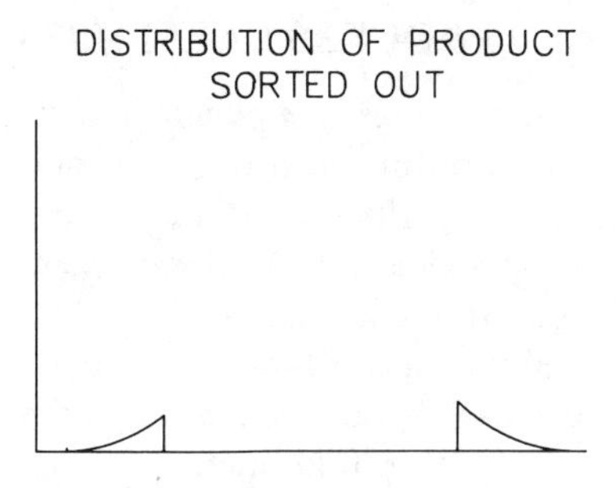

FIGURE 4.15. Sorted out material distribution.

Some suppliers will sell the sorted product at an alleged discount to try to salvage that product. Companies receiving the sorted raw material (Figure 4.15) may incur unplanned, incremental costs trying to run it.

All of the distributions shown in this section have been for populations. As stated in Section 4.1, we rarely know all of the values in a population, so we use values from samples taken from the process to predict the distribution of the population. When we deal with samples, we must understand that just because all of the sample values meet specification, the entire population may not be in specification. This can be seen by drawing a continuous curve around a histogram of the sample values (Figure 4.16). Although all of the sample values meet the specification, the curve predicts that part of the population falls outside the specification.

This is not always easy to see from a histogram of sample values. Determining if the population meets the specification can be mathematically calculated. The mathematical concept for determining whether a process that is stable or in control meets specification is known as process capability. Process capability is discussed in the next section, Section 4.5.

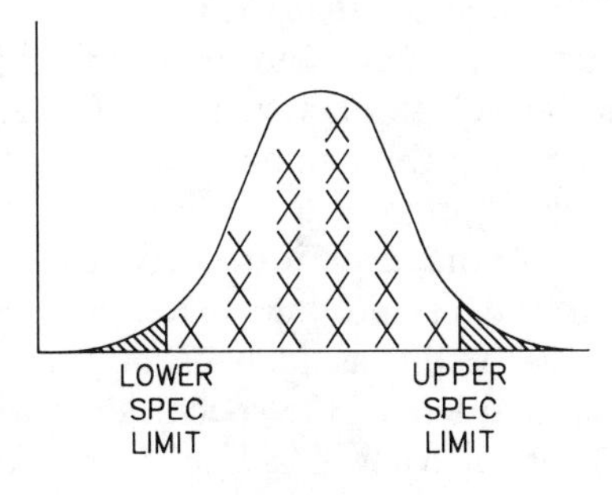

FIGURE 4.16. Sample in spec but process output not.

4.5 PROCESS CAPABILITY

Another basic SPC concept is *process capability*. Process capability is the measure of the total variation in the process output against the specifications. In plain English, it describes whether or not the process output is capable of meeting the specification. As discussed earlier, the total variation of a process output for a normally distributed process can be described by $6s$. That is, almost all of the data values (99.7%) of the output fall within $\pm 3s$ of the mean. Process capability compares the total variation, $6s$, against the specification tolerance.

It is important to know the capability of a process to produce materials that meet the specification statistically. A product sample may show all of the samples measured to be within specification, but the process is not capable of meeting specification. Despite being within specification, the statistics predict that some of the total population will be out of specification.

There are three measures of the process capability: two process capability indices C_p and C_{pk}, and the process capability ratio C_r. The process capability indices C_p and C_{pk} have become standards in industry. The choice of C_p versus C_{pk} depends on whether the process is centered (i.e., the mean is close to the midpoint of the specification) and on whether the specification is bilateral (two-sided) or unilateral (one-sided, for example having only a maximum or minimum). The process capability ratio C_r is presented because some companies continue to use it. (We don't recommend its use as a measure of process capability.)

The process capability index C_p should be used only if the process is centered and the specification is bilateral. C_p is defined by

$$C_p = \frac{\text{total specification tolerance}}{6s}$$

A value of 1 for C_p indicates that $6s$, or 99.7%, of the process output just meets the specification. There is no room for the process to move or it will be making some (more) product that is out of spec. For this reason, most companies require that the minimum C_p for their processes be 1.33. This means that the distribution of $6s$ of the process takes up 75% of the tolerance (Figure 4.17), so that even if the process changes slightly, it has a higher probability that it won't produce materials out of spec. The goal is to drive the process capability as high as possible so that the process variation takes up less and less of the specification tolerance.

A note of caution about using C_p: Use it only for processes that are centered. If the process is not centered, it can have a C_p greater than 1.33

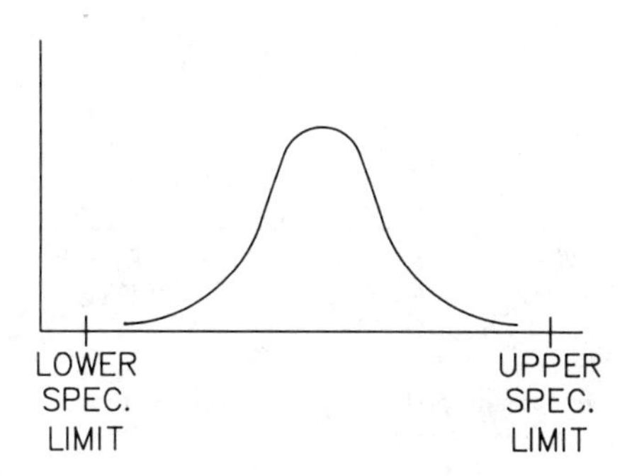

FIGURE 4.17. Process capability of 1.33—process centered.

and still manufacture product out of specification. Figure 4.18 presents such a case; the process is highly capable, but yet it shows product out of specification.

To combat this problem with the C_p statistic, a second process capability index C_{pk} was developed. C_{pk} divides the total variation of the process ($6s$) in half and compares each half to how far the process average is from the upper specification limit and then from the lower specification limit. Mathematically, this is represented by

$$C_{pk} = \min(C_{pu}, C_{pl})$$

where
$$C_{pu} = \frac{\text{upper spec. limit} - \text{process mean}}{3s}$$

$$C_{pl} = \frac{\text{process mean} - \text{lower spec. limit}}{3s}$$

If the process mean is centered in the specification, then $C_{pu} = C_{pl}$. The goal when using C_{pk} is the same as for C_p, a process capability index greater than 1.33. This again means that there is room in the specification for the process to move slightly without product being made out of spec. (Actually, even with a C_p or $C_{pk} > 1.33$, theoretically a small percent of the process output can be out of spec. Even if a process is operating at a C_{pk} of 2.0 (equivalent to Motorola's six-sigma effort), "only" 99.9999998% of the output will be in spec.)

C_{pk} can also be used with one-sided specifications. If the specification is a maximum, then use $C_{pk} = C_{pu}$. If the specification is a minimum, use

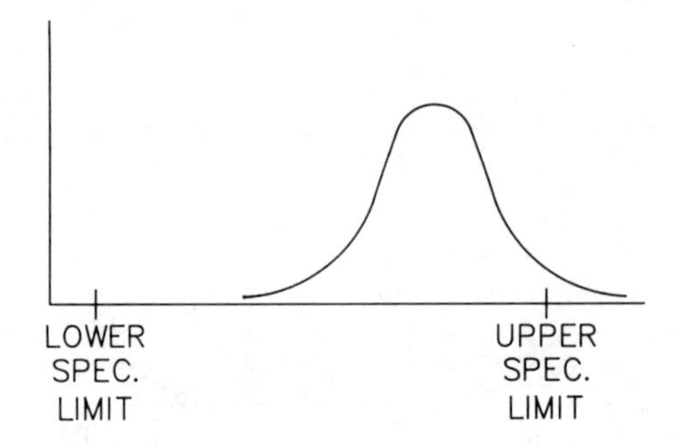

FIGURE 4.18. Process capability of 1.33—process not centered.

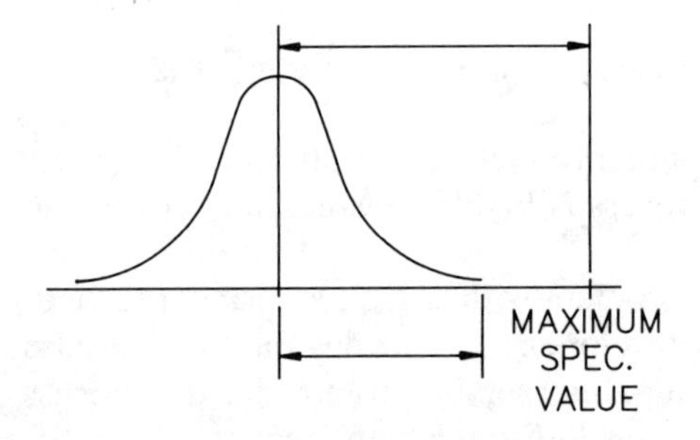

FIGURE 4.19. Process capability—maximum specification.

$C_{pk} = C_{pl}$. These are shown graphically in Figures 4.19 and 4.20, respectively.

Looking at the formula for C_{pk}, it can be seen that the C_{pk} will be a negative value if $\bar{x}$ lies outside the specification. In contrast, C_p will never be negative even if the entire distribution lies outside the specification. We could have a C_p of 1.33 and produce only noncomforming product. The

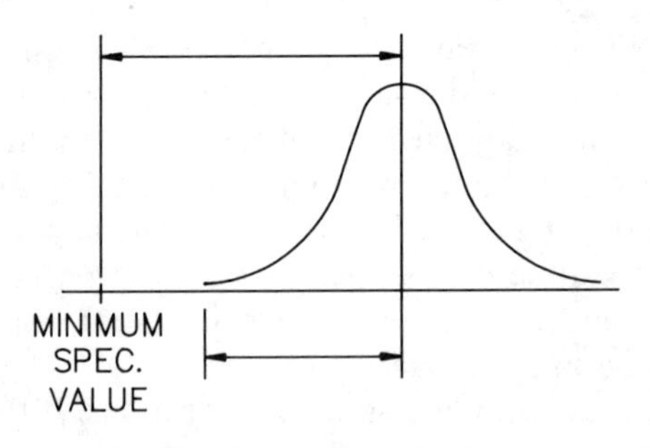

FIGURE 4.20. Process capability—minimum specification.

use of the C_{pk} instead of the C_p would keep us from being mislead by an uncentered process.

The process capability ratio C_r is simply the inverse of C_p:

$$C_r = \frac{6s}{\text{total tolerance}}$$

For those companies using C_r, their process capability objective is to drop their C_r below 0.75. (A C_r of 0.75 equals a C_p of 1.33.) Some companies use C_r because it tells exactly what proportion of the specification is taken up by the distribution of the process output. These companies try to improve their processes by moving the C_r toward zero.

The process capability ratio C_r has exactly the same limitations as C_p. It tells only that the process is capable of meeting specification, not that the process actually is meeting spec. A process can have a $C_r < 0.75$ and still produce significant quantities of materials out of specification.

Conducting a Process Capability Study

In order to conduct a proper process capability study, preparations must be made in advance. These preparations include:

1. Define the process to be studied. The process may be one machine or a group of machines that produce the product. The processing parameters should all be specified in advance. These should be the typical parameters the process is run at, unless the study is being made to see if a change in operating parameters can improve the process capability. If the process is operator dependent, select a typical operator. The study may be more of a case of operator variability than equipment variability.
2. Define the measurement system. What's going to be measured? What will it be measured with? Is the measurement equipment properly calibrated? Refer to Chapter 3 for a discussion on measurement system variation.
3. Select a sampling plan. A process capability study should typically be run with consecutive pieces. The most common capability study involves pulling 100 pieces from the process and putting them in subgroups of 5. That is, subgroup 1 is the first 5 pieces pulled; subgroup 2 contains pieces 6 to 10; and so on up to subgroup 20.

   Another approach is to pull 40 consecutive pieces without putting them into subgroups. Some texts refer to this as a short-term capability study, which it might be for a discrete part manufacturer. This approach may be most suitable for companies in the process industries. Capability studies conducted on a continuous flow process may be based on 40 batches or on 40 discrete samples pulled at regular time intervals from the process.

As a last resort, companies that face expensive destructive testing, or who don't run a product frequently enough to get 40 samples in a reasonable time can run a mini capability study. This involves taking a sample of 10 consecutive pieces. This approach is not very accurate, but it is better than nothing.

4. Develop a data collection form. As described in Chapter 2, make the form simple to use for data collection and well organized to help facilitate the calculations.
5. Take the samples. If the process involves recycle loops or takes time to warm up, allow time for it to reach steady state before sampling. Be certain to number the pieces or the samples as they are taken from the process.
6. Measure the samples.
7. Calculate the process capability.
   a. For 20 subgroups of size 5, calculate the mean $\bar{x}$ and the range R for each subgroup. Calculate the mean of the means $\bar{\bar{x}}$ (also known as the average of the averages) and the mean of the ranges $\bar{R}$ from:

$$\bar{\bar{x}} = \frac{\Sigma \bar{x}}{n} \qquad \bar{R} = \frac{\Sigma R}{n}$$

In this case, $n = 20$. The value for $\bar{R}$ is used to estimate the standard deviation s:

$$s = \frac{\bar{R}}{d_2}$$

where d_2 is a constant value that depends on the sample size. For a sample size of 5, $d_2 = 2.326$. Other values of d_2 can be found in the Appendix.

Use the values for $\bar{\bar{x}}$ and s along with the specification tolerance to calculate the process capability index C_{pk}:

$$C_{pk} = \min(C_{pu}, C_{pl})$$

$$C_{pu} = \frac{\text{upper specification limit} - \bar{\bar{x}}}{3s}$$

$$C_{pl} = \frac{\bar{\bar{x}} - \text{lower specification limit}}{3s}$$

b. If the process capability study consisted of 40 consecutive pieces or batches, calculate the mean of the samples $\bar{x}$ and the sample standard deviation s directly from the data:

$$\bar{x} = \frac{x_1 + x_2 + x_3 + \cdots + x_n}{n}$$

$$s = \sqrt{\frac{\Sigma(x - \bar{x})^2}{n - 1}}$$

in this case, $n = 40$.

The simplest way to calculate $\bar{x}$ and s would be to use a scientific calculator. The values for $\bar{x}$ and s are used with the specification tolerance to calculate the C_{pk}:

$$C_{pk} = \min(C_{pu}, C_{pl})$$

$$C_{pu} = \frac{\text{upper specification limit} - \bar{x}}{3s}$$

$$C_{pl} = \frac{\bar{x} - \text{lower specification limit}}{3s}$$

c. For the mini capability study with a sample size of 10, calculate the mean of the samples $\bar{x}$ and their range R. The value for the standard deviation s is roughly approximated from R:

$$s = 2R$$

Calculate the value for C_{pk} as shown in step 7b.

Using the Process Capability

The major use of the process capability is to determine whether or not the process can manufacture product that can meet the specification. If the value for C_{pk} is greater than 1.33, the process is capable of meeting the specification. If the value for C_{pk} is less than 1.33, we consider the process not capable from a practical standpoint.

If the process is not capable, there are three alternatives:

- Improve the process (see Chapter 5 for improvement techniques).
- Renegotiate the specification with the customer (is that tight a spec really needed?).
- Buy a new process.

Buying a new process should not be done until work to improve the process and renegotiating the specification have already been done. Until we reach an understanding of why our present process is deficient, we will have a difficult time selecting something better. Additionally, as we study and learn about our existing process, often we find and can then unlock the process "secrets", and render the existing process capable.

Managers are often approached with requests for capital for a new, "better" process. No manager should approve such a request without having data on the process capability of the existing process, and if possible, on the proposed process. Manufacturers of equipment for the process industries normally have pilot plant facilities where process capability studies can be conducted on the proposed process or the key equipment in the proposed process. We further recommend that not only the C_{pk} be shown to be better for the new process, but that the use of the improvement techniques in Chapter 5 be required prior to approval of a new process.

The capability of a process should not be kept just to the engineers, manufacturing personnel, and their managers. It is important that the sales and marketing people in the organization know the process capability also. If the C_{pk}, is high, they may be able to use it as a competitive advantage. If it is lower than competitors, they can provide feedback that process improvements are needed. Their knowledge of the C_{pk} and what it means will prevent many problems with customers if the sales and marketing personnel use it in establishing realistic specifications up front with the customers.

Process Capability for Nonnormal Distribution

The preceding measures for the process capability are all based on the process output being normally distributed. As mentioned previously, some processes do not produce output with normal distributions. There are two approaches that can be used with nonnormal distributions to measure the process capability. However, with either approach, the process output must be stable, that is, the distribution should not be changing over time.

The first approach is a simple graphical technique. It may not give an exact process capability, but it should be adequate. Follow steps 1 to 6, described previously, to collect data on the process output. Then plot the data on a histogram and draw a smooth curve around it. Be sure not to be too abrupt when drawing the tails of the curve. Approximately where the two tails hit the x axis, read their values. Refer to Figure 4.21, which shows the tails at values of 1.59 and 1.73. The process capability would be the total tolerance of the specification divided by the difference between the two tails:

$$\text{Process capability} = \frac{\text{upper spec. limit—lower spec. limit}}{\text{value of upper tail—value of lower tail}}$$

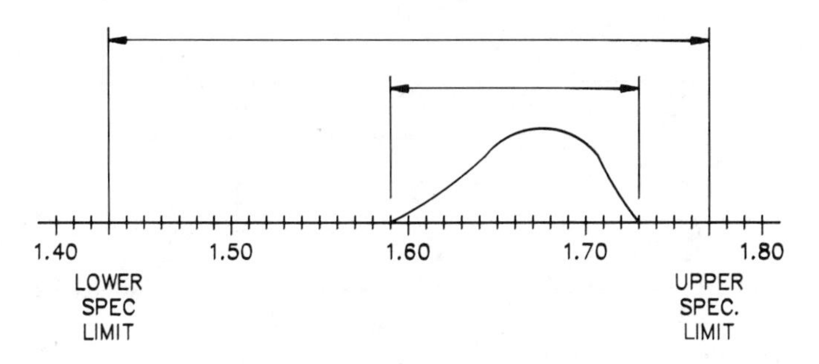

FIGURE 4.21. Process capability—nonnormal distribution.

In the figure shown, the process capability would be

$$\text{Process capability} = \frac{1.76 - 1.43}{1.73 - 1.59} = 2.36$$

Like the process capability index C_p, this measure of process capability does not tell whether or not the process output is within specification; it only tells that the process is capable of producing output within the specification provided that it is targeted correctly.

If the process is stable but skewed, an equation similar to the process capability index C_p can be developed using the mode to describe the location and the differences between the mode and the tails to describe the process variation:

$$\text{Process capability} = \min\left(\frac{\text{upper spec. limit—mode}}{\text{value of upper tail—mode}}, \frac{\text{mode—lower spec. limit}}{\text{mode—lower tail value}}\right)$$

The sample shown in Figure 4.22 has a mode of 1.68. The process capability would be

$$\begin{aligned}\text{Process capability} &= \min\left(\frac{1.76 - 1.68}{1.73 - 1.68}, \frac{1.68 - 1.43}{1.68 - 1.59}\right) \\ &= \min(1.60, 2.78) \\ &= 1.60\end{aligned}$$

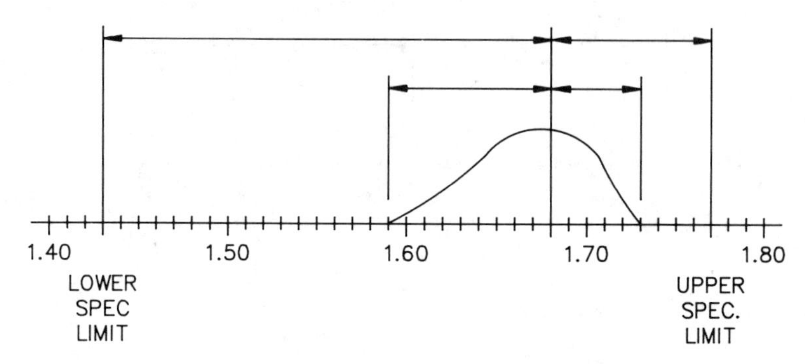

FIGURE 4.22. Using the mode to calculate process capability for a skewed distribution.

These two measures described do not have the same statistical basis as the process capability calculations discussed earlier for a normally distributed process. They are only "quick" approximations based on a best guess at the shape of the curve. A more precise approach to determining the process capability of a nonnormal distribution would be with a computer software. (Several are on the market.) The computer program would itself generate the process capability after fitting the inputted data to a curve.

Long Term Process Capability

A process capability study is normally done under controlled conditions. It is overseen by an engineer, technician, or manufacturing supervisor, and it is done with the process shown in the best light. The process capability, then, would not be taking into account things such as ambient condition effects, the effects of operators tweaking the process, and the effects of the process not being set up exactly the same, among others. In order to measure the process capability taking all of the causes of variation inherent in the process into account, a long term process capability study should be made.

A long term process capability study is conducted by using data from control charts (described in subsequent Sections 4.5 to 4.9) as input. The length of time for the capability study can be as long as needed to account for various sources of variation. For example, in a normal process capability study, the process is not shut down and started up repeatedly or set up by various operators. This type of activity may be an additional source of variation. They're not really special causes because they are part of the normal operation of the process. Because they are basically common causes of variation, it may be desirable to conduct a capability study over a longer period of time to take all common causes of variation into account.

To conduct a long term capability study, normally, more than 20 subgroups of size 5 would be used. A capability study over a month or more (to take into account multiple setups and start-ups) may use more than 100 subgroups. The procedures for calculating the process capability would be exactly the same as outlined for 20 subgroups of 5.

Summary

The process capability study ties the process variation to the customer's specification. The appropriate type of study depends on its goal and the distribution of the output. The standard capability study should be the most frequent technique used, although there are places to use the capability study for nonnormal distributions and the long term capability study.

4.6 CONTROL CHARTS FOR VARIABLE DATA

The control chart was developed by Dr. Walter Shewhart of Bell Laboratories as a tool to distinguish controlled from uncontrolled variation in a process. As covered earlier, histograms can also describe the variation present and whether it stems from common causes (in control) or special causes (out of control). However, control charts view the process in real time; they provide the immediate feedback on a process output that histograms do not.

The control chart graphically presents the output of a process over time and compares this output to statistical limits using statistically based interpretation techniques. They tell when the operation is in a state of statistical control and they can quickly direct attention to the process when special causes of variation appear. This allows the process to be adjusted to bring it back into control. Control charts can prevent the occurrence of problems with the process output because patterns of output can be interpreted and a decision can be made to shut down or adjust the process before a defect occurs.

Control chart theory is based on the normal distribution. As such, control charts should be used with processes with normally distributed outputs either individual or sample averages (following the central limit theorem). The power of the control chart derives from the probabilities

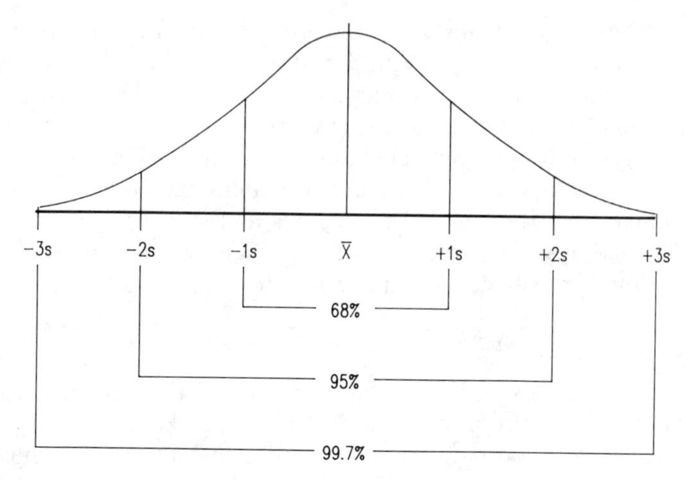

FIGURE 4.23. Distribution of the normal curve.

associated with the normal curve:

- That 50% of the distribution lies above the center of the distribution and 50% below.
- That most of the data points lie close to the center
- And that 99.73% of all the data lie within $\pm 3s$ (refer to Figure 4.23).

The control chart also relies on the randomness of the points under the normal curve. If the process fails to follow these probabilities, or is not random, the process is not stable and a special cause of variation is present that must be reacted to. For example, the probability of 7 points in a row falling on one side of the center line is $(0.50)^7 = 0.78\%$. There is less than 1 chance in 100 that this will occur, so it's likely that something has changed on the process and a special cause of variation is present. If we set limits to the process at $\pm 3s$, the chance of a point falling outside these limits is 27 in 10,000; again, a special cause is likely to be present when this occurs. Section 4.6.4 (Interpreting Control Charts) covers more on the probabilities of control charts.

The four major types of control charts for variable data are:

1. Range	R chart
2. Average	$\bar{x}$ chart
3. Sample standard deviation	s chart
4. Individuals	x chart

The R and the s charts monitor the variation in the process output whereas the $\bar{x}$ and the x charts monitor the location of the process output. Because both the variation and the location should be known and tracked, these charts are combined into an $\bar{x}$ & R chart, $\bar{x}$ & s chart, or x & R chart. Variations-to-variables data control charts can increase their functionality. Techniques such as control charts using normalized data, charts with modified limits, charts for batch operations and short runs, and precontrol charts are presented in Section 4.9.

Attribute data can also be control charted as percentages or counts. Section 4.7 presents p charts (percentages or proportion of nonconforming units), np charts (number of nonconforming units), u charts (percentage or proportion of nonconformities), and c charts (number of nonconformities).

4.6.1 $\bar{x}$ & R Charts

The $\bar{x}$ & R chart monitors the process location and the process variation as measured by the average and range of the output, respectively. Figure 4.24 shows a typical $\bar{x}$ & R control chart.

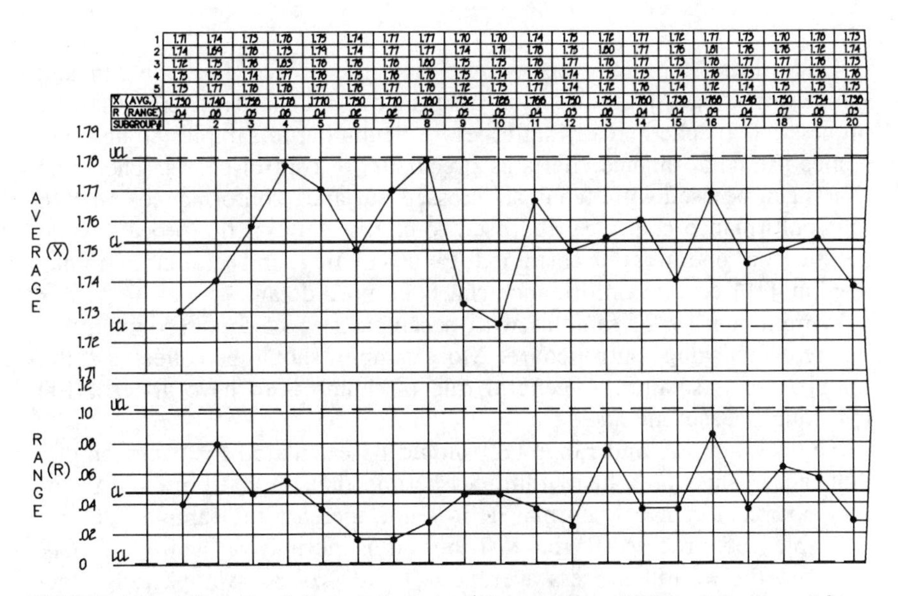

FIGURE 4.24. Typical x & R control chart. (*Reprinted from SPC In Action, copyright 1990, Resource Engineering, with the permission of the publisher, Quality Resources, White Plains, New York.*)

Before an $\bar{x}$ & R chart, or any other control chart, can be established for a process, several steps must be taken:

1. Define the process. The process may be a piece of equipment operating independently, or a group of equipment operating sequentially or interactively. The process may also be a procedure that yields an output that is not a manufacturing process, such as the hiring process or order entry process.
2. Define the variable(s) or characteristic(s) to be monitored. Frequently, these are defined by the requirements or specifications for a product as agreed on with the customer. These may also be defined by problems (e.g., yield losses) or areas that are desirable to maintain control over (e.g., indirect labor costs or on-time shipments).
3. Define the measurement system. The measurement system includes the test equipment, method, tester, and sample. The measurement system must be repeatable and reproducible. Without this, the control chart may actually be monitoring the variation in the measurement system rather than the variation in the process output. Chapter 3 covers this subject more completely and the next section, section 4.6.2, discusses sampling techniques for control charts.

Once these steps have been completed, the process must be run and sampled to collect data for establishing the control charts. Be sure that the process has reached steady state before beginning to sample. Some companies use conformance charts as precursors to control charts. The same charts can be used with "control" crossed out and "conformance" written in. Conformance charts can be used to collect data on the process while getting the operators accustomed to calculating and graphing points. Section 4.9.1 covers conformance charts in more detail.

A minimum of 20 samples with a sample size of 5 is recommended before establishing control charts. More samples should be collected if the sample size is smaller. A general rule of thumb is to have at least 100 individual measurements.

The average ($\bar{x}$) and range (R) should be calculated for each sample. Plot these values on a histogram to ascertain that the pattern of distribution is normal or nearly normal. If it is not, increase the sample size and resample the process. If the distribution is normal or nearly normal, calculate the overall process average ($\bar{\bar{x}}$) and process average range ($\bar{R}$) from the sample averages and ranges. These values will serve as the center lines, or "targets" for the $\bar{x}$ and R charts. To calculate the center lines:

$$CL_x = \bar{\bar{x}} = \frac{\Sigma \bar{x}}{n} \qquad CL_R = \bar{R} = \frac{\Sigma R}{n}$$

where $\bar{x}$ = sample average
R = sample range
n = number of samples

In order to know whether or not the process is stable or in control, control limits must be established for both the $\bar{x}$ and R charts. Each chart has an upper control limit (UCL) and lower control limit (LCL), which are symmetrical about the center lines (although for smaller sample sizes, there will be no LCL for the R chart). A stable process runs within the control limits (99.73% of the time) with no nonrandom patterns of variation. The control limits are set at $\pm 3s$ from the center lines using R to estimate the standard deviation:

$$UCL_{\bar{x}} = \bar{\bar{x}} + A_2\bar{R} \qquad UCL_R = D_4\bar{R}$$

$$LCL_{\bar{x}} = \bar{\bar{x}} - A_2\bar{R} \qquad LCL_R = D_3\bar{R}$$

where A_2, D_4, and D_3 are constants whose values vary with the sample size. Tables with these values can be found in the Appendix.

Once the center lines and control limits are calculated, they should be drawn on the control charts (this is explained in Section 4.6.3). The data from the measured samples should be entered onto the control chart and plotted. If the points are all in control (interpreting control charts is explained in Section 4.6.4), the control chart can be used. If there is an indication that the process is out of control, an investigation must be made into the cause of the out-of-control condition. Once this cause is eliminated, the process can be resampled, new center lines and control limits calculated, and the control chart put into use.

The values for the center lines and control limits should be reviewed on a regular basis (e.g., quarterly) similar to the calibration frequency for measurement equipment. If the most recent data show that the control limits are wider, it should cause an alarm. The control limits should not be widened—instead, the causes of this increased variation should immediately be investigated. In an operation where we are making small, steady improvements, we would expect the process to be operating closer to the target and with less and less variation over time.

Control limits should also be recalculated after any improvement to the process, such as an equipment change or a procedure change.

4.6.2 Control Chart Sampling Techniques

Much of industry uses a sample size of 5 for control charts. This stems from the early days of control chart use, because the average can easily be calculated by doubling the sum of the sample values and moving the decimal point one place to the left. With the advent of low cost calculators, there is no need to continue this convention. The sample size for control charts should depend on the pattern of distribution for the individual values, the cost per sample, and the frequency with which the process changes. The more skewed the distribution of the individuals, the greater the sample size should be (central limit theorem). If the distribution is nearly normal, the sample size should be small. The goal is to select a sample size so that the distribution of the sample averages ($\bar{x}$) is normal. A general rule of thumb:

Distribution of Individuals	Sample Size
Heavily skewed	10 or greater
Moderately skewed	5
Nearly normal	2

These samples must be rational. A rational sample is one where all of the pieces are taken under the same processing conditions. This ensures that the variation within the sample arises from common causes and that special causes of variation are not present. For this reason, samples taken for control charts should be set aside sequentially until the sample size is reached. Then, all of the sample parts should be measured at the same time. Parts put out by different machines or operators should not be sampled together unless t tests, F tests, or χ^2 tests have previously assured that the outputs from the different machines or operators are statistically the same. If rational samples cannot be obtained (e.g., if the process is shifting while the sample is being taken), control charts should not be used. Work on reducing the variation in the process until a rational sample is obtainable.

There are times when larger samples are desirable even though the process is not heavily skewed. The larger the sample size, the narrower the control limits are on the $\bar{x}$ chart and the easier to detect smaller changes in the process. It may be important to use large sample sizes if the ability of the process to meet the specification is questionable. This will help prevent defects. If the sample size is greater than 12, the R chart should be converted to an s chart.

The preceding sample sizes do not take into account the cost of sampling. Economically, it may not be possible to take a large sample size of an expensive item if testing is destructive. Where this is a concern, the reader should refer to *Introduction to Statistical Quality Control* by Douglas Montgomery (1985). This presents an entire chapter on the economic design of control charts, which outlines an optimization approach to minimize overall costs, sampling and measurement costs, and the costs of failure to detect a process change.

The frequency of sampling depends on the frequency with which the process changes. The more frequently the process shifts, the more frequently the process should be sampled, because the goal of control charts is to detect process changes. The frequency may be either time based (e.g., 1 sample every hour) or count based (e.g., 1 sample every 100 parts). The initial capability study on the process may give an indication of how frequently the process changes. In some cases, it may be possible to use the time it took to manufacture the parts for the capability study as the initial sampling frequency. This is not likely to be practical for high volume production.

If the frequency cannot be determined from the process history, start with a sampling frequency of every 15 or 30 min. Adjust the frequency based on the process performance. If a sample is taken every 30 min and process changes are not being caught, increase the frequency (or the sample size as an alternative). As reductions in the process variation are made, the frequency can be reduced. A reasonable goal may be to reach a sampling frequency of once or twice per shift.

4.6.3 Establishing an $\bar{x}$ & R Control Chart

The control chart is broken up into four portions:

- An information portion at the top.
- The data collection portion (in the middle).
- The plotting portion (also in the middle).
- A comment area for change control at the bottom.

Filling out each of these sections is critical to the successful use of the chart. The graphs for plotting may seem to be the most important, but without the data and information, investigations into special causes of variation will be more difficult to complete. Similarly, without the comments, an event that created a special cause may remain hidden.

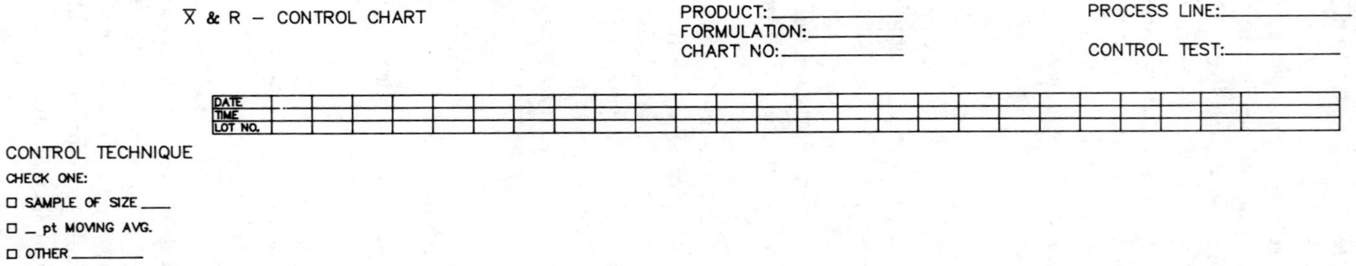

$\overline{X}$ & R – CONTROL CHART

PRODUCT: ____________
FORMULATION: ________
CHART NO: ____________

PROCESS LINE: ____________
CONTROL TEST: ____________

DATE																									
TIME																									
LOT NO.																									

CONTROL TECHNIQUE

CHECK ONE:

☐ SAMPLE OF SIZE ___

☐ _ pt MOVING AVG.

☐ OTHER ________

FIGURE 4.25. Information portion of an x & R chart.

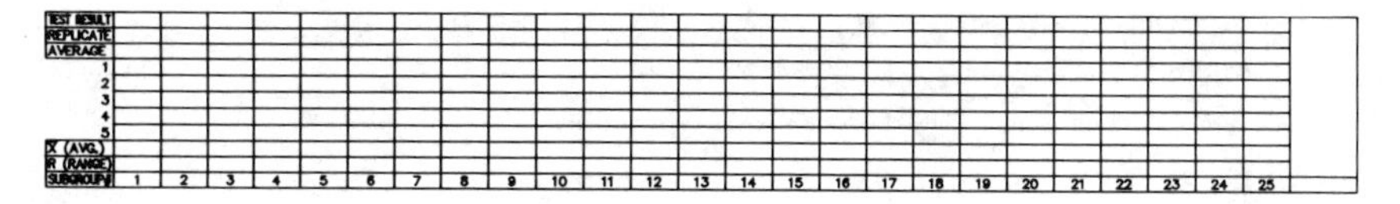

FIGURE 4.26. Data portion of an x & R chart.

The information portion (Figure 4.25) normally contains basic information on the product and the chart: product name, formulation of part number, variable measured, measurement technique, and chart number. Charts should always be modified to collect specific information needed by that particular process or plant.

The data portion (Figure 4.26) records some information on the data, the actual measurements made, and the calculations for $\bar{x}$ and R performed on those measurements. The sampling technique refers to how samples of the product were taken from the production run. Time is when the sample was taken. This may be the actual chronological time (e.g., 2:00 pm or 14:00) or the point in the run (e.g., part 100).

In the data portion, the operator should also record his/her initials or clock number. This will aid in investigating changes in the process should it go out of control. Enter the raw data in the blanks provided. The average ($\bar{x}$) and range (R) are calculated from the raw data and entered onto the chart.

In the graph portion (Figure 4.27), select the scales based on the values of the data. A general rule of thumb for the range chart is to set the lower value at 0 and the upper value at 1.5 to 2 times the largest range calculated during start-up. The scale for the average chart is often set at one-half the scale for the range chart. These scales should provide enough

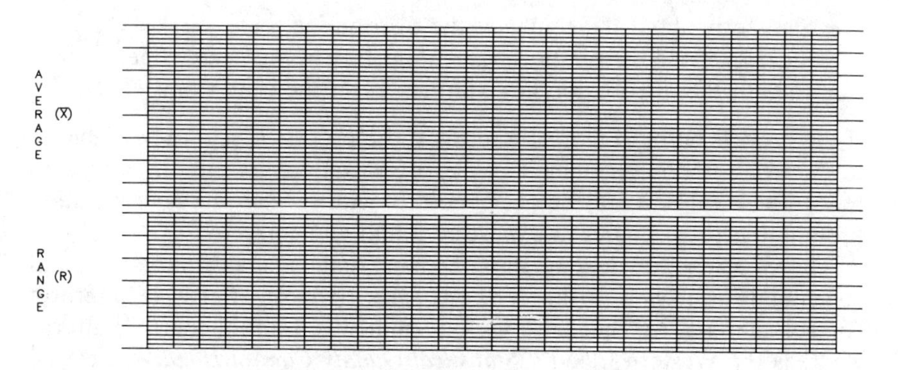

FIGURE 4.27. Graph portion of an x & R chart.

NUMBER ANY CORRECTIVE ACTION FOR OUT OF CONTROL CONDITIONS DIRECTLY ON THE CHARTS ABOVE, EXPLAIN BELOW IN COMMENT SECTION.

ALSO ANY CHANGE IN PEOPLE, MATERIAL, EQUIPMENT, METHOD OR ENVIRONMENT MUST BE NOTED BELOW. THE NOTES WILL HELP YOU TAKE CORRECTIVE ACTION AND IMPROVE THE PROCESS IN THE FUTURE.

DATE	TIME	NO.	COMMENT	DATE	TIME	NO.	COMMENT	DATE	TIME	NO.	COMMENT
		1				3				5	
		2				4				6	

FIGURE 4.28. Comments portion of an x & R chart.

room so that changes can be seen. If the plotted data later look like a straight line or there are points off the graph, the scales should be reset.

Once the scales have been set, the center lines should be entered on both the $\bar{x}$ and R charts as solid lines, and the control limits as dashed lines. The average and range values from the data portion are then plotted on the graphs. Solid lines are used to connect successive points. If the process or product runs intermittently, use a dashed line or no line to connect the last point of the previous run with the first point from the next run.

The comment portion (Figure 4.28) is much more important than most people treat it as being. All changes in the process equipment, manpower, methods, or materials should be recorded here. Whenever the process is adjusted, it should be noted. Comments help direct investigations of the process variation. Without the comments, special causes of variation may remain undetected or the problem-solving may take much longer than need be.

4.6.4 Interpreting Control Charts

Because the basis for control chart theory follows the normal distribution, the same rules that govern the normal distribution can be used to interpret control charts. These rules include:

- Randomness.
- Symmetry about the center of the distribution.
- 99.73% of the distribution lies within $\pm 3s$ of the center line.
- 95% of the distribution lies within $\pm 2s$ of the center line.

If the process output follows these rules, the process is said to be stable or in control with only common causes of variation present. If it fails to follow these rules, it may be out of control with special causes of variation present. These special causes must be found and corrected.

There is a great deal of disagreement among quality literature and consultants over how to apply these rules to most effectively interpret control charts. Dr. Wheeler (1984) recommends 3 out-of-control patterns whereas the Western Electric *Statistical Quality Control Handbook* (1956) describes more than 11 patterns showing when the process is out of

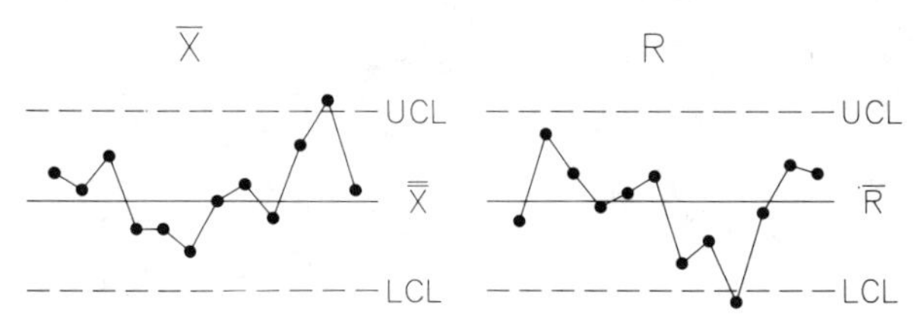

FIGURE 4.29. Single point above or below the control limits.

control. Actually there are nearly an infinite number of control chart out-of-control patterns because any nonrandom pattern shows the process to be out of control. There are six patterns we recommend concentrating on:

1. A single point above or below the control limits (Figure 4.29). The control limits are set at $\pm 3s$ so the probability of a point falling outside when the process is in control is less than 0.14%. This pattern may indicate:
   - A special cause of variation from a material, equipment method, operator, or measurement system change.
   - Mismeasurement of a part or parts.
   - Miscalculated or misplotted data points.
   - Miscalculated or misplotted control limits.
2. Seven points on one side of the center line (Figure 4.30). Because the probability of a point falling above or below the center line is 50–50, the probability of seven points in a row on one side is $0.5^7 = 0.78\%$. (Some literature suggests we use eight points on one side to decrease the probability to 0.39%.) This pattern indicates a shift in the process output from changes in the equipment, methods, or materials or a shift in the measurement system.
3. Two points in a row close to the UCL or LCL (Figure 4.31). In a normal distribution, the probability of two points in a row lying

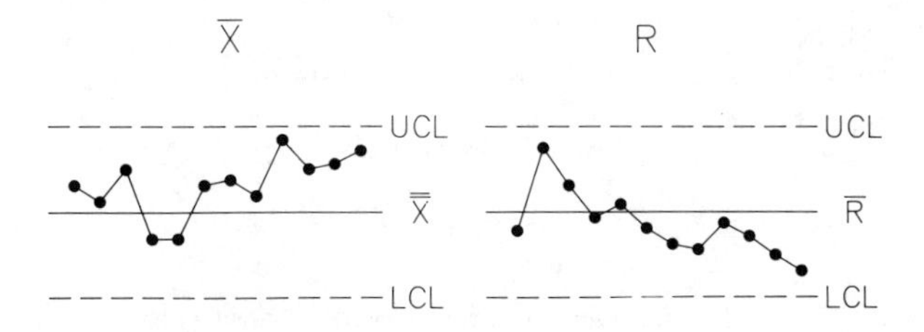

FIGURE 4.30. Seven points on one side or the center line.

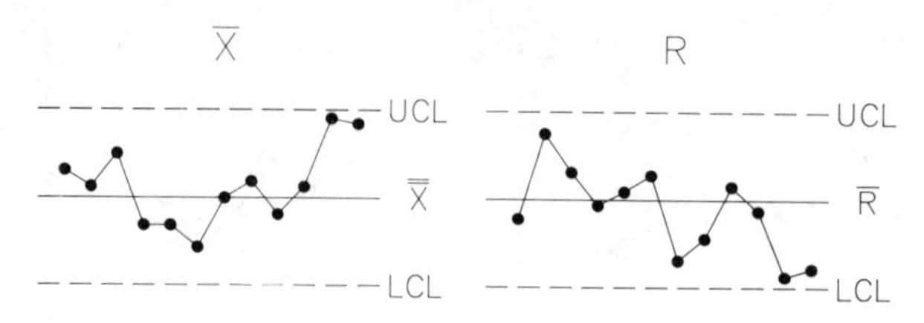

FIGURE 4.31. Two points in a row near the UCL or LCL.

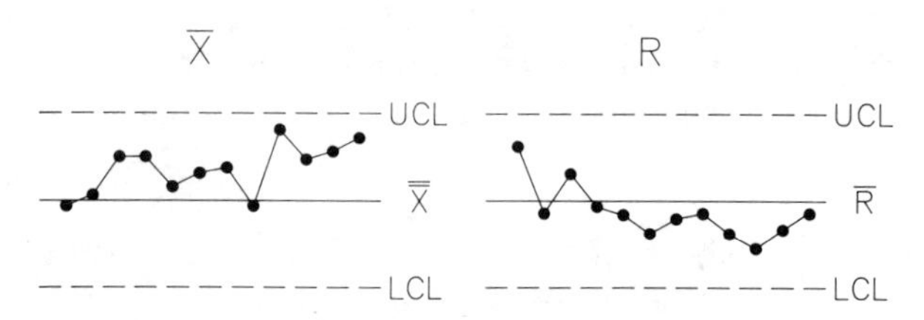

FIGURE 4.32. Ten of 11 points on one side of the center line.

between $\pm 2s$ and $\pm 3s$ from the center line is 0.05%. This may be the result of a large shift in the process in the equipment, methods, materials, or operator or a shift in the measurement system. It could also result from using the same chart to control the output from multiple process output streams or multiple processes.

4. Ten out of 11 points on one side of the center line (Figure 4.32). The probability of this occurring in a controlled process is 0.54%. This also indicates a shift in the process output.
5. A trend of seven points in a row upward or downward (Figure 4.33) demonstrates nonrandomness. This may show:
   - Gradual deterioration or wear in equipment.
   - Improvement or deterioration in technique.

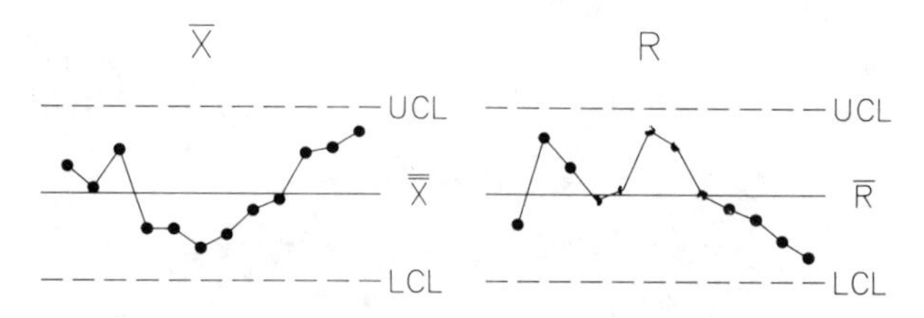

FIGURE 4.33. Trend of seven points in a row upward or downward.

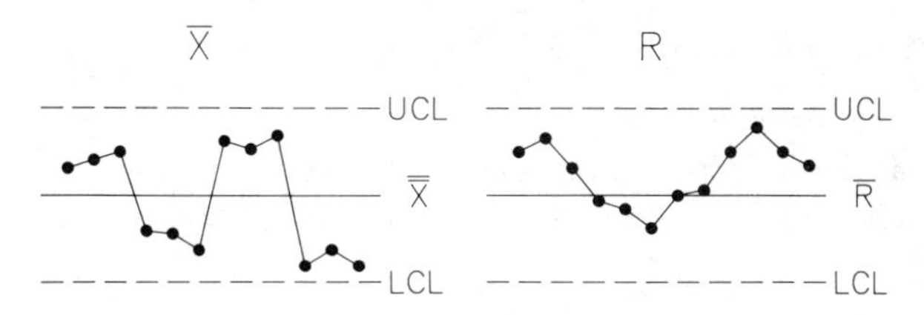

FIGURE 4.34. Cycling.

- Operator fatigue.

6. Cycling (Figure 4.34), also a demonstration of nonrandomness. This may be the result of:
   - Temperature or other recurring changes in the environment
   - Differences between operators or operator techniques.
   - Regular rotation of machines.
   - Differences in measuring or testing devices that are being used in order.

Engineers and managers must use some common sense when reviewing control charts. If the pattern shown on the control chart looks strange, but is not one of the preceding six patterns, it could still, of course, indicate a problem with the process. If in doubt, plot the control chart data on a histogram to see if the distribution is normal.

Control charts should not just be used for problem recognition. Some nonrandom patterns of variation may actually show that the process has improved. A process output with the average falling very close to the center line, or the range being close to 0 for several samples in a row may show a change in the process. Again, this is a sign to take action to identify the reason for the positive change. If it is actual improvement, then the special cause should be found so that it can be incorporated permanently into the process.

On the other hand, it may not be real improvement. Perhaps the operator is plotting using the wrong scale, or making a math error, or maybe the gauge is jammed. So regardless of whether the nonrandom pattern of variation shows improvement or deterioration, take action and identify the special cause.

Control charts with out-of-control conditions should be acted upon immediately. Many sources recommend shutting down the process when this occurs. This should not always be the first action, especially for a process that has run in control for a long time and has a single point that falls out. You should first:

- Check the calculations.
- Look at the individual values and remeasure any that appear strange.
- Investigate the process and the measurement system.

Normally, interpret the R chart first, because it is more sensitive to a change in the process, then, the $\bar{x}$ chart, and finally, look at the two charts together. Their interaction may assist in identifying the special cause of variation.

4.6.5 $\bar{x}$ & s Charts

When the sample size is greater than 12, the sample standard deviation (s) should be used rather than the range (R) to monitor the variation in the process. For larger sample sizes, the s chart is a better measure because it is more sensitive to a shift in the process. However, a word of caution: the

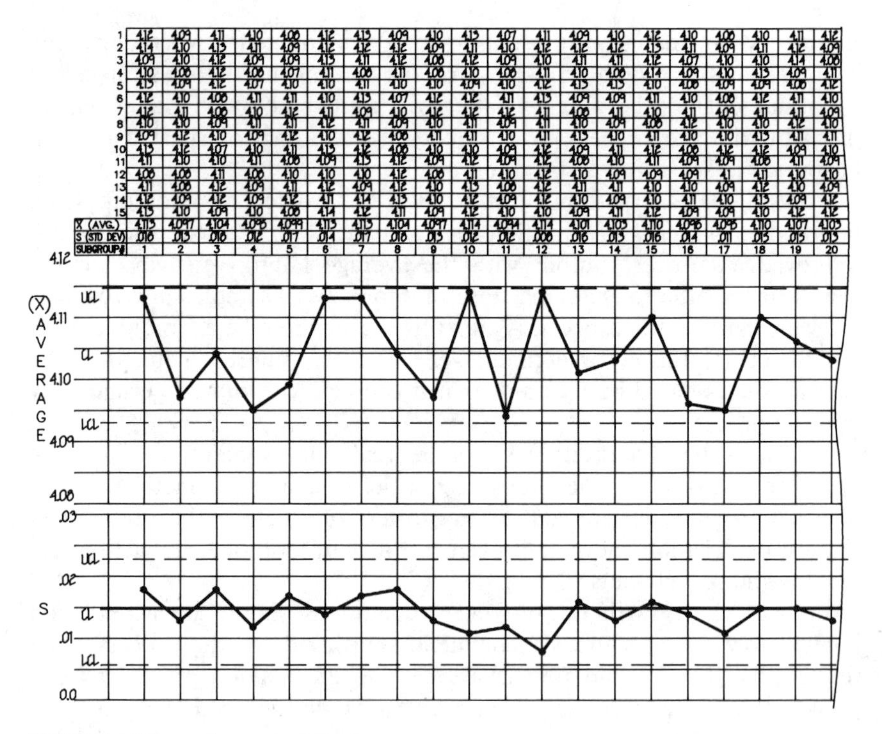

FIGURE 4.35. x & s control chart.

s chart is less sensitive in identifying nonnormal conditions that cause only a single value in a sample to be unusual. If this repeatedly occurs, the sampling techniques should be reexamined.

The center lines for $\bar{x}$ & s charts are $\bar{\bar{x}}$ and $\bar{s}$, respectively (see Figure 4.35). The calculation for $\bar{\bar{x}}$ has been shown previously in Section 4.6.1. The center line for the s chart is calculated by averaging the sample standard deviations of the individual sample groups:

$$\mathrm{CL}_s = \frac{s_1 + s_2 + \cdots + s_k}{k} = \bar{s}$$

where k = sample number and
s_k = sample standard deviation for sample k

The control limits are calculated from

$$\mathrm{UCL}_{\bar{x}} = \bar{\bar{x}} + A_3\bar{s} \qquad \mathrm{UCL}_s = B_4\bar{s}$$

$$\mathrm{LCL}_{\bar{x}} = \bar{\bar{x}} - A_3\bar{s} \qquad \mathrm{LCL}_s = B_3\bar{s}$$

The construction, use, and interpretation of $\bar{x}$ & s charts follow the same rules as for $\bar{x}$ & R charts.

4.7 CONTROL CHARTS FOR ATTRIBUTE DATA

Just like variable or measured data, control charts can be established for attribute or count data. It is just as important to know when a process is in control or out of control for an attribute as it is for a variable. In the process industries, yield is an important attribute over which to maintain control. In part manufacturing operations, cosmetic defects should be maintained in control. In administrative or service areas, the number of complaints can be monitored with attribute control charts.

There are four types of attribute control charts:

	Nonconforming Units	Nonconformities
Numbers	*np*	*c*
Proportion	*p*	*u*

The simplest to use are the number charts, *np* or *c*, because the total numbers of nonconforming units or nonconformities are charted. A drawback of these is that the sample size must remain constant. The pro-

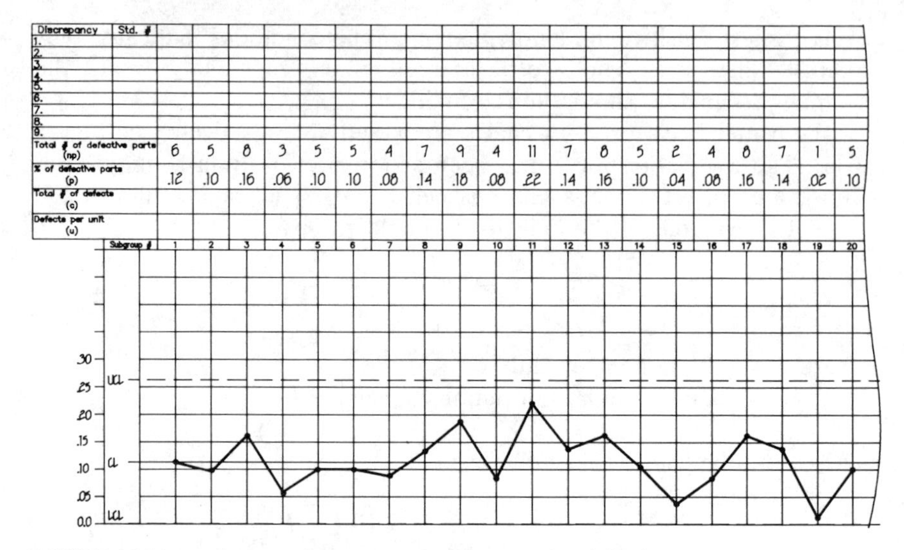

FIGURE 4.36. *p* chart monitors percentage or proportion defective.

portional charts, p or u, monitor the fraction or percentage of units nonconforming or nonconformities per unit. These are more complex and more difficult for operators to use. However, these charts can adjust for varying sample size.

4.7.1 *p* Charts

The p and np control charts are based on the binomial distribution, so they must follow the same rules as a binomial distribution. The attribute being looked at must have two mutually exclusive outcomes and must be independent from one part or event to the next. For example, the parts are either good or they are bad, or the shipment was made or it was not.

A p chart (Figure 4.36) is similar to an $\bar{x}$ or R chart in that is has a center line $\bar{p}$ and upper and lower control limits UCL_p and LCL_p.

The center line $\bar{p}$ is calculated from

$$\bar{p} = \frac{\text{number of nonconforming items or events}}{\text{number of items or events inspected}}$$

$$\bar{p} = \frac{(np)_1 + (np)_2 + \cdots + (np)_k}{n_1 + n_2 + \cdots + n_k}$$

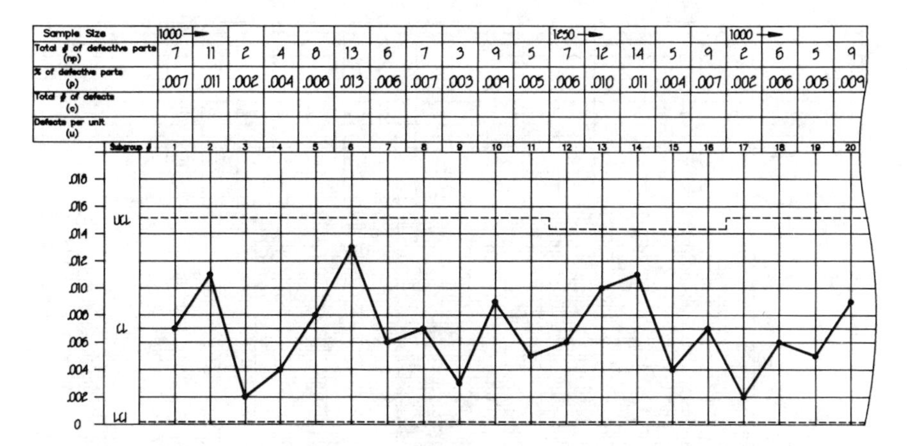

FIGURE 4.37. *p* chart with control limits changing as the sample size changes.

where k = sample number
n_k = number of items or events in sample k
$(np)_k$ = number of nonconforming items or events in sample k

The control limits are calculated from

$$\mathrm{UCL}_p = \bar{p} + 3\sqrt{\bar{p} + (1 - \bar{p})/n}$$

$$\mathrm{LCL}_p = \bar{p} - 3\sqrt{\bar{p}(1 - \bar{p})/n}$$

where n = sample size.

It is this relationship to n that allows the p chart to adjust to varying sample sizes (refer to Figure 4.37).

With a p chart and an np chart, it is possible that the LCL_p calculates to be negative. In this case, put a LCL_p line on the chart at a value of 0. Some recommend not putting a LCL_p line on at all because $p = 0$ falls within the random variation predicted for that process. However, we feel that a convention of including both the UCL and the LCL lines on the chart should be maintained.

A word of caution: If each sample is a different size, recalculating limits for each sample will be extremely ponderous. A more practical approach would be to calculate control limits based on the average sample size (n) and to recalculate limits only for those samples that exceed $n \pm 25\%$. Wherever possible, structure the control plan to collect samples with a constant size.

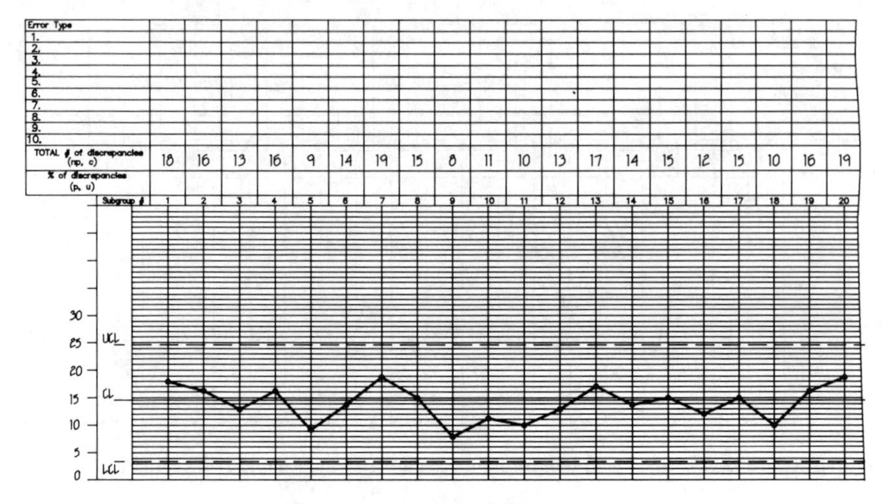

FIGURE 4.38. *np* chart monitors number defective.

4.7.2 *np* Charts

The *np* chart (Figure 4.38) looks at the number of nonconforming items or events in a sample. It uses the same basic information as a *p* chart, except that the total number of nonconforming items is charted rather than the proportion of nonconforming items. For this reason, it is simpler to use than a *p* chart. A drawback is that there must be a constant sample size. If this drawback can be overcome, many people prefer the *np* chart because of its simplicity.

The center line for an *np* chart is the average number nonconforming $(\overline{np})$:

$$\overline{np} = \frac{(np)_1 + (np)_2 + \cdots + (np)_k}{k}$$

where k = sample number
$(np)_k$ = number nonconforming in each sample

The upper and lower control limits are calculated from

$$\text{UCL}_{np} = \overline{np} + 3\sqrt{\overline{np}\left(1 - \left((\overline{np})/n\right)\right)}$$

$$\text{LCL}_{np} = \overline{np} - 3\sqrt{\overline{np}\left(1 - \left((\overline{np})/n\right)\right)}$$

where n = the sample size

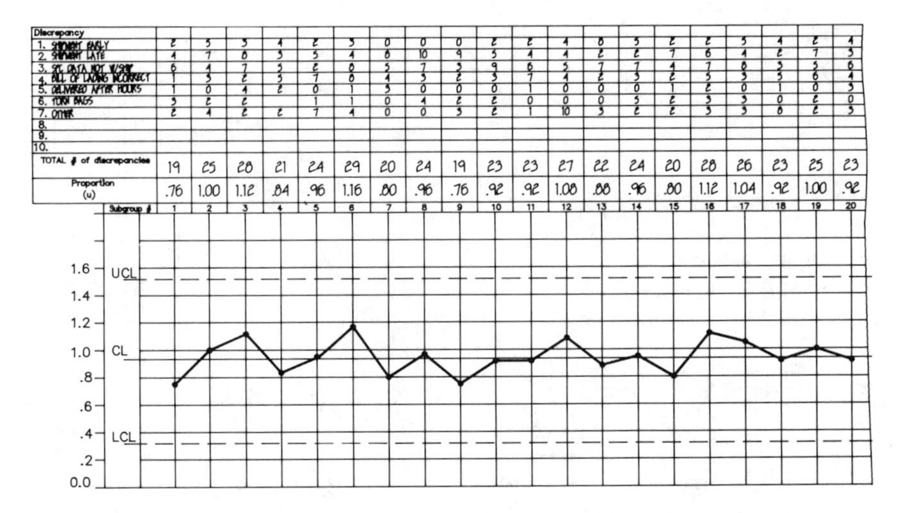

FIGURE 4.39. *u* chart monitors proportion of defects.

4.7.3 *u* Charts

The *u* chart and the *c* chart are based on the Poisson distribution. The *p* or *np* charts look at only two exclusive events: conforming versus nonconforming. More complex items or events, such as inspections of a continuous product or of television sets, require the use of *u* or *c* charts. The *u* chart looks at nonconformities or defects per unit. Like the *p* chart, it is a proportion and can accept a varying sample size. The *c* chart looks at total number of conformities. It is similar to the *np* chart and must also have a constant sample size.

The center line of a *u* chart (Figure 4.39) is the average nonconformities per unit ($\bar{u}$):

$$\bar{u} = \frac{c_1 + c_2 + \cdots + c_k}{n_1 + n_2 + \cdots + n_k}$$

where k = sample number
c_k = number of nonconformities in sample k
n_k = size of sample k

The control limits for a *u* chart are:

$$\text{UCL}_u = \bar{u} + 3\sqrt{\bar{u}/n}$$

$$\text{LCL}_u = \bar{u} - 3\sqrt{\bar{u}/n}$$

where n = sample size

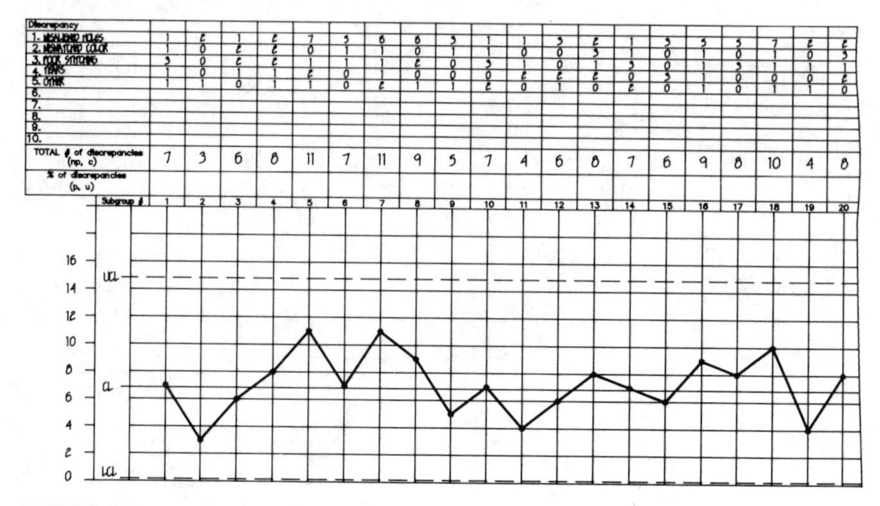

FIGURE 4.40. *c* chart monitors the number of defects.

The same problems exist with the *u* chart as with the *p* chart when the sample size varies constantly. Here also, many people construct the control limits based on the average sample size (n) and only recalculate limits if a particular sample size exceeds $n \pm 25\%$.

4.7.4 *c* Charts

The *c* chart (Figure 4.40) measures the number of nonconformities or defects in a sample.

The center line of the *c* chart is average number of nonconformities ($\bar{c}$):

$$\bar{c} = \frac{c_1 + c_2 + \cdots + c_k}{k}$$

where k = sample number
c_k = number of nonconformities in each sample

The upper and lower control limits are calculated from

$$\text{UCL}_c = \bar{c} + 3\sqrt{\bar{c}}$$

$$\text{LCL}_c = \bar{c} - 3\sqrt{\bar{c}}$$

4.7.5 Interpreting Attribute Charts

One concern with the u and c charts is that as u or c gets smaller, the potential of a false signal for an out-of-control conditions increases to 3 to 4%. This is because the Poisson distribution is skewed when the average is small. Replacing the $\pm 3s$ control limits with probability control limits will reduce this chance of a false signal to the level selected. A more detailed explanation is available in *Understanding Statistical Process Control* by Wheeler and Chambers (1986).

With all four of the attribute charts, there is the possibility that the LCL will calculate to be negative. This especially exists when the average is small. In the case where the LCL < 0, draw the lower control limit at a value of 0. Some feel that no lower control limits should be drawn in this case because the value of 0 falls within the normal probability of the distribution. However, we feel that control charts should be standardized to have both upper and lower control limits.

Interpretation of p, np, c, and u charts is done similarly to interpretation of $\bar{x}\,\&\,R$ charts. A point outside the control limits indicates an out-of-control process. Trends or patterns within control limits can be examined like an $\bar{x}\,\&\,R$ chart when np or c are greater than 9 because the distribution becomes nearly normal. When the average is less than 5, the distribution is not nearly normal, so the rules for interpretation for the normal curve-based $\bar{x}\,\&\,R$ charts cannot be strictly followed. This obviously reduces the power of these control charts. However, this should not totally dissuade their use, because the majority of out-of-control conditions are shown by a point falling outside a control limit. Those concerned about this basic weakness of attribute charts should use probability-based limits or individual–moving range ($x\,\&\,R_m$) charts.

4.8 SPECIAL CONTROL CHARTING TECHNIQUES

Some processes are difficult to apply standard control charts to. Standard control charts may not adequately monitor and control operations involving batch processes, short runs, and processes with multiple output streams. In cases such as these and others, special control charting techniques can be employed.

Control charting techniques for batch process depend on the homogeneity within the batch and between batches. If the concern is within-batch variation, then standard $\bar{x}\,\&\,R$ charts can be applied. However, most batch processes yield a product that is nearly homogeneous so that the within-batch measure of variation R is very small. This variation may actually be the variation in the measurement system rather than in the process. Using the R value to calculate control limits will give a control chart with tight limits that almost always shows the process to be out-of-control from batch to batch.

Several techniques can replace $\bar{x}$ & R charts, and the charts most often used for batch operations. An individual and moving range (x & R_m) chart, where the moving range monitors the variation between batches, a triple chart monitoring the average ($\bar{x}$), the within batch variation (R_w), the variation between batches (R_m), and the use of modified control limits are other tools. These will be discussed in Section 4.8.1 to 4.8.3.

Many job shops and some specialty chemical processing firms have short runs of a variety of products, rather than continuous runs of the same product. These short runs may not bè long enough to show that a special cause of variation is present in the process. Normalized control charts, discussed in Section 4.8.4, can pick up both special causes in the process and identify problem products.

A third area for special control charts is with processes that have multiple output streams. It is usually impractical to chart each stream of output. This is not a problem if the streams are related, such as the output streams from a simple distillation column. Here, a single output can be charted with the knowledge that if that stream goes out of control, there is a high probability that the other stream(s) has also gone out of control. If the streams are not related, group control charts can be applied to monitor the process. Group control charts monitor the overall process by monitoring the streams with the highest and lowest averages and the one with the greatest variation. Section 4.8.5 reviews this technique.

4.8.1 Individual and Moving Range Charts (x & R_m) Charts

The x & R_m chart can be applied in a variety of situations. It is most frequently used in batch operations to monitor the batch-to-batch variation when the within-batch variation is known to be nearly zero due to batch homogeneity. Some statisticians also recommend x & R_m charts in place of attribute charts. We agree; the reasoning is that the values plotted on attributes are really just individual values, so we can use the simpler x & R_m chart. Another use of these charts occurs when measurement costs prohibit larger sample sizes. This often happens when testing involves destructive testing on expensive products.

x & R_m charts (Figure 4.4.1) are constructed similarly to $\bar{x}$ & R charts. The difference is that the values are individuals and not subgroups. Therefore, the range must be calculated from batch to batch, not within the subgroup. The moving range is normally calculated from successive

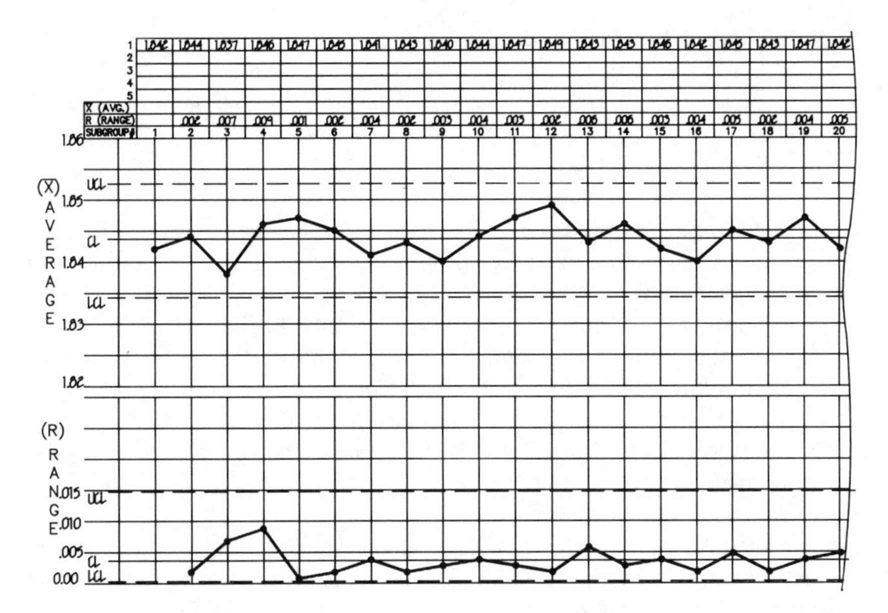

FIGURE 4.41. $x \& R_m$ chart.

pairs ($n = 2$) of data. However, it can also be calculated from three or more moving data points if the distribution of the moving range is not nearly normal with $n = 2$.

The steps for constructing an constructing an $x \& R_m$ control chart follow:

1. Take 25 individual readings over time and record them on a control chart.
2. Using successive pairs, calculate and plot the moving ranges. Because pairs of data are being used, there will be no range value to match the first individual reading. The first range value will be calculated between the first and second individual readings. The next range will use the second and third readings. (If the moving range is calculated from more than two successive data points, there will be no range value for the first $n - 1$ subgroups. The moving range will be the difference between the highest and the lowest values within the n subgroups.)
3. Calculate the center lines for the charts from:

$$CL_x = \bar{x} \qquad CL_{R_m} = \overline{R}_m$$

where $\bar{x}$ = the average of the individual readings
$\overline{R}_m$ = the average of the moving ranges

4. Calculate the control limits:

$$UCL_x = \bar{x} + E_2\bar{R}_m \qquad UCL_{R_m} = D_4\bar{R}_m$$

$$LCL_x = \bar{x} - E_2\bar{R}_m \qquad LCL_{R_m} = D_3\bar{R}_m$$

   where E_2, D_3, and D_4 are constants based on the value of n, the number of points used for the moving range. These can be found in the Appendix.
5. Plot the center lines and the control limits on the charts.
6. Plot the values for x and R_m on the charts. Interpret the control charts (see succeeding test). If the process is in control, deploy the charts on the floor. If it is not, work to identify special causes of variation. Once they have been eliminated, repeat the preceding steps.

As in an $\bar{x}$ & R chart, the x & R_m control chart is interpreted using the range first and then the individuals. Interpretation is done following the same rules as for x & R charts; however, care must be used. The range values are interrelated because they are calculated from a common data point or points. For this reason, trends are difficult to identify correctly. Additionally, because x & R_m charts are less sensitive to process change compared to $\bar{x}$ & R charts, there is a greater chance of not reacting fast enough to out-of-control conditions.

In operations where batches are later blended together to make a final product lot, or if the histogram of the individual values does not show a normal distribution, a moving average along with the moving range may prove useful. The sample size (in this case, the sample size is the number of individual values used to calculate the moving average) should be expanded until the histogram shows normality. The moving range should be calculated for the same sample size as the moving average.

4.8.2 $\bar{x}$, R_w, and R_m Charts for Batch Operations

This technique merely combines an $\bar{x}$ & R chart with an R_m, moving range, chart (see Figure 4.42). The $\bar{x}$ chart monitors the location of the process averages for the batches. The R chart is changed in name to an R_w chart and monitors the within-batch variations. The R_m chart monitors the variation between batches.

Because the R_w chart is actually a normal range chart, its control limits are calculated the same as for the R chart. In this technique, however, R_w

is not used to calculate the limits for the $\bar{x}$ chart. The average moving range R_m is used in its place. This is because the batch-to-batch variation is usually greater than the within-batch variation, $R_m > R_w$. The control limits for the $\bar{x}$ chart will better reflect what the overall process is doing using the greater (R_m) value.

A two-point moving average is normally used, so R_m is simply the difference between the measurements from successive batches. The average moving range $\bar{R}_m$ is calculated from

$$\bar{R}_m = \sum R_m/(n - 1)$$

where n = number of batches

The denominator is $(n - 1)$ because, by definition, there cannot be a moving range associated with the first sample. Once $\bar{R}_m$ is calculated, the center lines and control limits for the charts can be calculated:

Chart	CL	UCL	LCL
$\bar{x}$	$\bar{\bar{x}}$	$\bar{\bar{x}} + 3\bar{R}_m/d_2$	$\bar{\bar{x}} - 3\bar{R}_m/d_2$
R_m	$\bar{R}_m$	$D_4\bar{R}_m$	$D_3\bar{R}_m$
R_w	$\bar{R}_w$	$D_4\bar{R}_w$	$D_3\bar{R}_w$

Be careful in selecting the values for d_2, D_3, and D_4 from the Appendixes. For the $\bar{x}$ and R_m charts, the number of batches used for the moving range defines the values for d_2, D_3, and D_4 used in their control limit calculations. The values of D_3 and D_4 for the R_w chart are determined by the number of measurements made on each batch.

These charts are interpreted the same as directed for $\bar{x}$ & R charts in Section 4.6.4 and for R_m charts in Section 4.8.1. This technique can also be combined with the normalized, short-run chart approach, described in Section 4.8.4, should the situation warrant it.

4.8.3 Modified Control Limits

Charts with modified control limits are not true control charts because they incorporate the product specification limits. However, they are another tool that can be used for monitoring batch processes where the within-in batch variation is small and the batch-to-batch variation is much greater. They are also useful for processes where the within sample range is small, but the average varies significantly from sample to sample. In order to use modified control limits, the within-sample or within-batch range must be stable. Modified control limits are designed to keep the

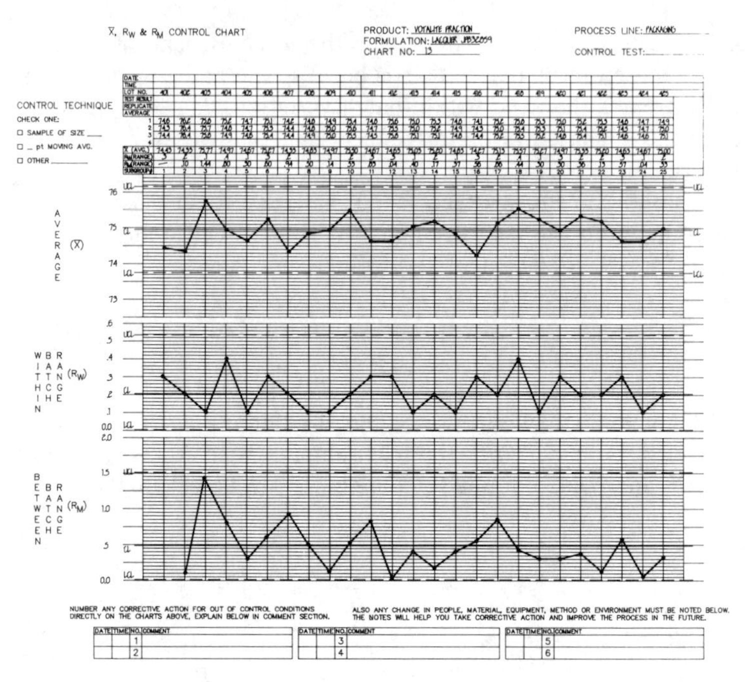

FIGURE 4.42. $\bar{x}$, R_w, and R_m chart.

process and product within specification. This technique should not be employed over the long term for processes that have shifts in their average, but should instead be employed while efforts are underway to reduce the excessive shifts in the process average. Modified limits are applied to the $\bar{x}$ chart only.

Charts with modified control limits (Figure 4.43) have two center lines plus upper and lower control limits. The limits are calculated from:

$$\text{upper center line} = \text{upper spec limit} - 3\overline{R}/d_2$$

$$\text{lower center line} = \text{lower spec limit} + 3\overline{R}/d_2$$

$$\text{UCL} = \text{upper center line} + A_2\overline{R}$$

$$\text{LCL} = \text{lower center line} - A_2\overline{R}$$

where $\overline{R}$ is the mean of the within-batch ranges. In interpreting $\bar{x}$ charts with modified control limits, points above or below the control limits indicate the process is out of control and should be investigated to find the

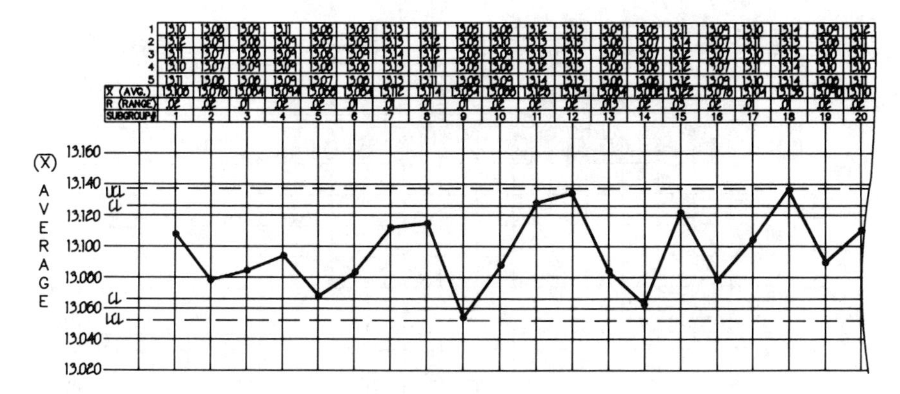

FIGURE 4.43. Control chart with modified control limits.

special cause. Patterns or trends between the center lines and their corresponding control limits are interpreted with the same rules as for $\bar{x}$ charts with standard limits. Patterns and trends between the two center lines are not interpreted and are not used as the basis for any action on the process.

4.8.4 $\bar{x}$ and R Control Charts for Short Runs

One of the most frequent problems faced in the process industries is how to control processes that have many short runs of many products. For example, a continuous reactor may produce several different types of resins and several grades of each type. The typical control chart approach would be to have a control chart for each grade of each resin. Although this approach is correct, it is *not practical* to ensure that the process is in a state of statistical control. Process cycles, trends, and other changes will not be noticed because the data will be spread across several control charts. Most of the control charts will be tucked in files waiting for the next run of that grade of resin. It is unlikely that the operator will recognize that the process has gone out of control until problems affecting the product quality occur.

There are several SPC approaches for short runs. The most common are nominal $\bar{x}$ and R charts, normalized $\bar{x}$ and R charts, and precontrol (or rainbow) charts. Nominal and normalized $\bar{x}$ and R charts will be discussed here. Nominal and normalized charts control all products made on a process with the *same* control chart. It should be noted here that neither of these techniques is "operator friendly". Both involve more complex calculations than standard $\bar{x}$ and R charts.

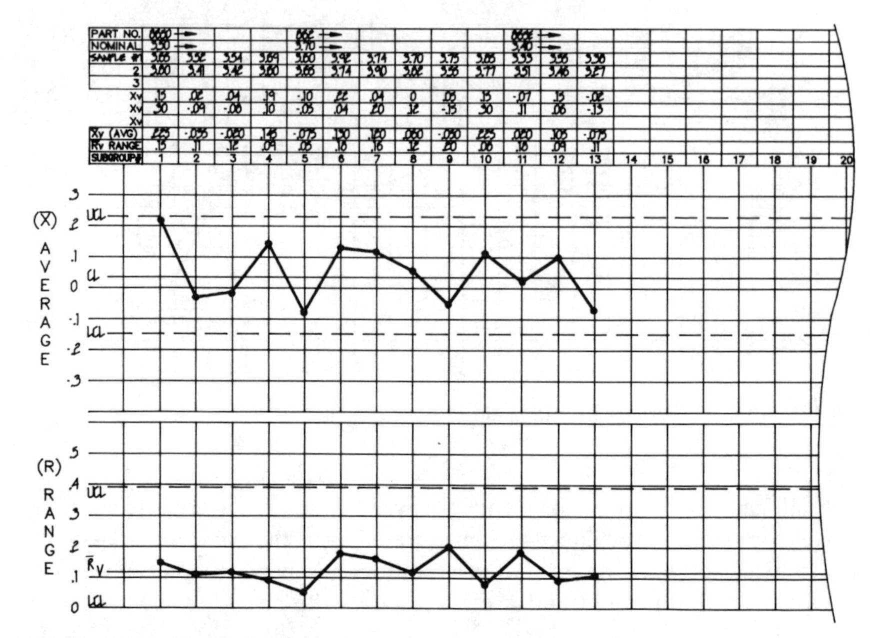

FIGURE 4.44. Nominal control chart.

Precontrol charts look at the process output versus the specification using individual charts. They may be suitable for some short run applications. Precontrol charts are discussed in Section 4.9.

4.8.4.1 Nominal $\bar{x}$ and R Charts

The nominal control chart (Figure 4.4.4) *uses the nominal value of the specification for each product as a reference point and looks at the sample measurements as variations from these reference points*. The control limits are calculated from these variations from the reference points. This technique is limited to products that have nearly the same standard deviations. Products that have different standard deviations will show out-of-control points each time they are run. In this case, use normalized $\bar{x}$ and R control charts as shown in Section 4.8.4.2.

To construct a nominal chart:

1. Measure samples from a run of product A. The sample values are denoted by x_1 to x_n, where n equals the sample size.
2. Determine the nominal value of the specification for product A. The nominal value is the central value of the specification range. If a specification has not been set, use the historical mean of the process as the nominal value.

3. Calculate the variation of sample 1 from the nominal value:

$$x_{v1} = x_1 - \text{nominal}$$

   Repeat this for samples 2 to n.
4. Calculate the average and range of variation for x_{v1} to x_{vn} for product A:

$$\bar{x}_v = \frac{\Sigma x v_i}{n}$$

$$R_v = \text{highest value of } \left[x_{v_1} \text{ to } x_{v_n}\right] - \text{lowest value of } \left[x_{v_1} \text{ to } x_{v_n}\right]$$

   (Use the approach in steps 1 through 4 to generate data during runs of product A.
5. When the process is switched to run product B, repeat steps 1 through 4 for using samples from product B and the nominal specification value of B. Use the same approach for products C, D, and so forth.
6. Once 25 sets of average and range values have been calculated, they are used to calculate the center lines and control limits for the nominal $\bar{x}_v$ and R_v charts for the process, regardless of the product. Standard formulas for $\bar{x}$ and R charts control limits are used:

	Nominal $\bar{x}_v$	Nominal R_v
Center line	$\bar{\bar{x}}_v = \frac{\Sigma \bar{x}_v}{n}$	$\bar{R}_v = \frac{\Sigma R_v}{n}$
UCL	$\bar{\bar{x}}_v + A_2 \bar{R}_v$	$D_4 \bar{R}_v$
LCL	$\bar{\bar{x}}_v - A_2 \bar{R}_v$	$D_3 \bar{R}_v$

   where $n = 25$ values in this case
7. Draw the center lines and control limits on the $\bar{x}$ and R charts. Plot the $\bar{x}_v$ and R_v values on the chart. Draw vertical lines on the chart to segregate products. If all of the points are in control, put the chart into use. If one or two parts are out of control, drop those points and recalculate the center lines and limits. If more than two points are out of control, a control chart cannot be used *unless* it is confirmed that one of the products yields all of the out-of-control points. If this is the case, the variation in that product differs significantly from the other products manufactured with that process. Then, that product and its measurements should be dropped from the nominal control

chart and a separate control scheme used for that product. Collect more data on the other products if necessary to replace the data from the product dropped.

4.8.4.2 Normalized $\bar{x}$ and R chart

The normalized control chart is used in similar situations as the nominal control chart. Multiple products running through a process can be tracked in order to determine if the process is in a state of statistical control. *The normalized chart goes one step* beyond the nominal chart because even *products with different standard deviations can be plotted on the same chart*.

In the normalized control charts, the values on the average and the range control charts are normalized by the variation in the process as measured by $\bar{R}$. The average chart plots $(\bar{x} - \bar{\bar{x}})/\bar{R}$ whereas the range chart plots $R/\bar{R}$, where $\bar{x}$ is the sample average for a given product, $\bar{\bar{x}}$ is the overall process average for the product, $\bar{R}$ is the average range for that product, and R is the range of the product sample.

The normalized control charts have the following parameters:

Normalized $\bar{x}$ Chart	Normalized R Chart
Center line = 0	Center line = 1
UCL = $+A_2$	UCL = D_4
LCL = $-A_2$	LCL = D_3

The formulas for plotting and the control chart parameters are derived from the acceptance regions for values of $\bar{x}$ and R for standard $\bar{x}$ and R charts. In an $\bar{x}$ chart, if $\bar{x}$ is within the region defined by the control limits, $\bar{\bar{x}} \pm A_2 R$, then the process location is in control. Likewise, for an R chart: If R is within the region bounded by $D_4\bar{R}$ and $D_3\bar{R}$, the variation is in control. The normalized control chart parameters are derived from these acceptance regions.

Normalized $\bar{x}$ Chart	
The acceptance region is	$\bar{\bar{x}} + A_2\bar{R} \leq \bar{x} \leq \bar{\bar{x}} - A_2\bar{R}$
Subtracting $\bar{\bar{x}}$ yields	$+A_2\bar{R} \leq \bar{x} - \bar{\bar{x}} \leq -A_2\bar{R}$
Dividing by $\bar{R}$ gives	$+A_2 \leq \dfrac{\bar{x} - \bar{\bar{x}}}{\bar{R}} \leq -A_2$

Normalized R Chart	
The acceptance region is	$D_4\bar{R} \leq R \leq D_3\bar{R}$
Dividing by $\bar{R}$ gives	$D_4 \leq \dfrac{R}{\bar{R}} \leq D_3$

To use normalized control charts, the overall process average $\bar{\bar{x}}$ and the average range $\bar{R}$ must be known for each product being manufactured with the process being controlled. It is suggested that these be determined from a *minimum of 100 data points*. That is, 20 samples of size 5 or the equivalent. With these known, the control chart is constructed in the same manner as the standard control chart:

1. Take samples from a run of product A and measure them. The measured values are denoted by x_1, to x_n, where n equals the sample size.
2. Calculate the sample mean $\bar{x}$ and range R:

$$\bar{x} = \frac{\Sigma x_i}{n}$$

$$R = \text{highest value of } [x_1, \ldots, x_n] - \text{lowest value of } [x_1, \ldots, x_n]$$

3. Calculate the normalized process average from $(\bar{x} - \bar{\bar{x}})/\bar{R}$ and the normalized range from $R/\bar{R}$, where $\bar{\bar{x}}$ and $\bar{R}$ are specific to product A.
4. Plot these results on the normalized $\bar{x}$ and R chart (see Figure 4.45) with its center lines and control limits as discussed earlier in this section. Repeat these calculations for the rest of the run for product A (and for subsequent runs of product A.)

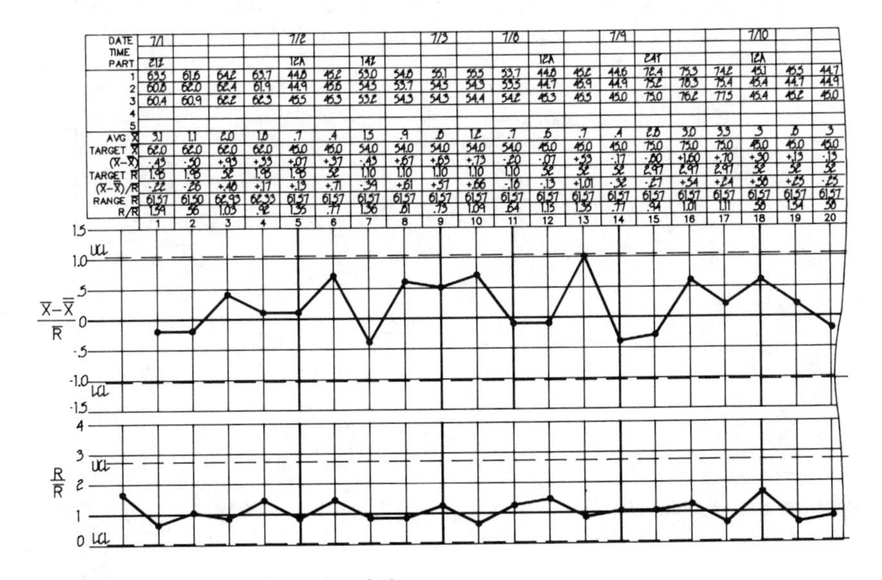

FIGURE 4.45. Normalized control chart.

ONE CHART ON A MACHINE

One Short Run $\overline{X}$, R chart at a machine can monitor all part numbers run over this machine (no matter what characteristic is tracked).

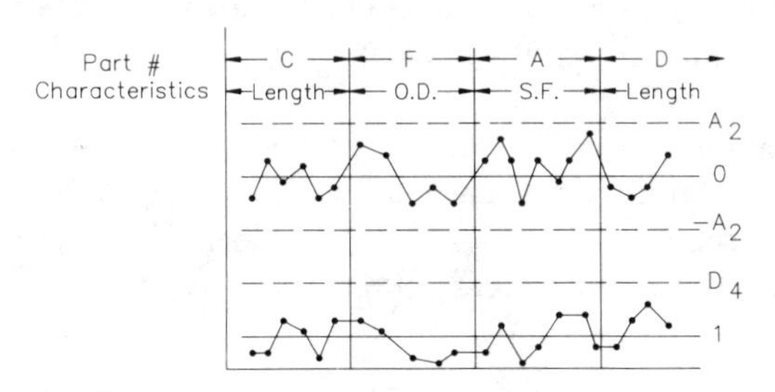

FIGURE 4.46. Nominal chart for various characteristics.

5. Repeat steps 1 to 4 for products B, C, and so forth using the values for $\bar{\bar{x}}$ and $\overline{R}$ associated with those products.

Normalized control charts for short runs can be used in many innovative ways.

- A single chart can be used to monitor various characteristics for the different products run on a machine (Figure 4.46).
- Multiple characteristics can be tracked on the same charts (Figure 4.47).
- A product may be followed from operation to operation through the process on the same control chart (Figure 4.48).

CHARTING DIFFERENT SUPPLIERS

Receiving inspection can track different suppliers (even for different characteristics) all on one short Run chart.

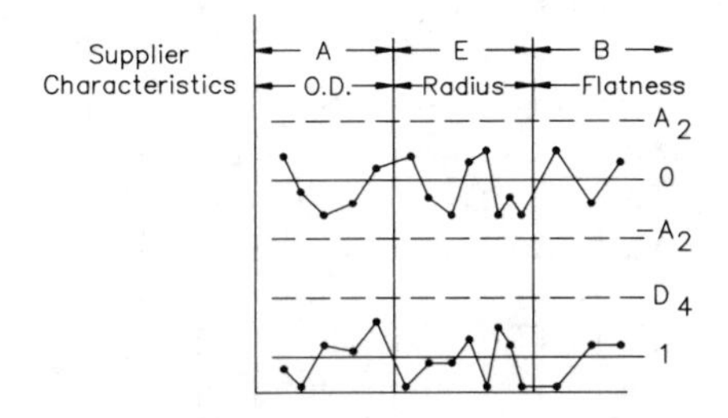

FIGURE 4.47. Multiple characteristics normalized chart. (*Courtesy of Rogers Corporation.*)

CHART FOLLOWS PART

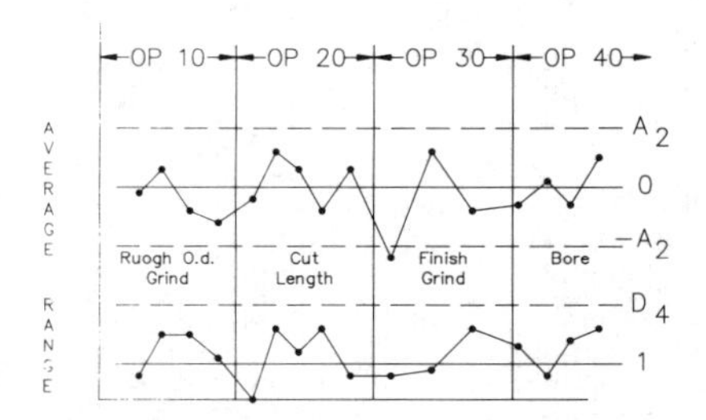

FIGURE 4.48. Normalized control chart for monitoring products from operation to operation. (*Courtesy of Rogers Corporation.*)

4.8.4.3 Interpretation of Nominal and Normalized Control Charts

There is nothing special about interpreting nominal and normalized control charts. Both are based on standard control chart theory so the same rules apply for determining if a process has gone out of control. Section 4.6.4 covers these rules.

One note of caution: Before shutting down a process for being out of control, check the calculations first. As stated earlier, nominal and normalized charts are *not operator friendly*. They need constant reinforcement and retraining to ensure they are being used correctly. Many out-of-control points, especially those outside the control limits, may be attributable to errors in calculating rather than the process actually going out of control.

4.8.5 GROUP CONTROL CHARTS

Group control charts are $\bar{x}$ & R charts used to monitor processes with multiple output streams. These streams may later be combined (for examples, a process with multiple lathes or screw machines manufacturing a single product or a molding operation using multicavity molds) or the streams may remain separate (as with the outputs from a chemical process such as a distillation column). Group control charts are much more efficient and practical in most cases than applying separate control charts to each output stream.

In the group control chart (Figure 4.49), only the output streams with the highest and lowest averages are plotted on the $\bar{x}$ chart and only the

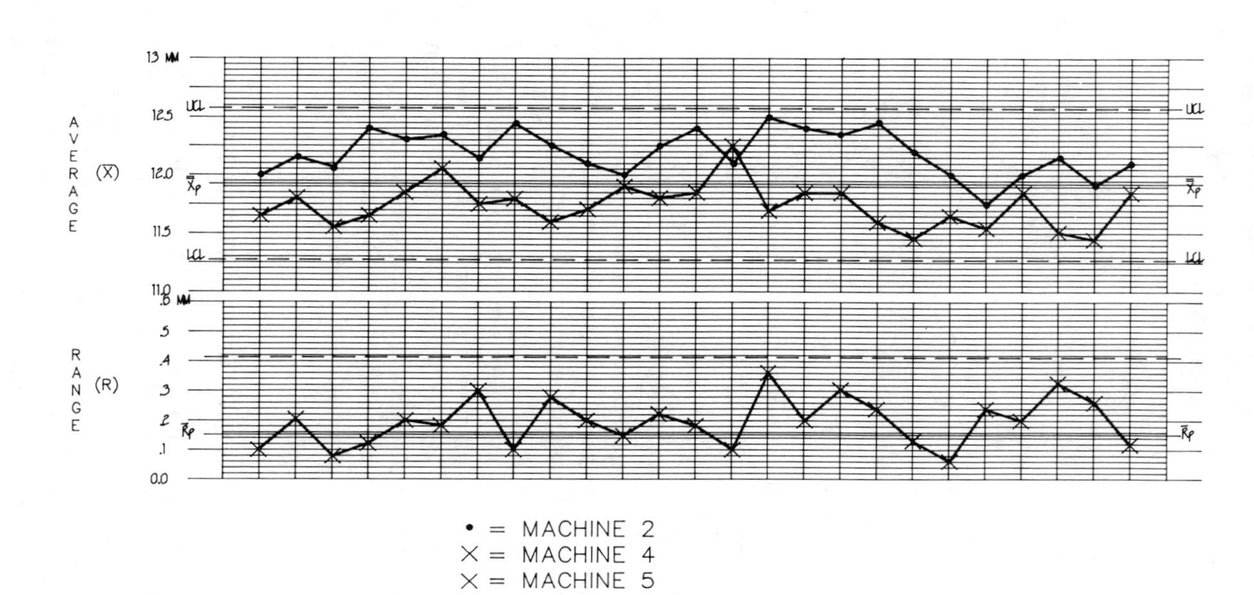

FIGURE 4.49. Group control Chart.

output stream with the highest range on the R chart. If the output streams are related as in a chemical process, as long as these streams remain in control so will the other streams normally. If the output streams are not related significantly, this SPC tool becomes limited to identifying special causes that relate to all streams. This could occur, for example, in machining or molding operations where only special causes associated with raw material or operator changes could be identified. Problems with individual lathes or individual cavities will not be identified unless the problem happens to occur in the two or three streams being monitored.

To establish a group control chart:

1. Take samples of each output stream as though a separate $\bar{x}$ & R chart were being set up for each stream. Use a constant sample size for all streams.
2. Calculate the average $\bar{x}$ and range R for each sample from each stream.
3. Use the $\bar{x}$ and R values to calculate the overall average $\bar{\bar{x}}$ and overall range $\bar{R}$ for each stream.
4. Calculate the process overall average $\bar{\bar{x}}_p$ and the process overall range $\bar{R}_p$ from $\bar{\bar{x}}$ and $\bar{R}$ for each stream. The center lines for the charts are:

$$\text{center line}_x = \bar{\bar{x}}_p \qquad \text{center line}_R = \bar{R}_p$$

5. The control limits are calculated identically to the standard $\bar{x}$ & R chart:

$$\text{UCL}_x = \bar{\bar{x}}_p + A_2\bar{R}_p \qquad \text{UCL}_R = D_4\bar{R}_p$$

$$\text{LCL}_x = \bar{\bar{x}}_p - A_2\bar{R}_p \qquad \text{LCL}_R = D_3\bar{R}_p$$

6. Identify the streams with the highest $\bar{x}$ value, lowest $\bar{x}$ value, and highest R value. These two or three streams are those that will be continually sampled, measured, and plotted to monitor the overall process.

This procedure should be repeated in part whenever a change or an improvement is made to the process. Frequently, the improvement process is applied to reduce the variation in the stream being plotted on the R chart because it has the most variation. Once changes are made to reduce the variation from that stream, it may no longer be the stream with the highest variation. If a change has been made, repeat steps 1 and 2 only for those streams affected by the change. Combine these new data with previous data from streams not changed and repeat steps 3 through 6.

In some cases, the variables being monitored on the output streams are not the same. Different streams from a process may have different concentrations or even materials. Group control charts can also be effective here by normalizing the measurements as shown in Section 4.8.4.

Group control charts are interpreted following the same rules as standard $\bar{x}$ & R charts. Whenever a special cause of variation has been identified after being highlighted by the chart, all of the output streams should be checked for that special cause. Do not check only the streams being monitored.

4.8.6 Gradual Process Changes

Processes that change gradually create special problems that standard control charts cannot overcome. Examples of such processes include polymerizations, blinding of particle separators, granulating equipment wear, and machine tool wear. Some people have suggested using control charts with slanting control limits, which follow the process change. However, these are impractical in most cases because they are difficult to implement, to maintain, and to interpret by operating personnel. There is also the problem that these charts treat gradual changes as all being linear. Many changes, especially chemical reactions, are not linear.

There are tools that can be used in cases where the process changes gradually. Two examples are:

1. Use of separate individual and moving range x & R_m control charts for specific points (elapsed time or quantity processed) in the process cycle. For example, if a reaction process normally takes 3 h, set up three x & R_m charts for $t_{30\ \text{min}}$, $t_{90\ \text{min}}$, and $t_{150\ \text{min}}$.
2. Use $\bar{x}$ & R charts with modified control limits on the $\bar{x}$ chart to halt the process before exceeding specification.

4.9 CONFORMANCE (OR RUN) CHARTS AND PRECONTROL CHARTS

Most companies are not ready to immediately jump into the use of control charts. Some companies are scared away from them by the statistics involved and because of the maintenance needs of control charts. For these companies, conformance and precontrol charts are two SPC tools

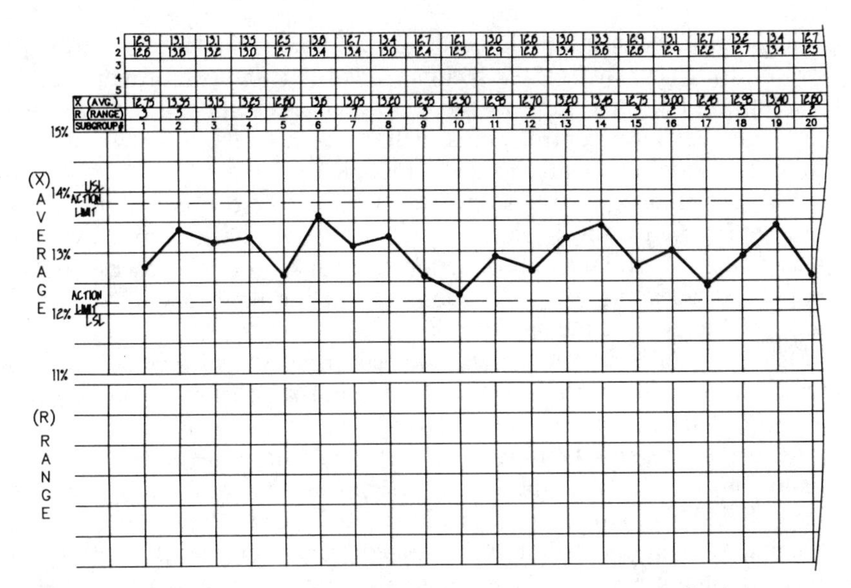

FIGURE 4.50. Conformance chart with action limits based on the specification.

that can be used to monitor the process against the specification. Conformance charts are often used as a precursor to control charts.

Dorian Shainin, a long-time statistical consultant, has been a leading proponent of the use of precontrol charts over control charts because of their simplicity.

4.9.1 Conformance Charts

When implementing control charts (Figure 4.50) in their operations, many firms start with conformance charts (also known as run or trend charts). Several reasons are behind this approach:

- Operators become accustomed to calculating and plotting their results.
- The local work force adjusts to viewing the real-time performance of the process.
- The process has special causes of variation, which have not yet been eliminated.
- Simultaneously with the preceding factors, the conformance chart serves to collect data on the process. These data can be used to verify that the sampling plan yields a normally distributed output and to calculate control limits once normality has been established.

Conformance charts can be used for the same variables and attributes as a control chart. In fact, it's recommended that the conformance chart be set up on the same form as the control chart except the conformance chart does not have control limits, it has action limits. Simply cross out "control" and write "conformance" in its place.

The conformance chart has no statistical basis, so probabilities don't come into play. However, action limits can be established based on the specification. These limits allow action to be taken on the process before the product goes out of specification. The action limits should be set to take up 60 to 80% of the specification.

Most companies set the action limits covering 80% of the specification range or at 10% above the lower spec limit and 10% below the upper spec limit. If the mean of the sample is outside the action limits, the process is shut down and reset to nominal. The parts made since the previous sample should be considered suspect and resampled or sorted. Conformance charts should not be used over the long term because both the α risk of adjusting the process when it's not necessary and the β risk of not adjusting when it is necessary are high. From time to time, parts outside of the specification may not be identified. A concerted effort should be made to bring the process into control so that the conformance chart can be converted to a control chart. If the range is in control but the process average is shifting, a control chart with modified control limits (refer to Section 4.8.3) could be used as an intermediate step.

4.9.2 Precontrol Charts

Precontrol charts are promoted by their developer, Dorian Shainin (1989), as an alternative to $\bar{x}$ and R charts for controlling a process. Although there are tradeoffs, they can offer a simpler and lower cost method to achieve the ultimate goals of both conventional control charts: preventing defective parts from being made and from reaching the customer. Precontrol charts require less in-process samples, so there is less measurement time and cost. In some operations, this also means less downtime. Precontrol charts do not require a large amount of sampling and calculating for control limits, nor the periodic effort to recalculate them.

There is a major disadvantage of precontrol charts, however. They do not supply information on whether the process is stable and in-control over time. Precontrol charts will not readily identify when new special causes appear. We recommend those that want to use precontrol introduce precontrol charts only AFTER the process is stable with no special causes of variation present and the C_{pk} is greater than 1.33.

Precontrol charts (Figure 4.51) use precontrol lines (PC Lines) set at $\frac{1}{4}$ and $\frac{3}{4}$ of the total specification tolerance. Although their positioning seems

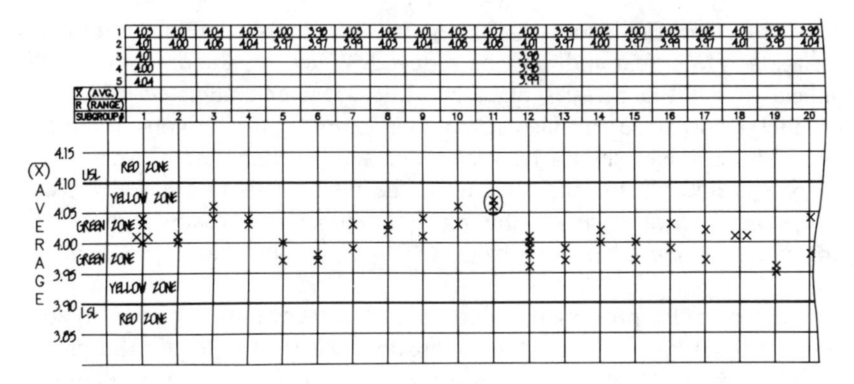

FIGURE 4.51. Precontrol charts.

to be arbitrary, it is not. Extensive Monte Carlo simulations were used to optimize the position of the lines and the sample size to minimize the risk for both α and β errors. The probability of an error depends on the relationship of the size and shape of the process output distribution to the total tolerance. For a normal distribution with a C_{pk} of 1, there is an α risk of 0.02 (a 2% chance of being signaled to adjust the process when there is actually no need to) and a β risk of 0.01 (a 1% chance of producing defective product by not being signaled to adjust the process when it should be).

Precontrol charts use the PC lines and the specification limits to establish zones. "Red" zones are outside the specification limits; they signal to stop the process. "Yellow" zones lie between the PC line and the spec limit; they indicate caution and a need to watch the process. A "green" zone lies between the PC lines and signals to keep going.

Adjustments are made in every process: changes in the equipment, methods, materials, or operators. The sampling plan for precontrol charts is to pull two consecutive pieces six times on average between process adjustments:

Avg. Time between Adjustments	Sample Interval
1 h	10 min
2 h	20 min
4 h	40 min
8 h	80 min

If either of the two pieces falls in a red zone or if both fall into yellow zones, stop the process and adjust; if both pieces fall into the same yellow zone, recenter the process; if they fall in opposite yellow zones, the variation is too great and must be reduced. Whenever the precontrol chart indicates that the process must be shut down and adjusted, all of the product made since the last acceptable sample must be considered suspect. This product should be rechecked and, in many cases, sorted. The process is not adjusted if just one of the pieces falls into a yellow zone. There is a 7% chance of one piece falling into a yellow zone (for a normally distributed process with $C_{pk} = 1$) but only a 0.5% chance of both.

Whenever the process is adjusted, the process must be resampled to verify it is set up correctly. After adjustment, it is okay to run the process once five consecutive prices are in the green zone. If a piece falls into a red zone or two pieces of the five into the yellow zones, the process is shut down, adjusted again, and resampled. With precontrol, this setup procedure should also be used whenever the process is started up, even if no adjustments were made.

4.10 CUSUM CHARTS

Cumulative sum control charts, or CUSUM charts for short, are another SPC tool used to control processes. CUSUM charts are more sensitive than the conventional $\bar{x}$-R chart for detecting small changes in the process output. $\bar{x}$-R charts are more effective for detecting process shifts of 2 standard deviations and greater. However, CUSUM charts will signal shift in the process mean of 0.5 to 1.5 standard deviations approximately twice as fast as conventional control charts.

CUSUM differs in construction from $\bar{x}$-R control charts in two major ways. First, CUSUM charts plot the sum of the deviation from the target of the statistic sample under study, whereas conventional control charts plot the value (or a nominalized value) of the sample statistic directly. Second, CUSUM charts don't have control limits in the same sense as those used for $\bar{x}$-R control charts; CUSUM charts use decision criteria to determine when a process shift has occurred.

The most common method for statistically determining whether there has been a shift in the process level is to use a mask on the CUSUM chart. The mask is placed over the last data point. If any of the previously plotted points are covered by the mask, then this indicates that the process shifted at the time associated with the covered point. Figure 4.52 shows this occurrence. In the case shown, the point is covered by the bottom of the mask. This means that the process average has increased. If a point were covered by the top of the mask, it would indicate that the process average had decreased.

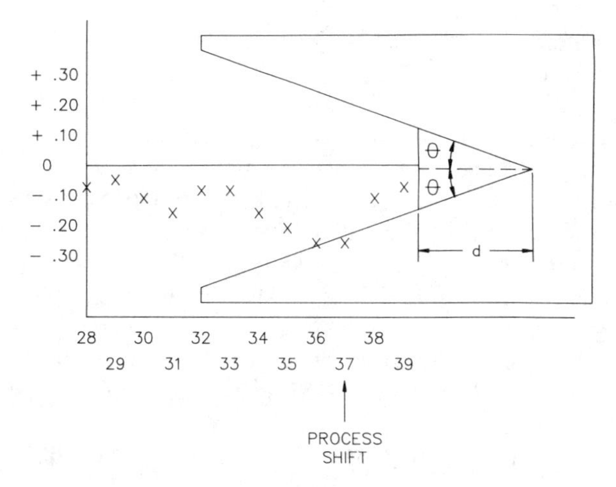

FIGURE 4.52. CUSUM chart.

The construction of the *V* mask depends on the standard deviation of the sample statistic, the level of shift to detect, the lead distance *d*, the mask angle Θ, and the risk chosen. Both Montgomery, in *Introduction to Statistical Quality Control* (1991), and Bicking and Gryna, in Juran's *Handbook of Quality Control* (4th ed., 1988), give a detailed explanation of the procedure for constructing a *V* mask. We won't duplicate their efforts. In fact, we don't recommend use of the *V* mask decision method for CUSUM charts. *V* masks are cumbersome to calculate and cumbersome to use as well. (A *V* mask is "another" piece of paper that can get lost in a manufacturing process, rendering the SPC tool inoperative.) We recommend the use of CUSUM signal charts instead.

CUSUM Signal Chart

The CUSUM signal chart is actually two parallel graphs of two modified cumulative sum values. The upper graph plots a cumulative upper signal value and monitors that value versus an upper signal alarm limit (USAL). The lower graph looks at a cumulative sum lower signal value versus a lower signal alarm limit (LSAL). The cumulative upper and lower signal values are calculated by adjusting the nominalized mean by a signal factor SF:

$$\text{Cumulative upper signal value} = \Sigma(\bar{x}_V - \text{SF})$$

$$\text{Cumulative lower signal value} = \Sigma(\bar{x}_V + \text{SF})$$

where $\bar{x}_V$ = nominalized mean = $(\bar{x} - \text{target})$
SF = signal factor

Both the cumulative upper and lower signal values have limits. The cumulative upper signal value can never be negative, so if the preceding equation for it calculates to be negative, the cumulative upper signal value is equal to zero. Similarly, the cumulative lower signal value can never be positive. If that equation calculates to be positive, the cumulative lower signal value is equal to zero. Values of 0 for either are not plotted on their respective graphs.

The signal factor is a function of the level of shift to be detected and the sample standard deviation. The SF is calculated as follows:

$$\text{SF} = (\text{shift to be detected}) \times (\text{sample standard deviation})/2$$

If a process shift of 1 standard deviation is to be signaled by the CUSUM chart, the SF is $s/2$.

On the graphs, the signal values are monitored against the upper signal alarm limit (USAL) and the lower signal alarm limit. The USAL and LSAL are simply $+10$ times SF and -10 times SF, respectively. When a data point falls outside the signal lines, the CUSUM chart has signaled a shift in the process mean and action should be taken to correct the process.

To use a CUSUM signal chart, the data from the process (the mean values) must be nominalized, then adjusted by the signal factor, and finally cumulatively summed. We recommend that this be done in a data table as shown in Table 4.1. For the process shown in this table, the sample standard deviation s equals 0.2 and it was desired to detect a $1s$ shift in the process mean so the signal factor calculates to be 0.1. The values for USAL and LSAL then calculate to be $+1.0$ and -1.0, respectively. The columns in Table 4.1 are described as follows.

Column 1 List the sample mean $\bar{x}$ data.
Column 2 Nominalize the data to a nominal or target value, showing the variation or deviation from nominal.
Column 3 Subtract the signal factor (SF) from the value in column 2.
Column 4 Calculate the cumulative values of column 3. However, if the value of the sum is less than zero, call it zero. These are the upper signal values.
Column 5 Add the SF to the value in column 2.
Column 6 Calculate the cumulative values of column 5. However, if the value of the calculation is greater than zero, call it zero. These are the lower signal values.

Table 4.1 Cumulative Signal Chart Data Table[a]

Sample Sequence	Column 1 Sample Mean $\bar{x}$	Column 2 Nominalized (Nominal = 5.0) $\bar{x}_V$	Column 3 Subtract SF[b] $\bar{x}_V - s/2$	Column 4 Cumulative US Value $\Sigma(\bar{x}_V - s/2)$	Column 5 Add SF[b] $\bar{x}_V + s/2$	Column 6 Cumulative LS Value $\Sigma(\bar{x}_V + s/2)$
1	4.8	−0.2	−0.3	0	−0.1	−0.1
2	5.3	+0.3	+0.2	+0.2	+0.2	0
3	4.9	−0.1	0	+0.2	0	0
4	5.2	+0.2	+0.1	+0.3	+0.3	0
5	5.4	+0.4	+0.3	+0.6	+0.5	0
6	5.1	+0.1	0	+0.6	+0.2	0
7	4.9	−0.1	−0.2	+0.4	0	0
8	5.5	+0.5	+0.4	+0.8	+0.6	0
9	5.4	+0.4	+0.3	(+1.1)[c]	+0.5	0
10	5.1	+0.1	0	+1.1	+0.2	0
11	4.8	−0.2	−0.3	+0.8	−0.2	−0.2
12	4.9	−0.1	−0.2	+0.6	−0.1	−0.3

[a] USAL = 10 SF = $5s$ = +1.0; LSAL = −10 SF = $-5s$ = −1.0.
[b] For our example, $s = 0.2$.
[c] Process exceeds upper signal alarm limit.

The upper signal values (column 4) and lower signal values (column 6) from Table 4.1 are plotted on the CUSUM signal chart as shown on Figure 4.53. Again, with CUSUM signal charts, values of zero are not plotted. Figure 4.53 shows that the process shifted at sample number 9. At this point, action should have been taken to find the special cause that created the shift and the process should have been reset.

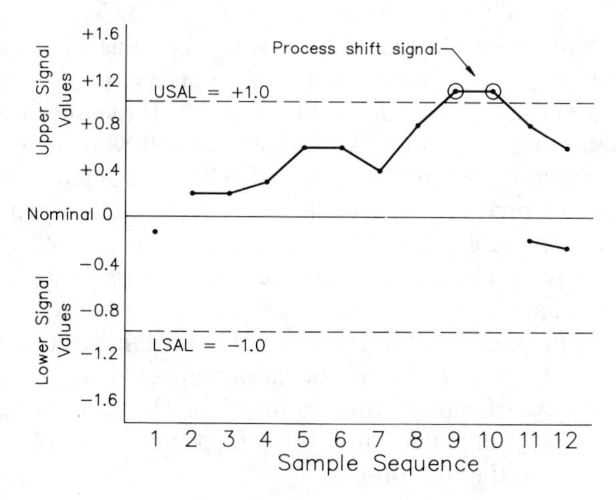

FIGURE 4.53. CUSUM chart mask showing process shift.

This shift would not have been picked up this early with an $\bar{x}$ and R control chart so the CUSUM signal chart gives us the chance to take action earlier. This minimizes the chance or the amount of bad product that the process might have produced before the $\bar{x}$ and R chart showed an out of control condition.

4.11 PROCESS TARGETING

One type recurring waste in industry is materials produced out of specification when a process starts up or after a planned change to the process has been made. This waste arises because few companies have established procedures to set up a process so that its aim, or mean value, lies close to the nominal value or center of the specification. When a process is not set up correctly, waste occurs not only from out-of-spec materials, but also from the costs to reset the process, sorting, added inspection, and engineering time, among others. A great effort could be made investigating the process for special causes of variation when the problem actually occurred in initially setting the process aim.

The best way to ensure that the process is aimed correctly is to mistake-proof the setup. That is, to ensure that the setup is the same each time it's done. This is relatively easy in an operation such as machining. The cutting tool and its positioning can be located the same repeatedly using rigid fixtures and stops. This is more difficult to do in the continuous flow industries, which generally don't have rigid fixtures and stops to rely on. These industries must rely on statistics to help ensure that their process is properly aimed.

We do not actually statistically set the aim of a process. We determine if our aim produces output that is statistically the same as prior production runs. The output at start-up is compared to an acceptance region that is calculated from the sample standard deviation s. If the output falls outside the acceptance region, there is a high probability that the process is incorrectly aimed. Although this approach will not eliminate the possibility of unacceptable product (i.e., waste) at start-up, it does minimize the amount of unacceptable product should our aim be way off.

As previously mentioned, the sample standard deviation s is used to calculate the acceptance region. In order to use s, the process must be in control. If the process is not in control, it will be difficult to determine if it is aimed correctly. A special cause of variation could be present at start-up that gives false signals on the process location. This would either result in the product being manufactured out of specification or in needlessly resetting and restarting the process.

For an existing process, the value of s is calculated from the sample range data from the control charts:

$$s = \overline{R}/d_2$$

where $\overline{R}$ = mean of the range values
d_2 = a constant that depends on the sample size

The range values will not be known if a new product is started-up on an existing process. For a new product that is similar to an existing product, set $s_{\text{new}} = s_{\text{existing}}$. If the new product is not similar to an existing product or if the critical properties are greatly different, estimate s from

$$s = (\text{specification range})/10$$

or

$$s = (\text{specification range})/8$$

These equations correspond to the equations for process capability. The equation with 10 in the denominator corresponds to a $C_{pk} = 1.67$. This is the more conservative approach because s will be smaller and a smaller s will give a smaller acceptance region. The equation with 8 in the denominator corresponds to a $C_{pk} = 1.33$ and gives a wider acceptance region. Neither of these two equations should be used except for process average setting. They are not statistically based, but are estimated to help avoid manufacturing a large amount of product out of specification.

Process Average Setting Procedure

There are several techniques for setting the process average. A technique developed by Dr. Donald J. Wheeler (1982) calculates a sample size based on the estimated process standard deviation, an α risk associated with the probability of needlessly adjusting the process, and a value, C, which defines how closely the process average needs to coincide with the nominal aim. The average of this sample then determines whether the process aim is acceptable or not. The basic procedure follows:

1. Use values from the R chart and determine if the range is in control. If it is not, this technique cannot be used.
2. Calculate the estimated standard deviation, $s = \overline{R}/d_2$. Select a value for α. This is the probability of needlessly adjusting the process. Most people select an α risk of 5 or 10%. It should be noted that the α risk of not adjusting the process when it should be is less than 0.5α.
3. Choose a C value. There are two approaches to selecting a C value. Both involve using a maximum nonconforming percentage that is

considered acceptable. Because few customers are willing to tolerate nonconforming materials, C should be based on the process capability. One recommendation:

For C_{pk}	Choose C
≤ 1	≤ 0.2
≤ 1.33	≤ 0.5
≤ 2	≤ 2
≤ 3	≤ 5

Obviously, if $C_{pk} < 1$, the process output will exceed the specifications, regardless of the C value chosen. However, the smaller the C value, the more there is assurance that the process output after adjustment is very close to the historical output.

4. Calculate the sample size:

$$N_s = \left[\frac{2(z_{\alpha/2})}{C} \right]^2$$

where $z = z$ table value for area under the normal curve (Appendix B.4).

5. Measure N_s pieces consecutively after the process is started up and reaches steady state.
6. Calculate $\bar{x}$ for the pieces measured.
7. Compare $\bar{x}$ to the acceptance interval. If nominal $- \frac{1}{2}Cs \leq \bar{x} \leq$ nominal $+ \frac{1}{2}Cs$, do not adjust the process. However, if it is not within this interval, readjust the process and return to step 5.
8. Proceed with the production run and monitor the process with $\bar{x}$ and R charts.

The techniques may not be practical to use if the sample is large and the product is expensive or the measurements are time-consuming. This is especially true if destructive testing must be performed. Additionally, some companies may want to use the same sample size for start-up as for their process control chart. To take this into account, Wheeler and Chambers (1986) further refined this technique:

For a sample size of 3, let $C = 2$. If the $C_{pk} \geq 1.67$, this gives a worst case of 0.5% nonconforming product.

For a sample size of 5, let $C = 1.5$. If the $C_{pk} \geq 1.33$, this also gives a worst case of 0.5% nonconforming product.

Table 4.2 Worksheet for Burr Procedure

Sample	L	Ls	x	x	Us	u
1	—					3.39
2	—					3.89
3	—					4.39
4	—					4.89
5	0.82					5.39
6	1.44					5.89
7	1.98					6.39
8	2.49					6.89
9	2.99					7.39
10	3.50					7.89
11	4.00					8.39
12	4.50					8.89
13	5.00					9.39
14	5.50					9.89
15	6.00					10.39
16	6.50					10.89
17	7.00					11.39
18	7.50					11.89
19	8.00					12.39
20	8.50					12.89

For a sample size of 10, let $C = 1.0$. If the $C_{pk} \geq 1.0$, this gives a worst case nonconforming of 2.4%.

Obviously, if the process capability is greater than shown in each of the three cases, the worst case for nonconforming product will be reduced.

Sequential Process Setting

Dr. Irving W. Burr (1949) developed a sequential approach for process setting.

This approach requires a minimum of five samples and is based on an α risk (of resetting the process unnecessarily) of 10% and a β risk (of failing to reset the process when needed) of 10%.

The Burr procedure

1. Calculate s from $\overline{R}/d_2$.
2. Set up a worksheet (see Table 4.2) for determining whether to approve the process or reset it.
3. Take a sample and measure it.
4. Calculate x_i, where x_i = (measurement − target), and Σx (equal to x_1 in this case).

5. If $|\Sigma x| > Us$, reset the process and repeat this procedure; if it is not, sample the process again.
6. Calculate x_2 and Σx_i (equal to $x_1 + x_2$ here).
7. If $|\Sigma x| > Us$, reset the process; if it is not, sample again. The value of Σx can be used to determine the amount to reset the process. Adjust the process by ($\Sigma x/n$ – target) upward if negative and downward if positive.
8. Repeat steps 6 and 7 as appropriate for samples 3 and 4.
9. For sample 5 and subsequent samples, calculate Σx. If $|\Sigma x| \leq Ls$, approve the process; if $Ls < |\Sigma x| < Us$, continue sampling; if $|\Sigma x| \geq Us$, reset the process.
10. If 20 samples have been taken and no decision made, Burr recommends adjusting the process by $\Sigma x/20$.

Using the *t*-Test to Set the Process Target

A third approach would be to use the t test. The mean of the population, μ, can be estimated from $\bar{\bar{x}}$. Then, the calculation for t would be

$$t = \frac{\bar{x}_{\text{sample}} - \bar{\bar{x}}}{s/\sqrt{n}}$$

where $\bar{x}$ and s are the sample mean and sample standard deviation, respectively. The acceptance region (for not adjusting the process) would be calculated as shown in Section 5.4.3 using the α risk and the degrees of freedom. This approach is not as accurate as the other methods because it only identifies that the sample is from the same population; it does not focus on how far off the sample is from the population mean. However, by using a very low α risk ($< 5\%$) this method may suffice.

Summary

To summarize, the best method for aiming the process correctly is to mistake-proof the process setup. However, statistical process targeting can be employed whenever mistake-proofing can't be done. With statistical process targeting, the risk of producing bad product and the risk of the process showing as out of control during process start-up are both minimized.

5

Using the Tools

5.1 FROM SPC TO SPI

The move from statistical process control (SPC) to statistical process improvement (SPI) does not signify a break with SPC. SPC will continue to be used as a statistical control tool; it will help maintain the status quo even as sequential improvements are made to the process. But once the process has been stabilized, improvement work must continue using the PDCA cycle. With a stable process, the statistical tools of improvement, primarily design of experiments, should be pulled from the statistical toolbox and put into use.

As Figure 5.1 demonstrates, there is a limit to quality and productivity improvement with SPC alone. SPC helps ferret out the special causes, those acute causes of variation that render processes unstable. But a stable process is not the goal: The process may be stable but at a mediocre level, and controlled mediocrity is still mediocrity. Actually, any level of performance, regardless of how high it may seem on a relative scale, will become mediocre or worse over time. In today's competitive world market, if every process within a business is not improving (at least as fast as the competition), it will look like the business is going backward. Continuous improvement must become a way of life, and SPI techniques are the improvement tools to use.

With SPI, engineers and managers can tackle the common causes, the chronic variation inherent in the process. By using the PDCA cycle over and over again, the hidden plant can be dismantled, block by block, and each time a piece of the hidden plant gets torn down, the real plant, the productive plant grows by the same amount.

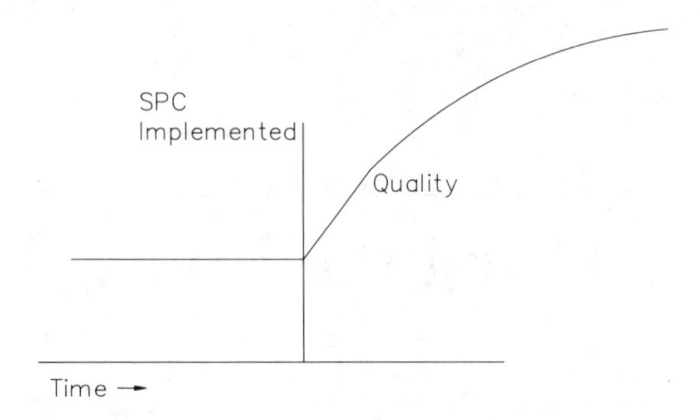

FIGURE 5.1. Quality and productivity with SPC alone.

In this chapter, a variety of SPI techniques will be discussed. These range from basic team problem-solving skills to advanced experimental designs for formulating materials. All of the techniques discussed are designed to give rapid, organized improvement. There are niches for each technique, even the "advanced" ones, that the engineer and manager must search for.

As mentioned in the preceding text, team problem-solving skills are covered in this chapter. These skills must be developed and used by every professional in the organization. For some people, the other techniques may be for informational and reference purposes only. However, it is recommended that all readers of this work pay special attention to the sections on screening experiments (Sections 5.8, 5.8.1, 5.8.2, and 5.8.3). In many cases, more than half of the improvements that can be made to a process can be accomplished just by using screening experiments on the process.

5.2 PROBLEM-SOLVING TECHNIQUES

In the text of Chapter 4 on control charts, it was repeatedly emphasized that special causes of variation must be identified and eliminated. This is easier said than done. Special causes of variation appear infrequently in a controlled process so that the trail to the underlying reason for that variation goes cold quickly. In order to identify the root cause so that it can be eliminated, a formal (and standard) problem-solving process should be adopted within each organization. The process should include training all employees in the same basic problem-solving techniques so that they can "speak the same language." The problem-solving process works best with a team approach. Many problems are not easily solved and having a team address them rather than an individual alone focuses a much greater amount of brainpower on the problem and will lead to a quicker solution.

The basic problem-solving process involves six steps:

1. Problem-solving team formation.
2. Identification and clarification of the objective for the problem-solving team.
3. Identification of the causes.
4. Identification of the solutions (to the causes).
5. Eliminating the causes (implementing the solution).
6. Verifying the changes made actually produced the desired results.

Problem identification is addressed in Sections 5.2.1 and 5.2.2. Techniques for identifying causes (and solutions) are covered in Sections 5.2.2 (brainstorming) and 5.2.3 (organizing problem-solving data). These techniques are often enough to identify the root cause of the variation thereby paving the way for basic manufacturing or engineering skills to eliminate the causes. In some cases where the special cause is rarely seen and not easily identified or where the common cause variation must be reduced, design of experiments (DOE) techniques need to be used to identify the root cause. DOE is discussed in detail later in this chapter.

After a problem has been eliminated through the problem-solving process, the manufacturing or service process should be rechecked to verify that the changes actually produced the desired results. This also serves to check that there are no side effects from the problem solution; that is, that the solution does not cause other problems. Hypothesis testing (Section 5.4) can be used to check the results, if the desired problem solution or process improvement is a measurable parameter. For example, if the problem was that a physical property of a material was too low, hypothesis testing can be used to determine if the solution actually produced a statistically valid improvement in that property. Process audits are used to verify that changes made are held in place. They are discussed in Section 5.13.1.

5.2.1 Problem Identification

As stated in the introduction to this chapter, identifying the right problem to solve is the key to the problem-solving process. Failure at this, the problem identification step, creates a great deal of wasted effort, time, and money because the attack is focused on the wrong problem. Most managers and engineers underestimate the need to proceed through a planned approach to identifying the problem. To them, the problems are all too

Table 5.1 Using the "5 Whys"
Problem—the yields have dropped

1. Why have the yields dropped?	The screens in the classifier have blinded
2. Why did the screens blind?	The relative humidity in the room reached 75%
3. Why did the relative humidity hit 75%?	The desiccator malfunctioned
4. Why did the desiccator malfunction?	The regeneration did not take place
5. Why didn't the regeneration take place?	The timer works erratically

apparent: the yields have dropped; the crystallizer shut down; the spot welds are cracking. Unfortunately, these do not get at the real problems nor the root causes that underlie these general problems. The causes of the yields dropping or the crystallizer shutting down are the real problems that must be identified. Once these causes are known, the problems may be easy to solve. The problem solution may be as straight forward as replacing an aging, open cage motor on the crystallizer, or the solution to the "real" problem may be complex, requiring additional problem-solving steps.

There are many methods to identify the true, root-cause problem(s). One of the simplest is the "5 whys". Another method is "what it is versus what it is not" analysis. In cases where the overall problem is complex, brainstorming should be used to start the problem-solving process.

The "5-whys" approach involves asking "why" at least five times. The goal is to reach the underlying problem (the root cause) in five steps. Although the root cause can usually be found within the five questions, don't treat the "five" in the "5 whys" literally. Five questions is normally enough but may not be in all cases. Tables 5.1 shows an example of asking the "5 whys."

After the fifth question in the example, the root cause of the problem is clear. The real problem underlying the yield problem is the timer. Replacing it will improve the yields. If the engineer had stopped after the second "why", the solution may have been to install a humidity monitor, which would have been a "band-aid" solution; it doesn't solve the real problem that caused the humidity to rise.

Problems that have multiple causes may not be so straightforward. A modification of the "5 whys" can be used where the problem is exploded into bits small enough that each bit can be solved. At each step, the investigator asks, "why?" until the underlying problems are uncovered. These smaller, underlying problems can then be corrected.

Table 5.2 What It Is/Is Not Analysis

Write down the answers to the following questions. Compare the answers to each pair of questions to narrow down the problem scope.

1. What is the specific problem? What other similar problem does not exist?
2. What part or product has the defect? What similar part or product could have the problem but does not?
3. What percentage of the parts or product has the defect? What percentage does not?
4. Where in the process does the defect first appear? Where else in the process could it appear but is not seen?
5. Where on the part does the defect appear? Where does it not appear?
6. Who normally detects the problem or defect first? Who else is in a similar position to detect the problem or defect but does not?
7. Who operates the process where the defect or problem occurs? Who does not?
8. When did the defect or problem first appear? When was it definitely known not to be present?
9. Is the problem or defect continuously present once it appears? Does the problem or defect come and go?
10. Does the problem appear at regular intervals? Does the problem appear on an erratic basis?

Another technique useful in identifying a problem is "what it is/is not" analysis. This technique establishes boundaries on the problem to more thoroughly focus the investigation. It prevents the investigation from getting bogged down with irrelevant information and from going off in the wrong direction. Table 5.2 shows a simple analysis. To identify the real problem, we use the "what it is/is not" analysis and ask "what causes these differences or changes?"

Many engineers in the chemical process industries are familiar with hazardous operations (HAZOP) analysis for process safety. The same techniques used in HAZOP can be used to identify the root cause of a quality problem or to identify an opportunity for improvement. Because of its familiarity in the industry and its treatment in many safety texts, HAZOP will not be covered in this work.

Sometimes, the underlying problem remains an unknown even after using the preceding techniques. When this occurs, then a brainstorming session should be conducted to start a formal problem-finding, problem-solving process.

Brainstorming uses a group process to generate a large number of ideas. The brainstorming process is covered in Section 5.2.2, which includes a discussion on brainstorming for solutions to problems. To use brainstorming for identification of the actual problem, simply change the brainstorming problem statement.

5.2.2 Brainstorming

Brainstorming is a powerful technique for generating a large number of ideas, to help solve a problem. It generates ideas (potential solutions) that may lead to the elimination of the underlying cause. Brainstorming can also be used to identify the actual problem cause when other problem identification techniques fail. Brainstorming is a group activity that provides a broad base for ideas, stimulates creativity by ideas triggering new ideas from others, and provides a means for buy-in by the participants.

Brainstorming Procedure

1. Invite everyone who may have ideas to the brainstorming session. Include operators as well as management—cross functional lines and department boundaries. Issue a clear statement of the problem, the objective of the brainstorming session, and any background information, well in advance of the session.
2. The brainstorming session can be set up to be round-robin or freewheeling. Round-robin, which means going around the group giving each person a turn in order, is normally the best approach because no one person or group of persons can dominate the session. We recommend round-robin brainstorming over the freewheeling approach.
3. Before beginning to brainstorm, the leader should review the rules of brainstorming. The rules for round-robin brainstorming are:
   - Each person is allowed one idea per turn.
   - People are allowed to "pass" on their turn, but may still add ideas on later turns. Someone else's idea may trigger new ideas after someone thinks they've run out of ideas.
   - Absolutely no evaluation of the ideas is allowed: No one should interrupt, censor, or criticize an idea. "Bad" ideas may stimulate ideas in others that lead to the solution.
   - Questions are allowed but only to clarify the idea.
   - Every idea is acceptable, even ones that seem unrelated to the brainstorming objective. Again, they may stimulate an important, related idea.
4. The leader should number and write down each idea on a flip chart. (A PC with a video imager on an overhead projector may also be useful as a time saver. Ideas can be typed in as they are generated.)
5. There is no right number of ideas for a brainstorming session. Keep the session going as long as someone has ideas.

6. If, toward the end of the session, only a few people are adding ideas, it may be expeditious to stop round-robin brainstorming and go to freewheeling. Freewheeling means that anyone can speak up with an idea at any time. Once everyone passes, the brainstorming session is over.

The ideas generated may lead quickly to the identification of the root cause of the problem or even to a solution to the problem. Most likely, the ideas will need to be organized into categories prior to action. Methods for organizing brainstorming ideas will be discussed in the following section, Section 5.2.3.

5.2.3 Organizing Problem-Solving Data

Once data have been collected on the process and ideas for identifying or solving the problem have been generated by brainstorming, the data and ideas must be organized. Organizing problem-solving data and ideas is crucial to solving the problem successfully. Organization directs our efforts to where they would be most beneficial. Without organization, improvement activities could quickly turn helter-skelter and luck would play a major role in finding a solution.

Techniques for organizing process data were described in Section 2.6. Histograms, Pareto diagrams, and scatter diagrams can all play important roles in the problem-solving process. Their visual presentation of data can quickly direct attention to where it is needed or, equally importantly, away from where it is not.

Brainstorming ideas must also be organized and prioritized in order to direct efforts toward maximum efficiency. The cause and effect diagram (Figure 5.2) is the best method for organizing brainstorming ideas because it shows the relationship of each idea (or, cause) to other ideas and their contribution to the problem (or, effect). If data can be collected on the ideas, then the Pareto diagram should be used to prioritize the ideas. If it is impractical or not possible to collect data on the ideas; then voting and ranking techniques should be used to prioritize the ideas for action. Discussions on both cause and effect diagrams and voting and ranking techniques follow.

Cause and Effect Diagrams

The cause and effect diagram was developed by Dr. Kaoru Ishikawa (1976) by applying Shewhart's concept of a process. Shewhart stated that a process can be broken down into four major categories: materials, methods, manpower, and equipment. Ishikawa organized ideas or elements of

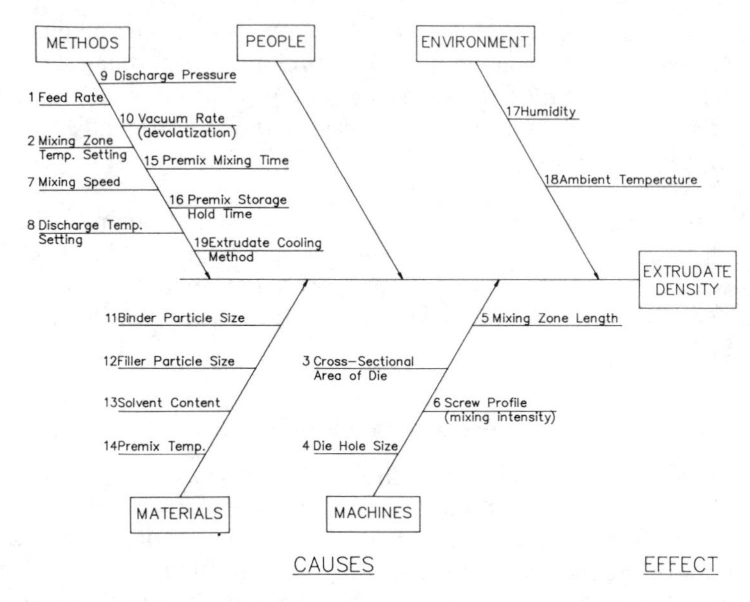

FIGURE 5.2. CEDiagram branches.

the process directly onto the Shewhart model. It is generally accepted today that a fifth category, the environment in which the process operates, be included on the cause and effect diagram. A sixth category, the measurement system, which actually filters the process output, can also be included.

The advantage of the cause and effect diagram is in its visual representation of the process and its elements. It shows how ideas fit together and aids in identification of those ideas that might interact. Constructing a good cause and effect diagram is often a major contributor to the proper selection of factors for an experimental design.

To construct a cause and effect diagram, use the template shown in Figure 5.3. Write the effect to be studied (eg. the problem) in the box to the right. Classify each of the brainstorming ideas into one of the five categories and write them on the diagram as branches. Some ideas are subsets of others so they are added as smaller branches (Figure 5.4); some ideas may seem to fit into more than one category. Resist the temptation to put an idea into more than one category. This would defeat one of the benefits of a CE Diagram—breaking a mountain of ideas into manageable piles. If the basic template does not seem to apply to the particular problem being solved, change the categories to ones more appropriate.

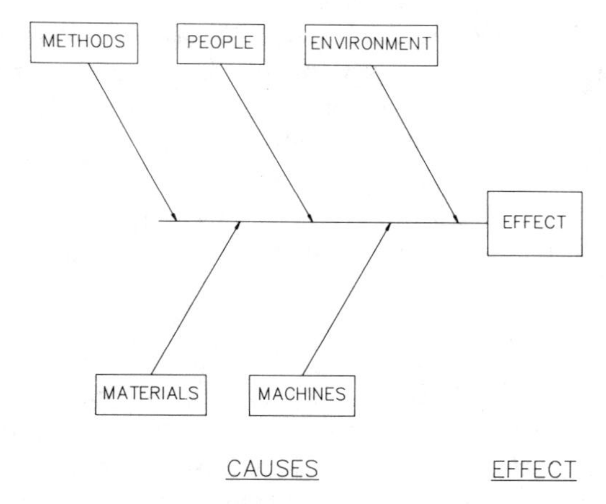

FIGURE 5.3. Template for CEDiagram.

A useful variation of the cause and effect diagram is the cause and effect diagram employing cards (CEDEC). This technique is useful with longstanding or complex problems because it allows people time to think about the problem and to manipulate their ideas within or between categories to see if there are any relationships between the ideas. Identifying these relationships may simplify solving the problem.

CEDEC is started by drawing a large template for a cause and effect diagram. The template should be drawn on paper at least 24 by 30 in. —the larger, the better. The template should have the problem to be solved written in the right-hand box and the categories labeled. Post this template in an area of high traffic near the problem; for instance, or on an aisle wall leading to the problem department. Provide a stack of self-stick

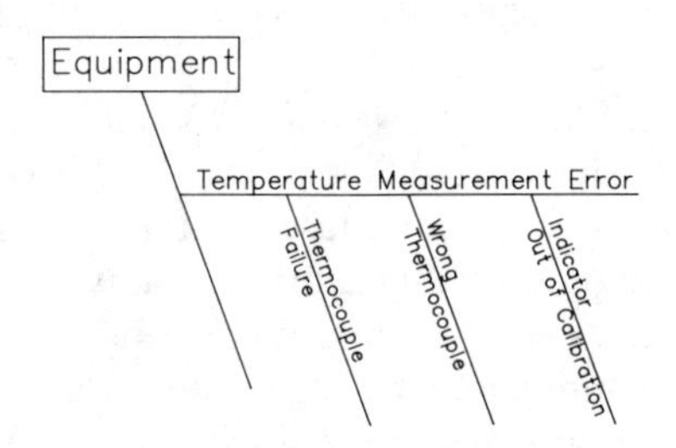

FIGURE 5.4. CEDiagram branch showing subbranches.

notes (or index cards with some tape) and encourage employees to write down their ideas for solving the problem. They should post their idea in the category they feel is most appropriate. Employees should also be encouraged to move the ideas around into other categories or into groupings of related ideas within a category. Leave the CEDEC up for a week or two and publicize it widely during that time period. Be careful not to leave it up too long or it will get stale.

Voting and Ranking

Whenever possible, facts should be used to determine which of the brainstorming ideas should be investigated or experimented with. Pareto diagrams and cause and effect diagrams are useful tools for focusing the problem-solving process. However, in some cases, a point is reached where these tools cannot be used further and facts (data) are unavailable. Here, experiential-based reasoning may be employed to determine the direction to take or the ideas to be experimented with. Unfortunately, this often leads to conflicts in team problem-solving activities because each person has their own experiences and perceptions. Voting and ranking techniques are useful to help put aside the emotion and to help the team find a consensus.

Voting and ranking is a subjective technique. It is organizing ideas based on perception. However, when used properly, it can remove most of the conflicts that arise within the team regarding which ideas to work on first because it allows everyone on the team to vote for the ideas he or she thinks is best. Even though it is a subjective technique, voting and ranking will nearly always identify ideas that lead to improvement.

Normally, the same group involved in the brainstorming session conducts the voting and ranking. The goal of voting and ranking is to break down the large number of ideas from brainstorming into a smaller number for initial action or investigation. As the name implies, each person votes for the ideas they feel will solve the problem. The ideas are then ranked by the number of points they receive. Those with the most points receive the initial effort.

Before voting, review the brainstorming list and combine similar ideas. The group may also agree that some of the ideas are not practical, feasible, or cost effective. These ideas should remain documented for future reference but should be omitted from the voting and ranking.

The number of votes each person can cast and the point value of each vote depends on the number of brainstorming ideas and the number of ideas to be investigated or experimented with. Typically, the initial experimentation will involve seven or less ideas. We recommend the following matrix for the number of votes and the point values for each vote:

Number of ideas	16–20	11–15	7–10	4–6
Votes	5	4	3	2
First choice	9	7	5	3
Second choice	7	5	3	1
Third choice	5	3	1	0
Fourth choice	3	1	0	0
Fifth choice	1	0	0	0
All others	0	0	0	0

After everyone votes, total the points for each idea. The ideas with the most points will be the ones investigated or experimented on first. If the brainstorming ideas were organized on a cause and effect diagram before voting and ranking, each of the categories should be voted on separately. This is needed especially when different functional groups within the organization have responsibility for different categories. For example, a materials engineering group may be responsible for acting on all materials-related ideas and the manufacturing engineering group may be responsible for the methods, equipment, and environment. Priorities should be set for both groups so their categories should each be voted on separately and the ideas ranked for each category.

After each of the categories is voted on, it is important to vote once again on the top ideas from each category. This helps set the overall priorities. In many cases, one or two categories will have more ideas that should be acted on than the other categories. Using the preceding example, the priorities for several of the materials-related ideas may be higher overall than even the top vote-getters in the other categories.

5.3 EXPERIMENTING FOR IMPROVEMENT

There is no tool more powerful for the improvement process than design of experiments (DOE). DOE is a series of statistically based techniques to organize experimentation to obtain the maximum amount of information at the minimum cost and time expenditure. This enables the experimenter to select variables thought to affect the process output, to methodically change those variables, and then to observe if the changes actually did affect the process output.

DOE was developed in the 1920s and 1930s by Sir Ronald Fischer for agricultural yield improvement in the United Kingdom. The theories were advanced in the United States and the United Kingdom in the 1940s and 1950s; then the Japanese applied them to other industries in the 1960s. They are part of the "off-line" quality techniques touted by the Japanese. (They are "off-line" because the process is not making production parts because it is shut down for experimentation.) The success of DOE techniques not only contributed to improvement in agricultural yields but have also contributed greatly to the rise to manufacturing excellence that the world has seen from the Japanese in the last 20 years.

In order to learn and apply DOE, its terminology must be understood. The important terms are explained in the following section, Section 5.3.1. Section 5.3.2 will discuss how to design a good experiment. Further background on experimenting for improvement will be covered in Section 5.3.3, which will review orthogonal arrays. These are the bases for many of the DOE techniques.

5.3.1 The Language of DOE

Mathematicians and statisticians created designed experiments. As such, many terms used in DOEs are different than common terms used in the process industries to describe the same thing. DOE has a language of its own which must be understood to be used. This section describes the important terms in the language of DOE.

Factor: One of the independent variables being studied in the experiment. Factors can be qualitative (such as changes in equipment, methods, personnel, or materials used) or quantitative (such as temperature, pressure, or time).

Level: The values of a factor being studied. For example, if temperature was a factor, then a level would be 150°C.

Treatment Combination: A treatment is a single level of a single factor in an experimental run (trial). A treatment combination is a set of levels for all factors in an experimental run.

Response: The result of an experimental run with a given treatment combination. The response is a dependent variable such as material compressive strength or molding press cycle time.

Effect: The change in a response caused by a change in the level of a factor. For example, if the temperature in a plate evaporator was raised from a level of 125°C to 130°C and the moisture content changed from 0.24% to 0.19%, then the effect would be 0.05%.

Interaction: Two or more factors working together to produce an effect different than the effects produced by those factors individually.

Experimental Run: The processing of product at a given treatment combination.

Randomization: The assignment of a treatment combination to an experimental run purely by chance. Designed experiments should be randomized whenever possible to minimize or eliminate the effects of variables not included in the experiment. If an experiment is not randomized, some of the effects of the factors studied may be confounded (confused) with factors not studied. For instance, if all levels of treatment combinations with factor A at level X are run during the day and then all at level Y are run at night, the effect of factor A may be confounded with the effect of ambient temperature or the effect of the operator. Neither of these may have been included as a factor in the DOE.

Blocking: A planned grouping of factors or treatment combinations. This is done to minimize the effects of background variables (those not included in the DOE) or when costs or time constraints occur. For instance, an experiment with oven temperature as a factor may be blocked if it is expensive and time consuming to change the level of the oven temperature and wait for a steady state to be regained. The DOE may be blocked in this case so that all experimental runs with oven temperature at level A are made first and then all at level B.

Replication: Repeating an experimental run. This is done when possible to increase the precision of the experiment and to eliminate the possibility of measurement error biasing the experiment. The entire DOE can be replicated in order to better estimate experimental error.

Statisticians have developed a number of models for designing experiments. Table 5.3 lists some of these models. Factorial experiments with the factors studied at just two levels are the most common in industrial environments because they are generally the most cost effective. In general, use screening experiments first to identify the factors that significantly effect the response. Then, if further optimization is needed, progress to fractional and full factorial designs to verify that those factors did have significant effects and to check the effects of their interactions on the response.

Once the process is operating with the significant factors at the levels identified through screening and factorial experiments, there is still an opportunity to optimize the process using evolutionary operations or response surface methods. Evolutionary operations (EVOP) are DOE techniques designed to make small, gradual improvements while the

Table 5.3 Models for Designing Experiments

Design Type	Format	Use To
Completely randomized	Two or more levels of one factor studied	Determine if one level is better than others for that factor
Randomized block designs	One factor is studied in a number of different blocked treatments; each block contains all treatments	Determine the best level for the factor when environmental factors must be blocked
Full factorial designs	All levels of all factors are studied	Study effects of all factors and all interactions to identify the best levels; practical to study two to four factors
Fractional factorial designs	Modified full factorial where effects of some interactions are confounded with main factor effects to reduce the number of experimental runs	Study three to six factors when all higher-order interactions do not need to be studied
Screening experiments	A severe fractional factorial	Study many factors (seven or more) with few runs with the compromise that interactions are not directly studied
Response surface analysis	A series of full factorials	Map the responses over the experimental region to determine the optimum levels; usually two factors at two levels
EVOP	Similar to response surface	Slow changes designed not to produce "bad" product
Mixture experiments	Often similar to screening experiments with two universal compositional constraints	Study product formulations to optimize component levels

process is on-line without causing any rejections. EVOP presents a low risk because it slowly makes changes so that no bad product is made and the process is not being shut down for the experimentation. This appeals to manufacturing personnel. If there is a need to optimize the process rapidly or if the response does not behave linearly, response surface methods should be employed. EVOP is covered in Section 5.11 and response surface methods in Section 5.10.

5.3.2 Designing a Good Experiment

As with most things in life, most of the time and effort involved in an experiment is spent in the preparation steps. (A *minimum* of 70% of the total resources spent on an experiment should be spent on the planning phase.) Why? Because the results you get out of an experiment are only as good as the:

- adequateness of the design
- method of data collection
- measurement system selected
- factor/levels studied

and the list goes on. If the effort is spent *up front* to insure that everything is in order, the data collection, analysis, and interpretation will be much easier. What types of things do the designer of the experiment have to do to make sure the experiment will provide useful, valid information? Figure 5.5 is a flow chart to offer some guidelines. A brief discussion about each point follows.

Define and Communicate Objectives Without clearly stated objectives or goals, it is difficult to determine how well you did when the experiment is over and the data are analyzed. State the objective to direct efforts to the root of the problem. For example, if the process in question produces salable product and a useless by-product, and the goal is to increase yield, try to focus on the real problem. The objective probably should be written accordingly:

NOT	Increase the yield of the process
BUT MAYBE	Reduce the amount of useless by-product generated

By focusing on the real problem, it will help others focus their efforts accordingly. Don't be afraid to set bold numerical targets (percent improvement or percent defective). Finally, communicate the experimental objectives to *everyone* involved. This is critical for getting everyone working toward a common goal.

Define the Process Once the goal is established, clearly define the process that will be used. This may seem like an unnecessary step but it is extremely *important* and will be needed in later steps of the planning process. Use the experimental objective to help here. Don't confine yourself to just the "process" or equipment at your facility. Always consider the extended process that could include suppliers and customers. For example, if the output being measured could be affected by the

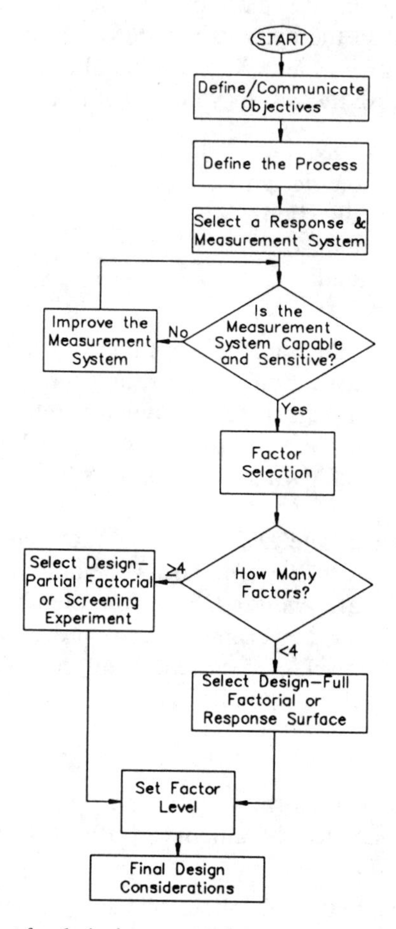

FIGURE 5.5. Flow chart for designing an experiment.

feedstock, consider including a supplier's process as part of "the process" used in the experiment.

A flowchart of the process is *always* useful and should be drawn once the process is established.

Select a Response and Measurement System This step ties indirectly with the experiment objectives. Determine (with help from others that are familiar with the process) the process outputs that will be measured, and the methods that will be used to measure them. Sometimes there are already capable measurement systems in place. However, often measurement systems will have to be developed for the experiment. If this is the case, additional work must be done to develop measurement systems that are stable, capable, and sensitive enough to detect a change in the process outputs. This step is *key* to insuring that the experimental results are useful and meaningful.

Factor Selection Get as many people that are involved with the process involved with the factor selection as possible. (If an extended process is under study, your suppliers and/or customers need to be involved, too!) Start with brainstorming (everyone in the room together) and encourage the free flow of ideas. Post the objectives of the experiment and the process flowchart to help spark ideas.

With the master list of potential factors generated, start a discussion with the same group to review the ideas. For each item, have everyone ask themselves the questions:

1. Is it practical?
2. Is it feasible?
3. Is it cost effective?

This exercise may rule out a few items but many others will remain in consideration. Group the remaining items into categories using a cause and effect diagram. Some factors may be able to be combined as they are categorized. Finally, use hard data or voting and ranking to "prioritize" the factors that will be studied in the experimentation.

Why so many steps and why so many people involved? To make sure that all the factors that could affect the process output are identified (two heads are better than one, four are better than two, etc.). Effective brainstorming needs representation from all corners of the process. Involvement of a group of people helps to make sure the experiment is free from one individual's personal bias or preconceived ideas (old theories die hard). Voting and ranking in a large group helps to temper individual bias as well.

Finally, make sure that the factors are independent of one another. It doesn't make sense to study the effect of material temperature on your output if material temperature is a function of equipment temperature and process time. It may make more sense to study the latter two independent factors than a dependent factor. Cause and effect diagrams help to identify independent factors by categorizing the ideas. In those cases where the effect of a factor may be confounded or confused with the effects of other factors, make sure that everyone understands how and why the factors were selected so they are aware of it when drawing conclusions about the results.

At this point, look at the list of prioritized items. If voting and ranking was used, there will probably be a natural break between high vote-getters and low vote-getters (remember the 80–20 rule: 20% of the ideas will probably get 80% of the votes). Try to include all of the high vote-getters in the initial experiment.

Design Selection Once the number of factors that will be included in the experiment is determined, the design can be selected. If there are a lot of factors, more than four (which there will be in most cases), select a partial factorial or screening experiment so that the main effects, those factors that control the process can be identified in a few experimental runs.

If there are only two or three factors, then choose a full-factorial or response surface experiment. CAUTION, if this is the first experiment being run on the process for a given objective, and there are only two to three factors selected, the group probably wasn't thorough enough in identifying potential factors!

Setting Factor Levels This is where the process experts get involved. [Again, if an extended process is being used, make sure to include experts from your supplier(s) and/or customer(s).] For screening experiments, set the levels as far apart as possible while still being able to produce the process output that can be measured. Run the process on the edges of the operating window. BE BOLD! If there is uncertainty whether or not certain combinations are possible, run some preliminary trials to confirm the processing window before running the full design.

Why is it important to set the levels as far apart as possible? It is important to make sure that it is possible to measure changes in the process that are caused by changing the factor levels. If they are set too close together, it probably won't be possible to pick up subtle changes in the output.

For factorial experiments, response surface designs, or EVOP, levels can be chosen using previous experimental results or cost constraint guidelines. Make sure, in general, that the levels selected pass the three point test from before:

- Is it practical?
- Is it feasible?
- Is it cost effective?

Final Design Considerations Once the type of design has been selected and the factors and levels that will be used have been determined, there still are a few final details to iron out before the experiment can be performed.

First, determine how an estimate of the experimental error will be calculated. (The experimental error is an estimate of the normal variation that exists in the process including the measurement system. We need to understand this to determine if changes in the process output are due to changes in the factor levels or are due to normal variation.) Will the entire experiment be replicated or just portions of it? How many samples will be tested at each set of conditions? Some of the answers to these questions will depend on how good the measurement system is. The cost of additional runs will also come into play.

Next, how will effects of external sources of variation be eliminated? Will the experiment be completely randomized? Will blocking be used? Again, cost and time are important things to take into considerations here. (When possible, completely randomize the experiment.)

With these questions answered, the final design can be completed and data collection can begin.

There is a great deal of work that goes into designing and planning experiments. Don't try shortcuts! They may come back and haunt you in the form of meaningless, incomplete, and misleading results!

5.3.3 Orthogonal Arrays

Many designed experiments use matrices called orthogonal arrays for determining which combinations of factors and levels to use for each experimental run, and for analyzing the data. What is an orthogonal array?

> An orthogonal array is a matrix of numbers arranged in rows and columns. Each row represents the state or level of the experimental factors. Each column represents a specific factor or condition that can be changed from experimental run to experimental run. The array is called orthogonal because the effects of various factors in the experiment can be separated from each other.

An orthogonal array is one form of matrix math. Let's look at a simple example that will illustrate why they are used (DeLassus 1990).

Suppose some of your relatives had to move to another country. They find that fruits and vegetables are sold only in outdoor markets where none of the prices are posted. The grocers total the amount in their heads and then tell them how much money they owe. Your relatives want to find our how much bananas, apples, and oranges cost so that they can compare the prices to those back home. They decide to go to the same market stall on three consecutive days and buy varying quantities of each fruit.

Day 1:

3 apples
2 bananas
4 oranges

Total cost was \$1.40

Day 2:

1 apple
5 bananas
2 oranges

Total cost was \$0.85

Day 3:

5 apples
3 bananas
1 orange

Total cost was \$0.90

Now, how can they find out what the cost of each individual fruit was? They have three equations and three unknowns, so the problem can be easily solved mathematically.

Apples (A)	Bananas (B)	Oranges (C)	Total
3	2	4	1.40
1	5	2	0.85
5	3	1	0.90

$$3A + 2B + 4C = 1.40$$

$$1A + 5B + 2C = 0.85$$

$$5A + 3B + 1C = 0.90$$

When they worked through the calculations, they found that

apples cost	\$0.10
bananas cost	\$0.05
oranges cost	\$0.25

The situation is the same for any situation where there are many unknowns. If there are enough independent equations, the individual contributions from each unknown can be determined. That is the whole idea behind orthogonal arrays: They help the experimenter make sure that the correct experiments are performed to allow separation and solution for the individual factor effects. In the case of the mystery of the prices, the proper orthogonal array would help avoid a situation where the wrong fruits were purchased or too many trips were made to the market. The purchases could have been preplanned so that our relatives could have solved for the maximum number of individual fruit prices in the fewest trips to the market. It turns out that they didn't do too badly without the help of an orthogonal array, but most experiments are not so straightforward.

If we talk about orthogonal arrays in terms of designed experiments language, the types of fruit would be the factors, the number of pieces of fruit that were purchased would be the factor levels, the trips to the market would be the experimental runs, and the price paid each day would be the response. This is a simplified example. We know that, unless the grocer is cheating, the same cost per piece of fruit would be charged each day; therefore, there would be no variation in the cost from day to day. In a designed experiment, when measuring the effect that a factor has on a process output as the factor is changed, the output is not a finite value. The output takes the form of a distribution (hopefully a normal distribution) that can be predicted by the samples measured.

Many designed experiments use matrices that will solve for f factors in $f+1$ (f unknowns in $f+1$ equations). How can a matrix be tested for orthogonality? Let's use the matrix for a 2^3 factorial experiment (the construction of this matrix is described in Section 5.6.2.):

	A	B	AB	C	AC	BC	ABC
(1)	−	−	+	−	+	+	−
a	+	−	−	−	−	+	+
b	−	+	−	−	+	−	+
ab	+	+	+	−	−	−	−
c	−	−	+	+	−	−	+
ac	+	−	−	+	+	−	−
bc	−	+	−	+	−	+	−
abc	+	+	+	+	+	+	+

This matrix can study seven factors ($A, B, AB, \ldots$) with eight runs ($(1), a, b, \ldots$). The rows are called treatment combinations. The matrix has some interesting characteristics that are easy to see. In each column there are an equal number of pluses and minuses. When comparing two columns, it is easy to see that when the sign in the first column is plus, then half the time the sign in the second column will be plus and half the time it will be minus. As an example, look at columns AB and C:

AB	C
+	−
−	−
−	−
+	−
+	+
−	+
−	+
+	+

This is true for any two columns. It is easy to test a matrix for orthogonality. Turn the pluses into values of +1 and the minuses into values of −1. Then, multiply any two columns together and add up the results.

AB	C	
+1	−1	−1
−1	−1	+1
−1	−1	+1
+1	−1	−1
+1	+1	+1
−1	+1	−1
−1	+1	−1
+1	+1	+1
	Total	0

If the sum of the products adds up to zero, the two columns are considered to be orthonormal. If all of the columns are orthonormal with respect to each other, the array is considered to be orthogonal.

Orthogonal arrays form the basis for many experimental designs, although not all designs use the same array. That is why it is important to understand how to recognize (mathematically) when an array is orthogonal. The use of orthogonal arrays is discussed in more detail in the following sections: Section 5.6.2, 2^f Factorial Designs, Section 5.7, Fractional Factorial Designs, Section 5.8.1, Plackett-Burman Screening Designs, and Section 5.8.2, Taguchi Designs.

5.4 TESTS OF SIGNIFICANCE

Every day comparisons are made between two or more items, and from these comparisons decisions are made about which is better: What mixing method produces the best product; what order entry system is the fastest; and what machine setting gives the dimensions closest to the target specifications are several examples. These decisions, which are often based on only a limited amount of data, can lead to procedural changes, capital investment, and other commitments. Fortunately, there are methods, based on statistics, to assist in the decision making. These methods provide a confidence level that the conclusions are correct. These methods are often referred to as tests of significance because they indicate whether differences measured between two or more samples or populations are due to normal (common cause) variation in the system or if they are the result of a planned or unplanned change in the process.

Ideally, it would be preferable to test the entire population (*everything produced*) from a process and compare that to the population produced from another process. In most cases this is not practical, so samples taken from each population are relied on to provide an estimate of the population. This is called statistical inference, drawing conclusions about a population based on information contained in a sample. Inferential statistics provide the basis for hypothesis testing and designed experimentation.

The foundation of the statistics involved in tests of significance is the normal distribution. This is a symmetrical, unimodal curve that has its mean, mode, and median all equal. The equation for the normal distribution is

$$f(x) = \frac{1}{\sigma\sqrt{2\pi}} \exp\left(- \frac{(x - \mu)^2}{2\sigma^2} \right)$$

The notations used for describing the normal distribution are listed in Table 5.4. The equations that go along with the normal distribution are given in Table 5.5. Note that there are subtle differences between those for populations and those for samples. Because it is rare that the entire population can be measured, most often, the equations for samples are the ones used in tests of significance.

Before a test of significance can be performed, the experimenter must decide what items are going to be compared. The comparison to be tested is called the hypothesis. (Hypothesis testing is covered in Section 5.4.1.) With the hypothesis in mind, the appropriate test statistic is chosen

Table 5.4 Symbols for Statistical Parameters

	Population	Sample
Mean	μ	$\bar{x}$
Variance	σ^2	s^2
Standard deviation	σ	s

Table 5.5 Mathematical Definitions of Statistical Parameters

	Population	Sample
Central tendency (mean)	$\mu = \dfrac{x_1 + x_2 + \cdots + x_N}{N}$	$\bar{x} = \dfrac{x_1 + x_2 + \cdots + x_n}{n}$
Variance	$\sigma^2 = \dfrac{\Sigma(x - \mu)^2}{N}$	$s^2 = \dfrac{\Sigma(x - \bar{x})^2}{n - 1}$
Standard deviation	$\sigma = \sqrt{\dfrac{\Sigma(x - \mu)^2}{N}}$	$s = \sqrt{\dfrac{\Sigma(x - \bar{x})^2}{n - 1}}$

depending on the nature of the test (whether it involves comparing means or comparing variances).

Comparisons between means (Z-test statistic and t-test statistic) are based on the normal distribution. The normal distribution has some unique properties, which are shown graphically in Figure 5.6. If the mean values and the variance or standard deviation values are known for a set of samples or populations, then the characteristics of the normal distribution can be used to examine the samples for differences. From this examination, conclusions can be drawn that are based on a statistical certainty.

If two samples predicted the distributions shown in Figure 5.7, then there would be a high degree of certainty that the samples are from different populations. However, if the distributions overlapped as in Figure 5.8, then there is a degree of uncertainty that the samples are from different populations.

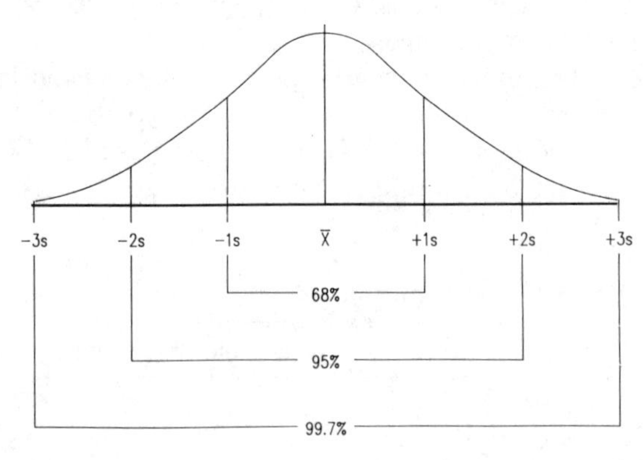

FIGURE 5.6. Area under the normal curve.

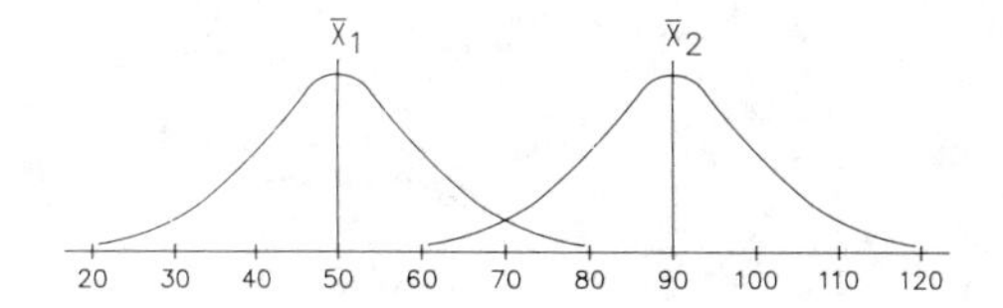

FIGURE 5.7. Two district distributions.

There is always a risk of making an error in determining if a set of samples is statistically different. Sample values are only predictions of the distributions of populations, so there is some uncertainty in whether the prediction is correct. There are also risks associated with how much overlap is allowed before it is reasonably certain that the samples or populations are different.

There are two types of risks or errors, the type I or α risk and type II or β risk.

Type I (α) error:	The error of saying that two items are not the same when they are.
Type II (β) error:	The error of saying two things are equal when they really aren't.

Whenever a test of significance is performed on a set of samples or populations, an acceptable α level must be chosen by the experimenter. Usually a value of $\alpha = 0.05$ is chosen. This means that 1 time in 20 the experimenter will be in error in stating that two distributions are not equal when they are. If a greater risk is acceptable, then experimenters may use an α value of 0.10. If a lower risk is required, then the value of α could be set at 0.005 or even lower.

The α risks are shown graphically in Figures 5.9 and 5.10. The α risk defines the amount of the tail of the normal distribution that will be allowed to overlap. Figure 5.9 shows a test of significance on only one tail. This would be used when comparing a set of samples or populations to

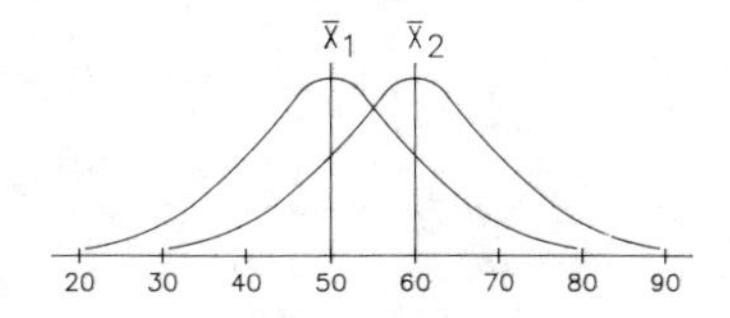

FIGURE 5.8. Overlapping distributions.

FIGURE 5.9. α risk for one-sided test.

determine if an improvement was made. A two-tailed test of significance is shown in Figure 5.10. This would be used when simply comparing the samples to see if they are different. Similar methods are used for tests of significance for the comparison of variance (F test and χ^2). Although the distributions involved here are not normal, the principles are the same and α levels must be chosen.

The t-test, Z-test, F-test, and χ^2-test statistics are described and applied in Sections 5.4.3 through 5.4.6.

5.4.1 Hypothesis Testing

Hypothesis testing involves the use of a sample, drawn from a population, to determine the validity of a theory posed about the population. Because an entire population cannot be tested, statistics must be used to draw conclusions about the distribution of the population from a sample.

The hypothesis itself is straightforward, but there are a number of different hypotheses that can be stated. The null hypothesis, or the hypothesis that will be tested, is designated by H_0. The alternate hypothesis is designated by H_a (some texts use H_1). Suppose it is necessary to know whether the mean of a population equals 4.3. The corresponding hypothesis would be

$$H_0: \mu = 4.3$$

where μ = population mean. The alternative hypothesis would be H_a: $\mu \neq 4.3$.

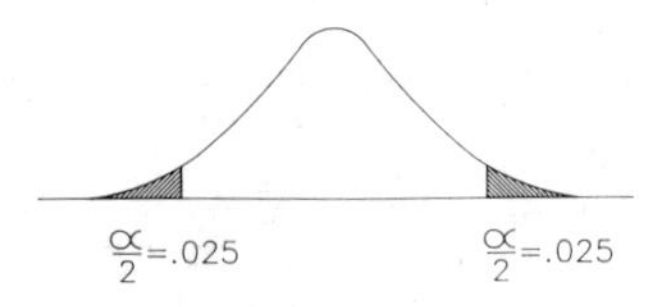

FIGURE 5.10. α risk for two-sided test.

It is also common to hypothesize about population variances (σ^2) using the same format:

$$H_0: \sigma^2 = 25$$

The alternate hypothesis would be H_a: $\sigma^2 \neq 25$.

The null hypothesis only contains equals, less than, and greater than signs and only population notation (μ, σ). In a null hypothesis, s or $\bar{x}$ is never used nor is the not equal to sign.

There are a number of different types of hypotheses that are commonly used:

- test on a single mean (sample vs. population)
- test on two means (population vs. population)
- test on a single variance (sample vs. population)
- test on two variances (population vs. population)

There are others, but they are beyond the scope of this text. For each type of hypothesis, there is a corresponding test statistic that is used to determine if the hypothesis is valid. For each test statistic, there is an acceptance region that helps to determine whether the null hypothesis is accepted or rejected. The acceptance regions are partially defined by the occurrence of two types of error.

Type I (α) error:	The error of rejecting H_0 when it is in fact true.
Type II (β) error:	The error of accepting H_0 when H_0 is false.

Confidence levels of α are selected for each hypothesis test and are included in the calculation (typically α is chosen to be 0.05 corresponding to a 95% confidence level).

Once the hypothesis is written, the appropriate test statistic is performed:

Tests on single mean
- σ is known — Z-test statistic
  $n \geq 30$
- σ is unknown — t-test statistic
  $n < 30$
  normal population

Tests on two means (independent samples)
- $\mu_1 = \mu_2$ and σ is known — Z-test statistic
- $\mu_1 = \mu_2$ but σ is unknown — t-test statistic

Test on single variance — χ^2-test statistic

Tests on two variances — F-test statistic

Examples of each are given in the following sections. Once the appropriate test statistic is applied, the calculated value is compared to the corresponding table value (given in Appendix B). If the calculated test statistic is greater than the table value, the hypothesis is rejected and the assumption is made that the alternate hypothesis is true. If the calculated test statistic is less than or equal to the table value, then the null hypothesis is not rejected. The null hypothesis cannot *theoretically* be accepted; the most that the test of significance can do is show that evidence doesn't exist to reject the hypothesis. However, from a practical sense, the null hypothesis would be accepted as true.

Example Problems: Writing Hypotheses

Write the null hypothesis for each problem.

Problem

Two different diets are fed to pigs to see if there is a difference in weight gain over a month's time. Twenty pigs were weighed at the beginning of the month and at the end of the month. Is the weight gain the same for the two diets?

Solution H_0: $\mu_1 = \mu_2$. Comparing the average weight gains of two groups of pigs.

Problem

The Department of Commerce released a report that showed that hourly employees in the brewery industry earned an average of $9.30 per hour. A state survey of 100 employees showed a mean of $9.20 and a standard deviation of $0.05. Is the income in the state the same as the national average?

Solution H_0: μ = \$9.30. Comparing the average salary for the brewery workers to the national average (considered a population).

Problem

In a molding operation, the temperature control is very important for molding consistent parts. A new press was installed and two different control systems were tested before a decision was made as to which one to buy. In each case, the controllers were turned on and allowed to reach steady state. Temperature measurements were then taken every 10 min for

2 h to determine how well the controllers could hold the set temperature. Was there a difference in the abilities of the two controllers to hold the test temperature?

Solution H_0: $\sigma_1^2 = \sigma_2^2$. Comparing the variances of two different temperature controllers.

Problem

Independent random samples of two makes of four cylinder compact cars were driven 40 mi/h for 200 mi with a single brand of gasoline. Is there a difference in the gas mileage for the two makes of car?

Solution H_0: $\mu_1 = \mu_2$. Comparing the average mileage values for two different cars.

Problem

A process for making large tablets has historically produced tablet weights with a standard deviation of 0.45 g. A designed experiment was performed to identify the factors that affect tablet consistency, and some of the process parameters were adjusted. After the adjustment, the following data were collected on the weights of six tablets (in grams):

2.43 2.54 2.75 2.46 2.63 2.55

What hypothesis would be tested to see if the process adjustments reduced the variation in the tableting operation?

Solution H_0: $\sigma^2 = 0.20$. Comparing the tablet weight consistency (variance) of an improved process to *historical* data from the old process (considered a population).

Problem

Last year, the mean IQ of entering freshmen at a university was 118. A random sample of 60 entering freshmen this year shows a mean IQ of 119.7. Has the mean IQ of the entering freshmen increased?

Solution H_0: $\mu = 118$. Comparing the average IQ of a sample of freshmen to the average of the entire population from last year.

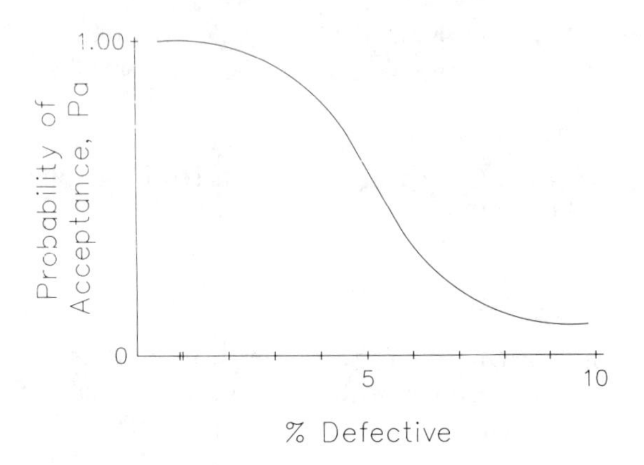

FIGURE 5.11. Operating characteristic curve.

5.4.2 Operating Characteristic Curves

When dealing with tests of hypotheses, there are two types of errors that come into play. Type I errors occur when the hypothesis is really true but it is rejected based on the results of the sample. The probability of a type I error is designated by α. Type II errors occur when the hypothesis is accepted based on the sample results when, in fact, the hypothesis is false. The probability of this occurring is represented by β.

α probabilities are commonly called producer's risks, because the producer is at risk of rejecting a batch of material that is acceptable.

β probabilities are commonly called consumer's risks, because the consumer is at risk of receiving a product that was considered acceptable based on sample inspection but was, in fact, not acceptable.

An operating characteristic curve (OC curve) is a plot of the probability of accepting a lot of material based on a sampling plan versus the actual fraction of defective parts in the lot. In a given sample plan with n equal to number of samples and c equal to acceptance number (number of defectives that will be allowed before the lot is rejected), there is a distinctive operating characteristic curve as shown in Figure 5.11.

OC curves indicate the ability of the sampling plan to discriminate between the good and bad lots. An ideal sampling plan would generate the "curve" shown in Figure 5.12, but this is not very realistic.

OC curves are constructed using an appropriate probability function. Typically, the binomial distribution is used. It is valid for all size lots. However, when dealing with large lot sizes, the Poisson distribution is

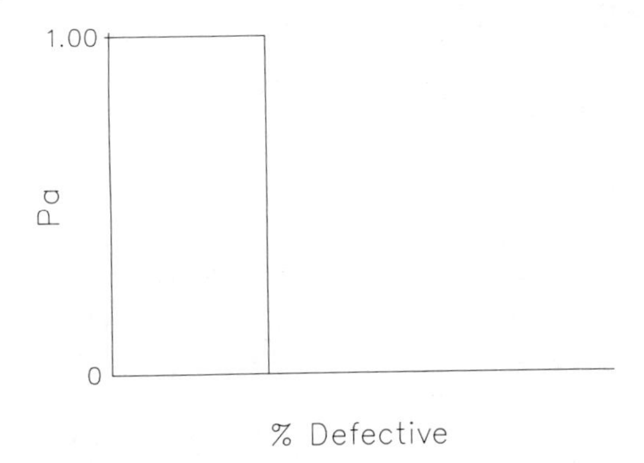

FIGURE 5.12. Ideal sampling plan operating characteristic curve.

sometimes used, and when dealing with small lots, the hypergeometric distribution can be used.

Binomial:
(all lot sizes)

$$P_a(n,c) = \binom{n}{c}_p^{\,c}(1-p)^{n-c}$$

Poisson:
(large lot sizes)

$$P_a(n,c) = \frac{np^c e^{-np}}{c!}$$

Hypergeometric:
(small lot sizes)

$$P_a(n,c) = \frac{\binom{np}{c}\binom{N-np}{n-c}}{\binom{N}{n}}$$

where P_a = probability of acceptance
n = sample size
c = number of defectives allowable
p = proportion defective in a lot
N = lot size

For a given sample size n and number of allowable defectives c, P_a can be calculated for the range of possible p values (proportion defective in the lot) to generate the OC curve. The level of consumer's risk ($\beta = P_a$) and producer's risk ($\alpha = 1 - P_a$) can then be interpreted from the graph.

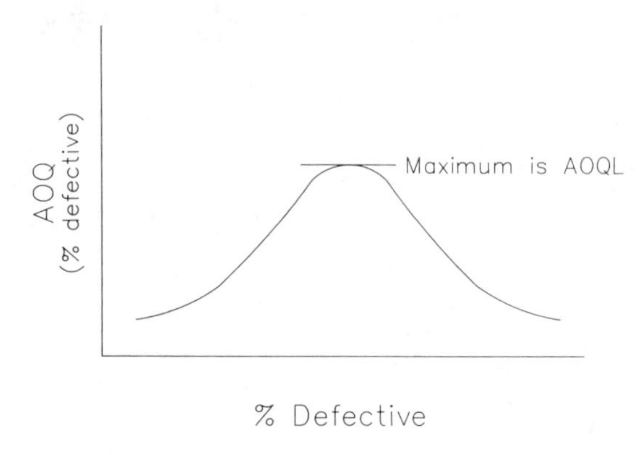

FIGURE 5.13. Average outgoing quality curve.

Some other terms commonly used with sample plans are:

AQL = acceptable quality level (acceptable percent defective for a lot)

LTPD = lot tolerances percent defective (lot percent defective considered to be unacceptable)

AOQL = average outgoing quality limit (maximum outgoing percent defective for a given plan)

AQL-based plans control the upper part of the curve and focus on producer's risks. High acceptance levels for low-percent-defective lots are the focus. LTPD plans control the area of low acceptance (consumer's risk) making sure that high-percent-defective lots are not accepted. AOQL plans control what is expected to be the worst average quality level that can exist in the product. An average outgoing quality (AOQ) curve (Figure 5.13) is constructed to determine the limit.

AQL and LTPD can both be identified on the OC curve as shown in Figure 5.14. Obviously, as the sample size increases, the curve will become steeper. (The steeper the curve, the better for both producers and consumers. Remember, ideally we want a vertical line at the level of acceptable percent defective.) If the level of allowable defectives is zero, then the curve will be exponential in shape (Figure 5.15).

5.4.3 *t* Test—Testing To See If a Change is "Real"

The *t* test, also known as the Student *t* test, examines either of two hypotheses: It is used to compare the mean of samples taken from two

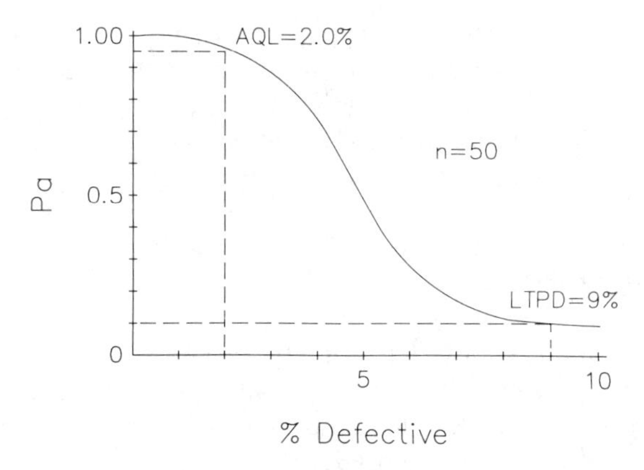

FIGURE 5.14. Using the operating characteristic curve.

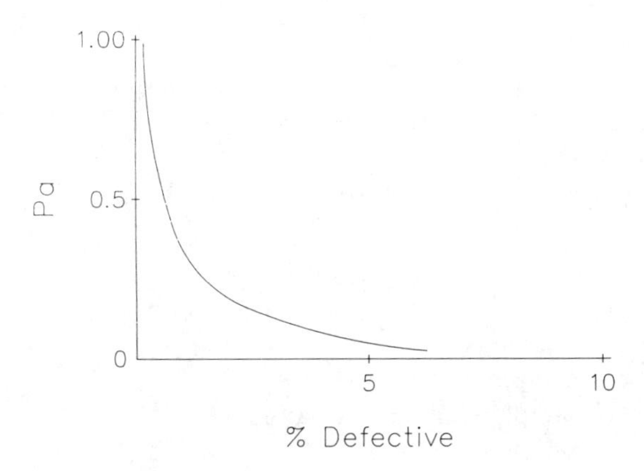

FIGURE 5.15. Operating characteristic curve for 0 defects acceptable.

populations (H_0: $\mu_1 = \mu_2$) or to compare the mean of a sample to the mean of a known population (H_0: $\mu = \mu_0$). The t test shows whether improvement has been made. To indicate improvement, the hypothesis would be rejected, showing that the means are not equal. This indicates that the change that occurred in the system had a significant effect on the output measured. The t test is normally used when the sample size is less than 30 and the value of the population standard deviation σ is unknown.

The t test uses a value t calculated from the sample mean(s), the sample standard deviation(s), and the sample size(s). This value is compared to an acceptance region. The acceptance region is found in Appendix B.3 using the appropriate α risk and degrees of freedom. The α

risk is the risk of rejecting the hypothesis when it is actually true. For example, if $\alpha = 0.05$ there is a 5% chance, or 1 out of 20, that the t test will reject the hypothesis when it is true. The degrees of freedom are based on the sample size(s). If t is within the acceptance region, the hypothesis is accepted and the means are equal; no change in the output is indicated. If t falls outside the acceptance region, the hypothesis is rejected because the means are not equal and a change in the process output is statistically verified.

Comparing a Sample to a Population. H_0: $\mu = \mu_0$
This hypothesis test uses the sample mean $\bar{x}$ to estimate the population mean μ. The hypothesis that the sample mean equals a given population is analyzed using the following procedure:

1. State the hypothesis.

$$H_0: \mu = \mu_0$$

$$H_a: \mu \neq \mu_0 \; [\text{or } \mu > \mu_0 \text{ or } \mu < \mu_0]$$

2. Calculate the value for t from

$$t = \frac{\bar{x} - \mu_0}{s/\sqrt{n}}$$

where $\bar{x}$ = sample mean
s = sample standard deviation
n = sample size
μ_0 = population mean

If the process generating the population has been control charted over time with no changes in the process, then the $\bar{x}$-chart centerline value $\bar{\bar{x}}$ can be used for μ_0.

3. Choose a value for the α risk. The values most commonly used are 0.01 (1 in 100 chance of rejecting the hypothesis when it is actually true), 0.05, and 0.10.
4. Determine whether the α is placed on one tail or both tails of the distribution. If a process improvement has been made so that it is suspected that the test value has improved or been reduced from the historical value, $\mu > \mu_0$, use a one-tailed test, which will use a t value using α. If the object is to determine whether the mean is different overall, $\mu = \mu_0$, use a two-tailed test, which will use a t value using $\alpha/2$.

5. Calculate the degrees of freedom where

$$\mathrm{DF} = n - 1$$

6. Determine the acceptance region from Appendix B.3.

   If the process was improved so that it is thought $\mu > \mu_0$, find t using α and DF;
   if the objective is to determine if the means are the same, $\mu = \mu_0$, find t using $\alpha/2$ and DF.

7. If the absolute value for t calculated in step 2 is less than the table value, accept the hypothesis that $\mu = \mu_0$; if it is greater than the table value, reject the hypothesis. The sample does not have the same mean as the population. This indicates that the process has changed.

Comparing Two Populations.

$$H_0: \mu_1 = \mu_2$$

$$H_a: \mu_1 \neq \mu_2 \; [\text{or } \mu_1 > \mu_2 \text{ or } \mu_1 < \mu_2]$$

The hypotheses test for comparing the means of two populations follow procedures similar to those given in the previous subsection. Again, the sample means are used to estimate the population means. The differences lie in the calculation for t and the degrees of freedom. In this case:

$$t = \frac{\bar{x}_1 - \bar{x}_2}{\sqrt{1/n_1 + 1/n_2}\sqrt{[(n_1 - 1)s_1^2 + (n_2 - 1)s_2^2]/(n_1 + n_2 - 2)}}$$

where $\bar{x}_1$ = average of sample 1
$\bar{x}_2$ = average of sample 2
s_1 = sample 1 standard deviation
s_2 = sample 2 standard deviation
n_1 = sample 1 size
n_2 = sample 2 size
$\mathrm{DF} = n_1 + n_2 - 2$

To simplify the calculations, this equation is often simplified to

$$t = \frac{\bar{x}_1 - \bar{x}_2}{s_p\sqrt{1/n_1 + 1/n_2}}$$

where

$$s_p = \sqrt{\frac{(n_1 - 1)s_1^2 + (n_2 - 1)s_2^2}{(n_1 + n_2 - 2)}}$$

Paired versus Unpaired *t* Test

There are actually two distinct ways to perform *t* tests when dealing with two samples. Previously, we discussed the *t* test for comparing two population means using their sample means $\bar{x}_1$ and $\bar{x}_2$. These two populations were independent of each other, so this was an unpaired test. There are times, however, when we want to look at dependent populations. For this we use a paired *t* test.

Paired *t* tests compare sets, or pairs, of data points. They are ideally set up for use when the *same* sample is used for both tests. This eliminates any effect due to lack of homogeneity between samples. For paired *t* tests, we look at the differences between the pairs of data. We could use this, for example, if we had 10 similar machining centers and wanted to see the effect of a change in the supplier of stock metal. We could sample a part from each machine made from the original stock metal and then sample a part from each machine with the new stock metal. We would calculate the difference in a key dimension(*s*) by machine and use the average difference in testing the hypothesis. Paired *t* tests are also useful in cases where measurement error or gauge differences are of interest.

For paired *t* tests, the hypothesis being tested is:

$$H_0: \mu_d = 0$$

$$H_a: \mu_d \neq 0$$

The calculation for *t* is

$$t_{\text{calc}} = \frac{\bar{d} - \mu_d}{s_d/\sqrt{n}}$$

where $d_i = x_{1i} - x_{2i}$ for 1 to i pairs of data

$\bar{d}$ = the average difference for 1 to i pairs of data

s_d = standard deviations of the differences

n = sample size

$\mu_d = \mu_1 - \mu_2$

Actually, the test being performed will confirm whether or not the difference between the means of the two samples is zero.

Example Problems: t tests

Problem 1

A manufacturer of needles has a new method of controlling a diameter dimension. From many measurements of the present method, the average diameter has been 0.076 cm. A sample of 25 needles from the new process shows the average to be 0.071 with a standard deviation of 0.010 cm. If a smaller diameter is desirable, should the new method be adopted?

Solution

$$H_0\colon \mu = 0.76.$$

$$H_a\colon \mu \neq 0.76.$$

Choose $\alpha = 0.05$; one-tailed test, so use $\alpha = 0.05$ and DF $= 24$ to determine t_{table}:

$$t_{\text{calc}} = \frac{0.071 - 0.076}{0.010\sqrt{25}} = -2.5$$

$$|t_{\text{calc}}| = 2.5$$

$$t_{\text{table}} = 1.711 \qquad (\alpha = 0.05, \text{DF} = 24)$$

$$|t_{\text{calc}}| \text{ of } 2.5 > t_{\text{table}} \text{ of } 1.711$$

Reject the null hypothesis. Adopt the new method.

Problem 2

In the garment industry, the breaking strength of cloth is important. A heavy cloth must have at least an average breaking strength of 200 lb/in.^{2} (the stronger the better). From one particular lot of this cloth, these five measurements were obtained (in pounds per square inch):

206 194 203 196 192

Does this lot meet the requirement of an average breaking strength of 200 lb/in.^{2}?

Solution

$$H_0: \mu \geq 200.$$

$$H_a: \mu < 200.$$

Choose $\alpha = 0.05$; one-tailed test, so use t_α and DF = 4 to determine t_{table}:

$$t_{\text{calc}} = \frac{198.2 - 200}{6.017\sqrt{5}} = -0.67$$

$$|t_{\text{calc}}| = 0.67$$

$$t_{\text{table}} = 2.132 \qquad (\alpha = 0.05, \text{DF} = 4)$$

$$t_{\text{calc}} \not> t_{\text{table}}$$

$$|t_{\text{calc}}| \text{ of } 0.67 \not> t_{\text{table}} \text{ of } 2.132$$

Therefore, we cannot reject the null hypothesis. We can conclude that for practical purposes the fabric does meet the requirement.

Problem 3

A random sample of eight parts produced on first and second shifts had the following weight in pounds:

First shift:	6.2	4.7	5.8	5.0	7.1	5.3	3.8	4.1
Second shift:	6.0	4.2	5.6	5.1	6.3	4.6	3.7	4.6

Test the hypothesis that there is no difference between the weight of the parts produced on first and second shifts. Perform both a paired and an unpaired t test.

Solution (Unpaired t test)

$$\bar{x}_{\text{first}} = 5.25, \qquad s_1 = 1.097, \qquad s_p = \sqrt{\frac{7(1.097)^2 + 7(0.903)^2}{8 + 8 - 2}}$$

$$\bar{x}_{\text{second}} = 5.01, \qquad s_2 = 0.903, \qquad s_p = 1.005$$

H_0: $\mu_1 = \mu_2$; H_a: $\mu_1 > \mu_2$:

$$t_{\text{calc}} = \frac{5.25 - 5.01}{1.005\sqrt{1/8 + 1/8}} = 0.478$$

$$t_{\text{table}} = 2.14 \quad \left(\text{two-tailed test so } \frac{\alpha}{2} = 0.025 \text{ and DF} = 14\right)$$

$$t_{\text{calc}} \not> t_{\text{table}}$$

Cannot reject H_0.

Solution (Paired t Test) H_0: $\mu_d = 0$; H_a: $\mu_d \neq 0$; d values: 0.2 0.5 0.2 −0.1 0.8 0.7 0.1 −0.5. *Note:* Signs matter!

$$\bar{d} = \frac{\Sigma d_i}{8} = \frac{1.9}{8} = 0.2375$$

$$s_d = 0.427$$

$$t_{\text{calc}} = \frac{.2375 - 0}{.427/\sqrt{8}} = 1.573$$

$$t_{\text{table}} = 2.36 \quad \left(\text{two-tailed test so } \frac{\alpha}{2} = 0.025 \text{ and DF} = 7\right)$$

$$t_{\text{calc}} \not> t_{\text{table}}$$

Cannot reject H_0.

5.4.4 Z Test—Testing for Normality

The Z-test statistic represents a standard normal distribution. When a normal distribution curve is drawn, the curve will reach a maximum at $Z = 0$ with a variance of 1. Figure 5.16 shows the normal distribution with values for Z ranging from approximately −3.5 to 3.5.

Because of the symmetry of the curve, there is an equal probability of achieving $Z = +2$ as $Z = -2$, based on the area under the curve. Z,

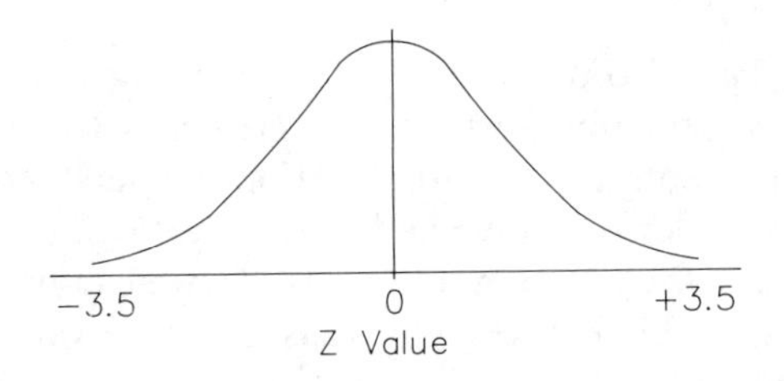

FIGURE 5.16. Normal curve showing $Z = 0$.

therefore, describes the distribution of a population and a sample using slightly different equations. For a sample:

$$Z = \frac{\bar{x} - \mu}{(\sigma/\sqrt{n})}$$

This is the calculated Z value used in the test of significance. Z tests are used to compare a sample mean to a known population mean to determine if the sample is from the same population. It is also used to compare two population means that are unknown but can be estimated from their sample means. The Z test is normally used with sample sizes greater than 30 and with the population (s) standard deviation known.

To show the use of the Z test, we'll use an example with a hypothesis to be tested of:

$$H_0: \mu = 2.5$$

Knowing that $\sigma = 0.8$ and using a sample size of $n = 30$, which produced a sample mean value of 2.2, Z can be calculated:

$$Z = \frac{\bar{x} - \mu}{(\sigma/\sqrt{n})}$$

$$= \frac{2.2 - 2.5}{(0.8/\sqrt{30})}$$

$$= -2.05$$

Because the normal curve is symmetrical around $Z = 0$, we can use the absolute value of the calculated Z to compare with the table value.

To find the Z table value, an α level must be chosen. Typically, $\alpha = 0.05$ is used. Because the hypothesis is H_0: $\mu = 2.5$, a two-sided test should be used and $\alpha/2$ will be used to find the table value. If the hypothesis was H_0: $\mu \geq 2.5$, a one-sided test would be used. Knowing that $\alpha/2 = 0.025$, it is possible to find the corresponding Z table value for 0.975, which is 1.96. Z_{calc} (2.05) > Z_{table} (1.96). Therefore, *reject* the null hypothesis.

The Z table can also be used to determine areas of the normal curve and probabilities. By definition, the curve is symmetric about $Z = 0$. The value $Z = 1$ represents 1 standard deviation. Understanding that $\pm 1\sigma =$ 68% of the area under the curve and $\pm 2\sigma = 95\%$ of the area, and so on, and by normalizing the notation to $Z = 1, 2, 3$, we can find areas under the normal curve using the Z table and translate those values into probabilities; see Figures 5.17, 5.18, and 5.19.

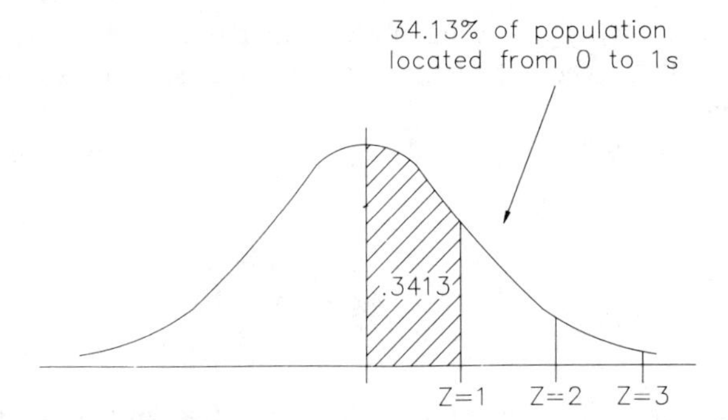

FIGURE 5.17. Area under the normal curve for $Z = 0$ to $Z = 1$.

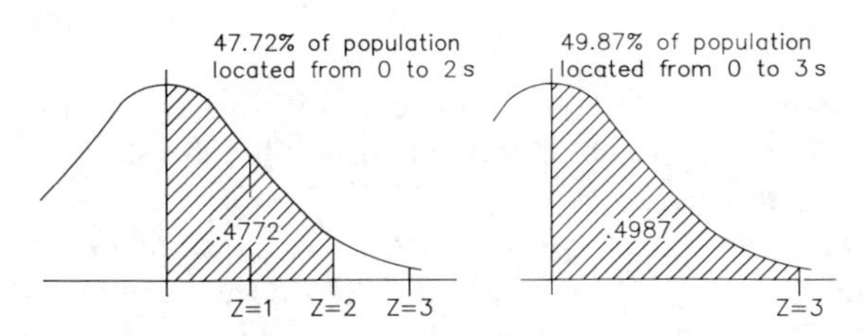

FIGURE 5.18. Areas under the normal curve for $Z = 0$ to $Z = 2$ and to $Z = 3$.

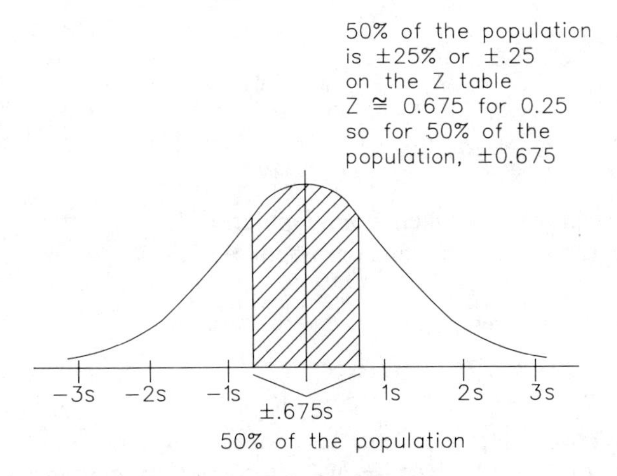

FIGURE 5.19. Normal curve showing 50% of the population.

Conversely, knowing the area under the curve (assuming the entire population is 100%), the corresponding standard deviations can be found from the table.

Confidence intervals (CI) for means can also be found using the Z statistic:

$$\mathrm{CI} = \bar{x} \pm Z_{\alpha/2}\left(\frac{\sigma}{\sqrt{n}}\right)$$

From the preceding example with $\bar{x} = 2.2$, $\sigma = 0.8$, and $n = 30$,

$$\mathrm{CI} = 2.2 \pm 1.96\left(\frac{0.8}{\sqrt{30}}\right) = 2.2 \pm 0.286$$

for $\alpha = 0.05$, which calculates to a range of 1.914 to 2.486. It is easy to see why the hypothesis was rejected when stating that the sample 2.2 ± 0.286 was equal to the mean of 2.5.

Just like with the t test, we can use the Z test to determine if two population means μ_1 and μ_2 are equal. We use the sample means $\bar{x}_1$ and $\bar{x}_2$ to estimate the population means. However, to use the Z test, we must know the population standard deviations σ_1 and σ_2. If we do not know the values for σ_1 and σ_2, then we would use the t test with the sample standard deviations. The equation for the Z test for two population means is:

$$Z = \frac{\bar{x}_1 - \bar{x}_2}{\sqrt{\sigma_1^2/n_1 + \sigma_2^2/n_2}}$$

where n_1 = sample size taken from population 1.
n_2 = sample size taken from population 2.

Example Problems: Z test

Problem 1

An automatic weighing operation has historically produced bags weighing 60.7 lb with a standard deviation of 0.4 lb. After an equipment repair, 20 bags were sampled to determine if the mean output had changed. The mean of the samples was 60.2 lb.

Solution H_0: $\mu_1 = \mu_2$; H_0: $\mu_1 \neq \mu_2$. Choose $\alpha = 0.05$; two-tailed test, so use $\alpha/2$.

$$Z_{\text{table}} = 1.96$$

$$Z_{\text{calc}} = \frac{60.2 - 60.7}{0.4/\sqrt{20}} = -5.59$$

$$|Z_{\text{calc}}| = 5.59$$

$$Z_{\text{calc}} > Z_{\text{table}}$$

We reject the null hypothesis. For practical purposes, the equipment repair has changed the mean output of the weighing system.

Problem 2

Two random samples of senior men and women students at a university gave the following grade point averages (GPA):

Group	Sample Size	Sample Mean	Population Standard Deviation
Women	35	2.70	0.63
Men	45	2.78	0.55

Is there a difference in the mean GPA between men and women at this university?

Solution H_0: $\mu_1 = \mu_2$; H_0: $\mu_1 \neq \mu_2$.

$$Z_{\text{calc}} = \frac{2.70 - 2.78}{\sqrt{(0.63)^2/35 + (0.55)^2/45}} = -0.011$$

$$|Z_{\text{calc}}| = 0.011$$

Choose $\alpha = 0.05$; two-sided test, so use $\alpha/2 = 0.025$.

$$Z_{\text{table}} = 1.96$$

$$Z_{\text{calc}} \not> Z_{\text{table}}$$

Because the Z value calculated falls within the acceptance region, the hypothesis cannot be rejected. There is no difference between the grade point averages of men and women.

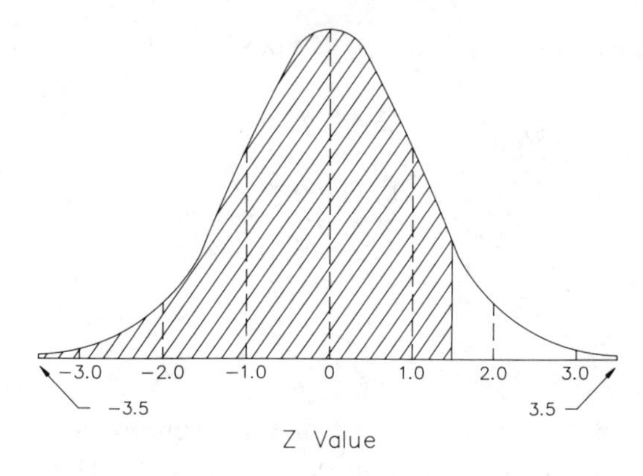

FIGURE 5.20. Total area under the normal curve for $Z = 1.5$.

Problem 3

At $Z = 1.5$, what percentage of the total area under the normal curve is represented?

Solution At $Z = 1.5$, the table value is 0.9332. Therefore, $100\% \times 0.9332 = 93.32\%$, meaning that 93.32% of the total area is represented as shown in Figure 5.20.

Problem 4

The lengths of a machined bushing are normally distributed around a mean of $\bar{x}$. How many standard deviation units symmetrical about $\bar{x}$ will include 80% of the lengths?

Solution $Z = 0$, the area = 50% or table = 0.500. The problem asks for 80% of the data, or 80% of the area centered around $Z = 0$. That means ± 0.40 on the table.

$$\text{At } 0.500 + 0.400 = 0.90,\ Z = 1.28$$

$$\text{At } 0.500 - 0.400 = 0.10,\ Z = -1.28$$

So, 80% of the data fall between -1.28 and 1.28 standard deviations as shown in Figure 5.21.

5.4.5 *F* Test—Testing To Determine If Two Variances Are Different

The F test is used to test the differences in the variation between two samples using the sample standard deviations. This test will indicate

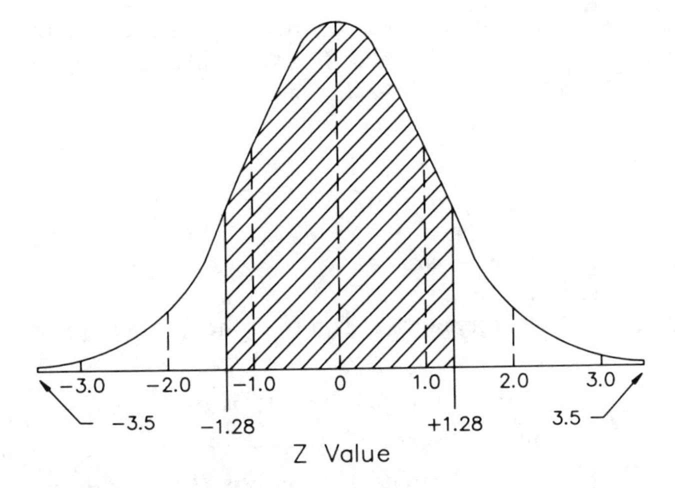

FIGURE 5.21. 80% of the area under the normal curve.

whether or not the improvements made to the process result in real improvement in reducing process variation. The hypothesis tested is that the variation between the samples is equal, H_0: $\sigma_1^2 = \sigma_2^2$. If improvement has resulted in real reduction in the variation, this hypothesis will be rejected. The procedure for completing the F test follows a procedure similar to the t test:

1. State the hypothesis.

$$H_0: \sigma_1^2 = \sigma_2^2$$

$$H_a: \sigma_1^2 \neq \sigma_2^2 \left[\text{or } \sigma_1^2 > \sigma_2^2 \text{ or } \sigma_1^2 < \sigma_2^2\right]$$

2. Calculate the value of F from

$$F = \frac{s_1^2}{s_2^2} \quad \text{put the largest of the two in the numerator}$$

where s_1 = sample 1 standard deviation
s_2 = sample 2 standard deviation

3. Choose the value for the α risk (that the hypothesis will be rejected when it is actually true). Again, use common values of 0.01, 0.05, or 0.10.

4. In the improvement process (H_a: $\sigma_1^2 > \sigma_2^2$), a one tailed test is used so an F based on α is used. If a two-tailed test (H_a: $\sigma_1^2 \neq \sigma_2^2$) is desired, use $\alpha/2$ to find the F value.
5. Calculate the degrees of freedom:

$$DF_1 = n_1 - 1 \qquad DF_2 = n_2 - 1$$

where n_1 = sample 1 size
n_2 = sample 2 size

6. Determine the acceptance region from the F table in Appendix B.6.

If the alternate hypothesis is H_a: $\sigma_1^2 \neq \sigma_2^2$,
find F using $\alpha/2$, DF_1, DF_2.

If the alternate hypothesis is H_a: $\sigma_1^2 > \sigma_2^2$,
find F using α, DF_1, DF_2.

7. If the value for F calculated in step 2 is less than or equal to the F table value within the acceptance region, accept the hypothesis that $\sigma_1^2 = \sigma_2^2$. If it is greater than the table value, then reject the hypothesis indicating that the two populations do not have the same variation. If the hypothesis is rejected, there has been real improvement in reducing the variation.

F tests are very useful for comparing the variances of two samples and measuring process improvement. F tests form the foundation of analysis of variation (ANOVA), which is a tool commonly used in the analysis of data generated during designed experiments. (ANOVA is discussed further in Section 5.4.8.)

Example Problems: F Test

Problem 1

Determine whether the following two types of rockets have significantly different variances in their nozzle velocities.

	Rocket 1	Rocket 2
Sample size	61 readings	31 readings
Variance	1346.89	2237.29

Solution H_0: $\sigma_1^2 = \sigma_2^2$; H_a: $\sigma_1^2 \neq \sigma_2^2$. Choose $\alpha = 0.05$; two-tailed test so use $\alpha/2$:

$$F_{\text{calc}} = \frac{s_2^2}{s_1^2} = \frac{2237.29}{1346.89} = 1.66$$

F_{table} with 30 DF (numerator), 60 DF (denominator), and $(1 - \alpha/2) = 0.975$.

$$F_{\text{table}} = 1.82$$

$$F_{\text{calc}} \not> F_{\text{table}}$$

so the null hypothesis cannot be rejected. There is no significant difference.

Problem 2

A manufacturer of a plastic molding materials is trying to decide which formulation to use for a particular customer. High tensile strength is desirable. Formulation 1 is less expensive to mix, but product engineering isn't sure if the strength is the same as formulation 2. Five batches of each were made by each formulation and tested for tensile strength.

Formulation 1	Formulation 2
3067	3200
2730	2777
2840	2623
2913	3044
2789	2834

Which formulation should be used?

Solution H_0: $\sigma_1^2 = \sigma_2^2$. H_a: $\sigma_1^2 \neq \sigma_2^2$. Choose $\alpha = 0.05$; two-tailed test, so use $\alpha/2$:

$$F_{\text{calc}} = \frac{s_2^2}{s_1^2} = \frac{51713.3}{16923.7} = 3.056$$

4 DF (numerator), 4 DF (denominator), $(1 - \alpha/2) = 0.975$.

$$F_{\text{table}} = 9.60$$

$$F_{\text{calc}} \not> F_{\text{table}}$$

The null hypothesis cannot be rejected, so the experimenters cannot say the formulations are different. Formulation 1 should be used because it is less expensive.

5.4.6 Chi-Square Distribution

The chi-square (χ^2) test statistic is used when testing a hypothesis involving one known population variance and one unknown population variance that can be estimated from its sample variance s^2. The hypothesis for a chi-square test is:

$$H_0: \sigma^2 = \sigma_0^2; \qquad H_a: \sigma^2 \neq \sigma_0^2 \left[\text{or } \sigma^2 < \sigma_0^2\right]$$

where σ_0^2 is the known population variance. It is a skewed distribution. The test statistic used is

$$\chi^2 = \frac{(n-1)s^2}{\sigma_0^2}$$

where n = sample size
s^2 = sample variance
σ_0^2 = population variance

To test the hypothesis, compare the table value for $\chi^2_{\alpha, n-1}$ to the calculated value.

If $\chi^2_{\text{calculated}} > \chi^2_{\text{table}}$, then the hypothesis is rejected. Again, use α for a one-tailed test and $\alpha/2$ for a two-tailed test.

For example, suppose that the population has a variance $\sigma_0^2 = 1.34$. A sample of 20 parts is selected and their dimensions are measured. A variance of 1.20 is calculated for the sample. The hypothesis is stated as:

$$H_0: \sigma^2 = 1.34; \qquad H_a: \sigma^2 \neq 1.34$$

From these data the value of χ^2 can be calculated:

$$\chi^2 = \frac{(19)(1.20)^2}{(1.34)^2} = 15.24$$

The chi-square Table (Appendix B.5) for $n - 1$ degrees of freedom (19 in this case) and $\alpha/2 = 0.025$, indicates the value of approximately 34. Because $15.25 < 34$, the null hypothesis is accepted. The variances are equal. Again, if $\chi^2 \geq \chi^2_{\alpha/2}$ or $\leq \chi^2_{1-\alpha/2}$, reject H_0. (Most tables list both $\chi^2_{\alpha/2}$ and $\chi^2_{1-\alpha/2}$.)

To set confidence limits, the following formula is used:

$$\frac{(n-1)s^2}{\chi^2_{\alpha/2}} \leq \sigma^2 \leq \frac{(n-1)s^2}{\chi^2_{1-\alpha/2}}$$

So for this example, the standard deviation of the process will fall somewhere between

$$\frac{(20-1)(1.2)^2}{32.85} \leq \sigma^2 \leq \frac{(20-1)(1.2)^2}{8.825}$$

or $0.833 \leq \sigma^2 \leq 3.10$. This is based on 95% confidence limits.

Example Problems: Chi-Square Test

Problem 1

In a drug firm, the variation in the weight of an antibiotic from batch to batch is important. With the present process, the standard deviation is 0.11 g. The research department has developed a new process that they believe will produce less variation. The following weight measurements were obtained using the new process:

7.47 7.49 7.64 7.59 7.55

Does the new process have less variation?

Solution H_0: $\sigma^2 = 0.0121$; H_a: $\sigma^2 < 0.0121$; $\sigma_0 = 0.11$; $s = 0.07$; choose $\alpha = 0.05$; one-tailed test:

$$\chi^2_{\text{calc}} = \frac{(4)(0.07)^2}{(0.11)^2} = 1.6198$$

$$\chi^2_{\text{table}} = 0.711$$

$$\chi^2_{\text{calc}} > \chi^2_{\text{table}}$$

So the null hypothesis is rejected. There is enough evidence to show less variation in the new process.

Problem 2

A paper manufacturer has a new method of coating paper. The less variation in the weight of the coating, the more uniform and better the product. The following 10 sample coatings were obtained by the new method:

223 215 220 238 230 234 229 223 235 227

If the standard deviation in the past has been 9.3, is the proposed method any better?

Solution H_0: $\sigma^2 = 9.3$; H_a: $\sigma^2 < 9.3$: choose $\alpha = 0.05$; one-tailed test:

$$\chi^2_{\text{calc}} = \frac{(9)(7.229)^2}{(9.3)^2} = 5.438$$

$$\chi^2_{\text{table}} = 3.33$$

$$\chi^2_{\text{calc}} > \chi^2_{\text{table}}$$

Again, the null hypothesis is rejected. There is evidence to conclude that there is less variation.

5.4.7 Regression Analysis

Regression analysis is the study of the relationship between two or more variables. In its most simple form, it is a graphical technique, known as a scatter diagram, used to visually interpret the relationship between two variables. Mostly, however, it is a statistical method for establishing an equation relating one variable to one or more other variables. This statistical method is also known as least squares or curve fitting.

There are many reasons for wanting to know the regression equation for a set of variables. The following list describes some of their uses:

- To locate optimal operating conditions. For example, a knowledge of granulator speed verses fines generation would enable the minimization of fines output by operating at the correct speed.
- To quantitatively understand the relationship. For example, measuring pressure drop across an orifice plate and the actual flow rate can establish their relationship so that the flow rate can be adjusted by adjusting the pressure drop, which is more easily measured. Another example is that the cure rate of an adhesive could be correlated with both time and temperature.
- To predict results. Establishing a regression equation would enable prediction of the dependent variable response at levels of the independent variable(s) that may not be able to be measured or that were not thought of originally.

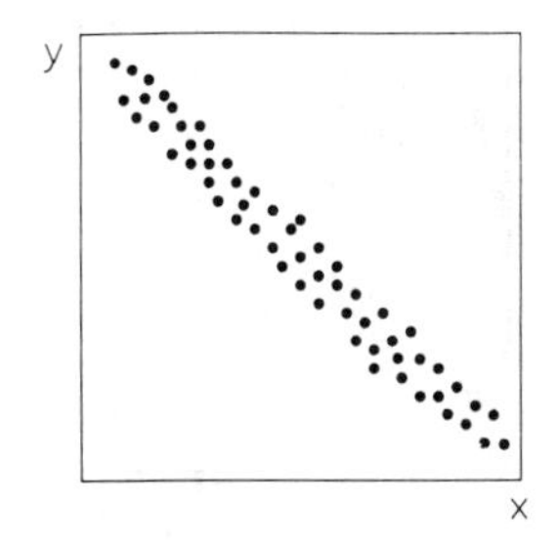

FIGURE 5.22. Linear correlation of two variables.

The basic steps in a regression analysis study are:

1. Define the objectives of the study. This is normally to study the response of a dependent variable to variations in the independent variable(s).
2. Define the measurement system. This includes the sampling technique, measurement equipment, and methods.
3. Collect data.
4. Plot the data on scatter diagrams. As described in Section 2.7.3, this plot may provide enough information on the relationship between the variables to draw a conclusion. Additionally, the shape of the points plotted on the scatter diagram can guide the calculation of the regression equation. The shape may indicate if the target equation is linear or a higher order polynomial as seen in Figures 5.22 and 5.23.
5. Calculate the regression equation. For linear relationships, this equation will take the form

   $$y = mx + b$$

   where m is the slope and b is the intercept. This equation does not take into account random errors such as measurement errors so y is actually an estimate based on sample data. The values for the slope a y intercept are calculated from

   $$m = \frac{\Sigma(x_i - \bar{x})(y_i - \bar{y})}{\Sigma(x_i - \bar{x})^2}$$

   $$b = \bar{y} - m\bar{x}$$

   where i is the sample number. These numbers are rarely calculated by hand now. Most hand calculators designed for engineers and

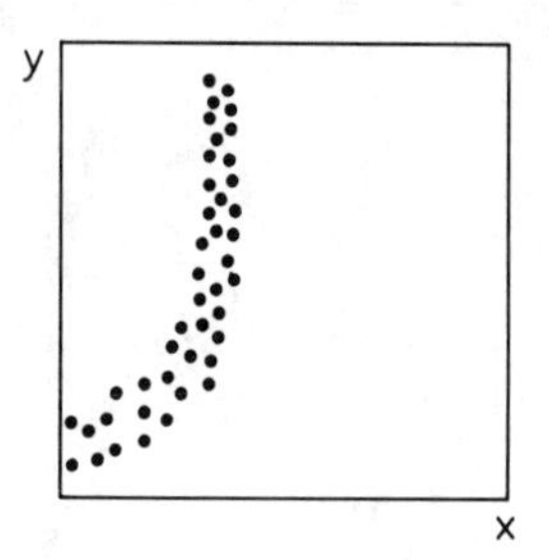

FIGURE 5.23. Nonlinear correlation of two variables.

scientists will calculate linear equations automatically from inputted data.

One method for checking the fit of the equation to the data is the correlation coefficient r:

$$r = m \Bigg/ \sqrt{\frac{\Sigma(y_n - \bar{y})^2}{\Sigma(x_n - \bar{x})^2}}$$

The value for r can be obtained along with the value for m and b with an appropriate scientific calculator. If $r = 1$, there is a perfect correlation so the equation is a perfect fit. This rarely occurs. Values of r less than 0.67 indicate the equation is not necessarily a good fit. This may indicate there is little correlation or that the regression equation is a higher-order polynomial.

Higher-order polynomials take the form

$$y = a_0 + a_1 x + a_2 x^2 + \cdots + a_k x^k$$

where a_k values are unknown coefficients. If the equation is a higher-order polynomial, most often these are second or third order. Reaction equations are often second- or third-order polynomials. These take the form:

Second order: $y = a + a_1 x + a_2 x^2$

Third order: $y = a + a_1 x + a_2 x^2 + a_3 x^3$

With sufficient data, these equations can be solved by hand using differential calculus. In some cases, these can also be solved by matrix algebra. Both of these approaches are extremely rigorous. PC-based software such as Statgraphics® or SAS® by SAS Inc. can quickly calculate equations for higher-order polynomials and determine the fit of the data to the curve.

Multiple regression involves identifying the equation relating the dependent variable y to more than one independent variable x_k. This may occur in plastic molding operations, for example, where the cycle time will be related to a number of the plastic characteristics and to the molding parameters. This equation takes the general form

$$y = a_0 + a_1x_1 + a_2x_2 + \cdots + a_kx_k$$

where the a_k values are unknown coefficients and x_k represents the independent variables. The a_k values can be calculated using matrix algebra or with the previously mentioned software. Those interested in the underlying principles and in calculating multiple regression coefficients by hand should refer to Juran's *Quality Control Handbook* or to works such as *Applied Regression Analysis* by N. P. Draper and H. Smith and *Fitting Equations to Data* by C. Daniel and F. S. Wood.

5.4.8 Fundamentals of ANOVA

Sections 5.4.1 through 5.4.6 discussed the concepts of hypothesis testing and the various test statistics for use when comparing two means, two variances, and sample means and variances to population means and variances. In other words, cases where two things are compared and tested for statistical similarity. Although there are many cases where there is only a need to compare two things, there are also many instances where more than two things must be compared simultaneously. The hypothesis may be for multiple means or variances:

$$H_0\colon \mu_1 = \mu_2 = \mu_3 = \mu_4 \quad \text{or} \quad H_0\colon \sigma_1^2 = \sigma_2^2 = \sigma_3^2 = \sigma_4^2$$

For these situations, a slightly different approach must be used: analysis of variance (ANOVA). This is a common tool for designed experimentation, where a number of different variables are being studied simultaneously. ANOVA allows the experimenter to quickly analyze the variances present in the experiment with the F test (see Section 5.4.5). The results of the F tests indicate whether there are differences in the means due to varying the test conditions.

ANOVA studies the overall variation along with the variation between treatments (variables) and within treatments (error). The F-test estimates the variances of all subgroups involved in the experiment. If the estimates are similar, the chances of the subgroup averages being detectably different are small. If the estimates are significantly different, then the subgroup

Table 5.6 Generic ANOVA Table[a]

Source of Variation	DF	SS	MS	F
Treatments Variation between (Variables)	$i - 1$	$\sum \frac{Y_{i\cdot}^2}{n_i} - \frac{(Y_{..})^2}{N}$	$\frac{SS_{treat}}{DF} = MS_{treat}$	$\frac{MS_{treat}}{MSE} = F_{calc}$
Error Variation within Unexplained	$(N - 1) - (i - 1)$	by subtraction = SSE	$\frac{SSE}{DF} = MSE$	
Total (Sum of all existing variation)	$N - 1$	$\sum \sum Y_{ij}^2 - \frac{(Y_{..})^2}{N}$		

[a] i = treatment; j = observation within each treatment; n = sample size; N = total number of observations. Dot notation represents a sum: $Y_{i\cdot}$= sum of all observations for treatment i; $Y_{..}$= sum of all observations.

averages may be significantly different. This is accomplished by comparing the between-subgroup estimate of variances to the within-group estimates of variance. The F test performed in an ANOVA is actually a ratio of the variation between treatments and the various treatment variations to the variation due to error. If the F value calculated based on the data is greater than the F value found in the table (as with any other F test), then the null hypothesis is rejected and the means are considered to be statistically different.

The generic format for an ANOVA table is shown in Table 5.6. The basic format can be modified depending on the experimental design and can be expanded to handle multiple treatments (variables) and interaction.

A one-way ANOVA refers to the situation where there is only one variable (treatment) involved. Similarly, a two-way ANOVA has two treatments and a three-way ANOVA handles three treatments. The next few sections will cover various types of designed experiments with examples of ANOVA applied to each.

5.5 RANDOMIZED DESIGNS

In the hierarchy of experimental designs (see Table 5.7) the simplest form of experimental design is where only one factor is to be studied (or varied). This is referred to as single-factor experimentation. To improve experimental techniques, the order of experimentation is randomized with regard to the different levels of the factor being studied. This eliminates bias by providing an equal chance of testing any of the levels at any time.

Table 5.7 Types of Randomized and Blocked Experimental Techniques

Design	Factors	Blocks	Output
Completely randomized	One	None	One factor effect Estimate variance
Randomized block	One	None	One factor effect One block effect . Estimate variance
Latin square	One	Two	One factor effect Two block effects Estimate variance
Graeco–Latin square	One	Three	One factor effect Three block effects

This technique is referred to as a *completely randomized design*. It allows the experimenter to compare treatment effects when only *one* factor has to be studied. (There is no blocking used in a completely randomized design.) Completely randomized designs will be covered in Section 5.5.1.

In situations where there are environmental or material effects that can be divided into groups or "blocks" of effects (for example, days, operators, piece of test equipment), it may be advantageous to introduce blocking. One factor can still be investigated at a number of different levels, as in a completely randomized design, and the effects of blocks on the response can be separated from experimental error and measured. This type of design is called a *randomized block design*: Each treatment appears in each block once. Randomized block designs are covered in Section 5.5.2.

In some extreme cases, it may even be desirable to add a second blocked variable (for example, operator and day of the week) while studying the effect of one factor at several levels. The effect of the treatments (the factor at its various levels) can be measured as well as the effects of the two blocks. This type of design is called a *Latin square*, which is also covered in Sections 5.5.2.

Going one step further, and adding a third blocking constraint is possible—one factor with three blocks. These designs are called *Graeco–Latin squares*. We do not recommend this design because of the high degree of experimental constraint imposed by having three blocks and the fact that the size of the error term is greatly reduced. This makes the experiment less sensitive to subtle factor effects. Graeco–Latin squares will not be covered in this text.

5.5.1 Completely Randomized Experiments

In most real-life situations, comparisons have to be made between many different combinations of process parameters. In Section 5.4.1, hypothesis

Table 5.8 Generic One-Factor ANOVA Table

	Method 1	Method 2	Method 3
	56	49	71
	59	52	72
	61	53	65
	60	49	65
	56	48	71
$\bar{x}$	58.4	50.2	68.8

testing was discussed, but only hypotheses involving two items were included; comparisons of two means or comparisons of two variances. Sections 5.4.3 through 5.4.6 discussed applicable test statistics where two items were being compared. Experimental design methods allow for the organization of experimentation to maximize the output for the minimum input by applying these test statistics to more complicated hypotheses.

The simplest form of experimentation design is a completely randomized design. This is where a variable or factor is being tested at more than two treatments (or levels). If one variable (manufacturing method, temperature, time) is tested at two treatments, then the normal test statistics, such as t test or F test, can be used to compare them. With more than two treatments (levels), an analysis of variance (ANOVA) is used.

Take for example, a process where there are three different mixing methods being used to load raw materials. An experiment was set up with its objective to compare the effect of the three methods on the finished product density. Five measurements were made for each mixing method. The density measurements for the samples for each method are shown in Table 5.8. In this example, there are three levels, Methods 1, 2, and 3 of the mixing method. This means we are looking at three treatments. The basic ANOVA table for a completely randomized experiment is shown in Table 5.9.

In the ANOVA table, the DF column and the SS column are always additive, so everything above the total line should add up to the total.

In the mixing experiment example, some of the equation variables are immediately evident:

$i = 3$ because three methods are being studied.
$N = 15$ because five data points for each method were collected for a total of 15 data points.

With that information, the first column of the ANOVA can be completed as shown in Table 5.10.

Table 5.9 Basic ANOVA Table for One-Factor[a]

Source	DF	SS	MS	F
Treatments	$i - 1$	$\sum \frac{Y_{i\cdot}^2}{n_i} - \frac{(Y_{..})^2}{N}$	$\frac{SS_{treat}}{DF}$	$\frac{MS_{treat}}{MS_{error}}$
Error	$N - i$	$SS_{total} - SS_{treat}$	$\frac{SS_{error}}{DF}$	
Total	$N - 1$	$\sum \sum Y_{ij}^2 - \frac{(Y_{..})^2}{N}$		

[a]DF = degrees of freedom; SS = sums of squares; MS = mean square; F = F-statistic comparisons; MSE = mean square for the error term; i = number of treatments (methods); N = total number of observations (data points); n_i = sample size for each treatment.

For the second column, more complicated mathematics are needed.

$\Sigma Y_{i\cdot}^2$ is the sum of the squares of all the observations for each method i divided by the number of samples for each.
$(Y_{..})^2/N$ is the sum of all observations, which is then squared and divided by the total number of observations.
$\Sigma\Sigma Y_{ij}^2$ is the sum of the squares of all of the observations.

Using data from Table 5.8, the calculations are

$$\sum \frac{Y_{i\cdot}^2}{n_i} = \frac{(56 + 59 + 61 + 60 + 56)^2}{5} + \frac{(49 + 52 + 53 + 49 + 48)^2}{5} + \frac{(71 + 72 + 65 + 65 + 71)^2}{5}$$

$$= 17052.8 + 12600.2 + 23667.2$$

$$= 53{,}320.2$$

$$\frac{(Y_{..})^2}{N} = \frac{(56 + 59 + 61 + 60 + 56 + 49 + 52 + 53 + 49 + 48 + 71 + 72 + 65 + 65 + 71)^2}{15}$$

$$= \frac{(887)^2}{15}$$

$$= 52{,}451$$

$$\sum \sum Y_{ij}^2 = 56^2 + 59^2 + 61^2 + 60^2 + 56^2 + 49^2 + 52^2 + 53^2 + 49^2 + 48^2 + 71^2 + 72^2 + 65^2 + 65^2 + 71^2$$

$$= 53{,}409$$

Table 5.10 Partial ANOVA Table Showing Degree of Freedom for 1-Factor Designs

Source	DF
Treatments (method)	2
Error	12 (by subtraction)
Total	14

Plugging the results from these calculations into the formulas in column 2, the values for SS_{treat}, SS_{error}, and SS_{total} can be found. Using the values from columns 1 and 2, the calculations shown in column 3 can be made. The values in column 3 are used to generate the F value in column 4. The completed ANOVA table appears in Table 5.11.

In column 4, the F-test statistic was calculated. This value must then be compared to the F-table value for $i - 1$, $N - i$ degrees of freedom in Appendix B.5. The F-table values for 95% and 99% ($\alpha = 0.05$ and 0.01), respectively, are 3.89 and 6.93. It is obvious that for either risk level:

$$F_{\text{calc}} \gg F_{\text{table}}$$

Therefore, the null hypothesis can be rejected. But, what is the null hypothesis?

For completely randomized designs, the mathematical model that applies is

$$Y_{ij} = \mu + \tau_i + \epsilon_{ij}$$

where Y_{ij} = the ith observation of the ith treatment
μ = common effect for all treatments
τ_i = treatment effect for treatment i
ϵ_{ij} = random error

This model is based on some assumptions: the error term is considered to be normally distributed around zero; μ is fixed; and τ_i is fixed for each

Table 5.11 Completed ANOVA Table for Density Experimentation

Source	DF	SS	MS	*F*
Treatment	2	869.2	434.6	58.7
Error	12	88.8	7.40	
Total	14	958.0		

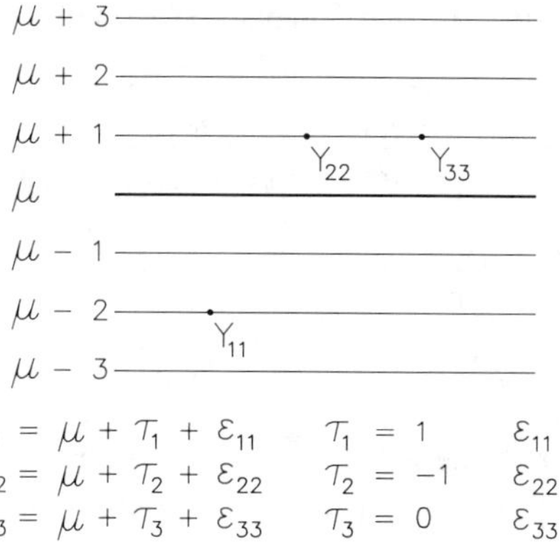

$\mu + 3$

$\mu + 2$

$\mu + 1$ — Y_{22} Y_{33}

μ

$\mu - 1$

$\mu - 2$ — Y_{11}

$\mu - 3$

$$Y_{11} = \mu + \tau_1 + \varepsilon_{11} \qquad \tau_1 = 1 \qquad \varepsilon_{11} = -3$$
$$Y_{22} = \mu + \tau_2 + \varepsilon_{22} \qquad \tau_2 = -1 \qquad \varepsilon_{22} = 2$$
$$Y_{33} = \mu + \tau_3 + \varepsilon_{33} \qquad \tau_3 = 0 \qquad \varepsilon_{33} = 1$$

FIGURE 5.24. Line graph.

treatment (i) and is normally distributed around zero. In order to visualize this, a line graph can be drawn. The baseline represents μ, which is fixed and the values of τ_i are positive or negative around μ and $\Sigma\tau_i = 0$. ϵ_{ij} is also additive or subtractive around the base line μ. This is shown in Figure 5.24.

An ANOVA for the completely randomized design tests the hypothesis

$$H_0\colon \tau_i = 0 \quad \text{for all treatments}$$

which means that there is no effect on the response due to the treatments.

In the preceding example of investigating the mixing method, the hypothesis was *rejected*, so there was an effect on the mean values due to the treatments. Once that is established, the individual treatment means should be tested to determine which are significantly different. There are many methods for accomplishing this. One effective way of comparing more than two means is the range test developed by Newman (1939) and modified by Keuls (1952). This test is now commonly known as the Newman–Keuls test. Its steps are:

1. List the means for each treatment from lowest to highest.

	$\bar{x}$
Method 2	50.2
Method 1	58.4
Method 3	68.8

2. Calculate the standard error for each treatment.

$$S_{\bar{Y}_{ij}} = \sqrt{\frac{\text{MS}_{\text{error}}}{\text{number of observations in } \bar{Y}_{ij}}}$$

$$= \sqrt{\frac{7.40}{5}}$$

$$= 1.22$$

3. From the table in Appendix C.8 for the upper 5% of the studentized range, find the p values up to one less than the number of treatments:

	$p = 2$	$p = 3$	
$\text{DF}_\text{E} = 12$	3.08	3.77	from the table

4. Multiply the table values by the standard error from step 2 to find the least significant ranges (LSR):

p	2	3
LSR	3.76	4.60

5. Compare the differences between means to the LSR values

Largest vs. smallest	Method 3 vs. 2	$18.6 > 4.60\ (\text{LSR}_3)$
Largest vs. next smallest	Method 3 vs. 1	$10.4 > 3.76\ (\text{LSR}_2)$
Second largest vs. smallest	Method 2 vs. 1	$8.2 > 3.76\ (\text{LSR}_2)$

(Note: If there had been six methods examined, for example, instead of three methods, the largest versus smallest would have been compared with LSR_6, the largest versus next smallest with LSR_5, and so on down to the largest versus the second largest with LSR_2. Then, the second largest versus the smallest would be compared with LSR_5, the second largest versus the next smallest with LSR_4, and so on. This would be cascaded until the last comparison of the next smallest versus the smallest was made with LSR_2.)

All three methods are statistically different. Therefore, for this example, the ANOVA showed that there was a significant effect on the response due to the different methods. The comparison of means indicates that all three methods have different effects on the response. By changing to method 3, the properties of the mix are significantly improved over both methods 1 and 2. If methods 3 and 1 had tested to show no statistical difference between the two, then either of these methods could be run rather than method 2.

There is a special notation that is often used to distinguish between treatments that are significantly different and those that are not. Capital letters ($A, B, C, \ldots$) are assigned to each treatment. Significantly different treatments have different letters. Treatments that are not found to be significantly different are assigned the same letters. To apply this procedure to the mixing method example, start by listing the treatments from highest to lowest:

Method 3
Method 1
Method 2

Because method 3 was significantly better than method 1, which was, in turn, significantly better than method 2, each treatment is assigned a different letter:

Method 3	*A*		
Method 1		*B*	
Method 2			*C*

That is a straightforward example because all of the treatments were significantly different from the others. It is usually more complicated. For example:

Treatment 1	*A*		
Treatment 2		*B*	
Treatment 3		*B*	
Treatment 4			*C*

This notation indicates that treatment 1 is significantly better than 2 through 4. Treatments 2 and 3 are not significantly different from each other, but both are significantly better than treatment 4. This notation allows the results of the experiment and the Newman–Keuls analysis to be presented in a format that is easy to interpret.

5.5.2 Randomized Block Designs and Latin Square Designs

One step beyond a completely randomized design is a randomized block design. Blocking, as defined in Section 5.3, is the use of homogeneous groups to investigate the effect of different treatments within strictly

Table 5.12 Table Representing a Three-Factor Three-Level Block Design

Mixing Method	Operators (Blocks) A	B	C
1	Y_{11}	Y_{21}	Y_{31}
2	Y_{12}	Y_{22}	Y_{32}
3	Y_{13}	Y_{23}	Y_{33}

controlled environments. In completely randomized designs, one factor is investigated as a number of different treatments in random order. In randomized block designs, the same experiment is performed, but the experiment is repeated with the treatments randomized within the different environments, or blocks.

For example, what if the mixing methods investigated in Section 5.5.1 (Completely Randomized Experiments) were thought to also be affected by the operators performing the tasks? Studying the mixing methods using a completely randomized design with operators assigned randomly may lead to conclusions that may not be completely valid. This type of situation is perfectly suited for a randomized block design. Suppose there are three trained operators in the department that makes the mixes. The best way to study the effect of mixing method is to design the experiment as in Table 5.12.

Within each block, the order that the mixing methods are performed is randomized, so the actual design might look something like Table 5.13.

One benefit of using the blocks is that the experimenter now has the ability to understand the effects of operator techniques on the product properties. Another benefit of using this method is that the error due to operator variability can be separated from the variation due to the treatment, so the results of the experiment are more sensitive to treatment effects.

The analysis of the results of a randomized block design is accomplished using a two-way ANOVA: it is two-way because both block effects and treatment effects are studied. The ANOVA table is given in Table 5.14.

Table 5.13 Randomization of the Mixing Methods within Each Block

	A	B	C
Mixing Method	1	3	2
	3	1	3
	2	2	1

Table 5.14 Two-Way ANOVA Table to Study Block and Treatment Effects[a]

Source	DF	SS	MSE	F ratio
Treatments	$I - 1$	$\frac{1}{J}\sum$	$\frac{SS_{treat}}{DF_{treat}}$	$\frac{MS_{treat}}{MSE}$
Blocks	$J - 1$	$\frac{1}{I}\sum Y_{.j}^2 - \frac{(Y..)^2}{N}$	$\frac{SS_{block}}{DF_{block}}$	$\frac{MS_{block}}{MSE}$
Error	$(N-1)-(I-1)$ $-(J-1)$ by subtraction	by subtraction = SSE	$\frac{SSE}{DF_{error}}$	
Total	$N - 1$	$\sum\sum Y_{ij}^2 - \frac{(Y..)^2}{N}$		

[a] $N = I \times J$ = total number of observations; I = number of treatments; J = number of blocks; $\Sigma Y_{i.}^2$ = sum of the squares of all of the treatments in each block; $\Sigma Y_{.j}^2$ = sum of the squares of all of the blocks in each treatment; $\Sigma\Sigma Y_{ij}^2$ = sum of the squares of all combinations of block and treatment; $(Y..)^2$ = the square of the sum of all observations.

To demonstrate this technique, the example involving both mixing methods and operators will be used. The response will be product density. The data collected are shown in Table 5.15. For this example,

$$I = \text{mixing methods} = 3$$

$$J = \text{operators} = 3$$

$$N = I \times J = 9$$

The degrees of freedom can be calculated and added to column 1 of the ANOVA table as shown in Table 5.16. Calculating the sum of squares

Table 5.15 Experimental Results

Mixing Method	Operators *A*	*B*	*C*
1	78	74	74
2	69	71	68
3	59	63	65

Table 5.16 Partial ANOVA Table Showing Degrees of Freedom for Three-Factor Block Designs

Source	DF
Mixing method	2
Operator	2
Error	4 (by subtraction)
Total	8

values for column 2:

$$\frac{1}{I}\sum Y_{.j}^2 = \frac{1}{3}\left[(78 + 69 + 59)^2 + (74 + 71 + 63)^2 + (74 + 68 + 65)^2\right]$$

$$= \frac{1}{3}(42{,}436 + 43{,}264 + 42{,}849)$$

$$= 42{,}849.7$$

$$\frac{1}{J}\sum Y_{i.}^2 = \frac{1}{3}\left[(78 + 74 + 74)^2 + (69 + 71 + 68) + (59 + 63 + 65)^2\right]$$

$$= \frac{1}{3}(51{,}076 + 43{,}264 + 34{,}969)$$

$$= 43{,}103$$

$$\sum\sum Y_{ij}^2 = 78^2 + 69^2 + 59^2 + 74^2 + 71^2 + 63^2 + 74^2 + 68^2 + 65^2$$

$$= 43{,}137$$

$$(Y_{..})^2 = (78 + 69 + 59 + 74 + 71 + 63 + 74 + 68 + 65)^2$$

$$= 385{,}641$$

So,

$$SS_{\text{block}} = \frac{1}{I}\sum Y_{.j}^2 - \frac{(Y_{..})^2}{N} = 42{,}849.7 - \frac{385{,}641}{9} = 0.7$$

$$SS_{\text{treat}} = \frac{1}{J}\sum Y_{i.}^2 - \frac{(Y_{..})^2}{N} = 43{,}103 - \frac{385{,}641}{9} = 254$$

$$SS_{\text{total}} = \sum\sum Y_{ij}^2 - \frac{(Y_{..})^2}{N} = 43{,}137 - \frac{385{,}641}{9} = 288$$

Table 5.17 Completed ANOVA Table for Three-Factor Three-Level Block Design

Source	DF	SS	MS	F
Mixing method	2	254	127	15.26
Operator	2	0.7	0.35	0.042
Error	4	33.3	8.325	
Total	8	288		

These values can be used to calculate the values for the mean squares (column 3), which in turn can be used to calculate the F values (column 4). Each F value is a ratio of the variance due to the factor (between groups) to the variance due to random error (within groups). The completed ANOVA table is shown in Table 5.17.

The F_{table} values are found using the corresponding degrees of freedom ($i - 1, N - i$ for mixing method and $j - 1, N - i$ for operator) and are listed at an α level of 0.05. For an α risk of 0.05, the F value in the table in Appendix B.5 is 6.94 for both.

Because F_{calc} for the mixing method is greater than F_{table}, *reject* the hypothesis. F_{calc} for the operators is less than F_{table}, so we accept the hypothesis. What are the hypotheses being tested? For a randomized block design, the mathematical model is:

$$Y_{ij} = \mu + \beta_j + \tau_i + \epsilon_{ij}$$

where β_j = block effect
τ_i = treatment effect
ϵ_{ij} = error term

The terms in the equation represent the effect that each fact or block has on the overall mean μ. The hypotheses tested in the ANOVA are that there is no effect on the mean due to the variation in treatments or blocks:

$$H_0: \tau_i = 0 \quad \text{and} \quad H_0: \beta_j = 0$$

In the example, the hypothesis for τ_i (mixing methods) was rejected. Therefore, there is an effect on the mean due to the different mixing method. The hypothesis for operator effects (blocks) was accepted, which means that variation due to the different operators was insignificant. With this information in hand, a simple comparison of means can be performed to determine which mixing methods are significantly different (better or worse) than the others as was done in Section 5.5.1.

Table 5.18 Randomized Block Design

	Operators		
	A	*B*	*C*
Mixing Methods	1	3	2
	3	1	3
	2	2	1

Latin-Square Designs

Whereas randomized designs introduce one constraint of blocking, Latin-square designs go one step further and introduce two blocking variables. In the randomized block design, the matrix in Table 5.18 was used.

Note that method 2 doesn't appear in row 2 and method 3 doesn't appear in row 3. This is because the order that the mixes were made randomized within each block (operator). With a Latin-square design, an additional blocking constraint (test order) can be added, which forces the position of the methods in each block. The design might look like Table 5.19.

With this matrix each mixing method appears in each column and each row only once. Remember that there is still only one factor being studied (mixing method), but that there are two blocking constraints: operators and test order.

Hicks (1982) presented a Latin-square design involving the study of tire wear. The first blocking constraint in this design is "car" and the second blocking constraint is "tire position". The result is a 4 × 4 Latin square (Table 5.20).

In this case, there are four types (brands) of tires being tested for tire wear. The two blocks are wheel position (there are four on each car) and cars (total of four cars). In order to meet the criteria of a Latin-square design, there must be equal numbers of block levels and treatments.

Table 5.19 Latin-Square Design for Three-Factor Three-Level Experiment

Order Tested	Operators *A*	*B*	*C*
First	1	2	3
Second	2	3	1
Third	3	1	2
	Mixing method = 1, 2, 3		

Table 5.20 4 × 4 Latin-Square Design for Tire Wear

	Car			
Position	I	II	III	IV
LF	*A*	*B*	*C*	*D*
RF	*D*	*A*	*B*	*C*
LR	*C*	*D*	*A*	*B*
RR	*B*	*C*	*D*	*A*
	Tires = *A*, *B*, *C*, *D*			

Designs for 3 × 3, 4 × 4, 5 × 5, and 6 × 6 Latin-squares are listed in Table 5.21.

The ANOVA format for a Latin-square design is given in Table 5.22. The procedure for its use follows the example for randomized block designs.

The mathematical model for Latin-square designs is

$$Y_{ijk} = \mu + \tau_i + \beta_j + \delta_k + \epsilon_{ijk}$$

where μ = overall mean
τ_i = treatment effect
β_j = block effect of block 1
δ_k = block effect of block 2
ϵ_{ijk} = error term

Table 5.21 List of Latin-Square Designs for 3 × 3, 4 × 4, 5 × 5, and 6 × 6 Experiments

3 × 3	*ABC*			
	BCA			
	CAB			
4 × 4	*ABCD*	*ABCD*	*ABCD*	*ABCD*
	BADC	*BADC*	*BDAC*	*BCDA*
	CDBA	*CDAB*	*CADB*	*CDAB*
	DCAB	*DCBA*	*DCBA*	*DABC*
5 × 5	*ABCDE*			
	BAECD			
	CDAEB			
	DEBAC			
	ECDBA			
6 × 6	*ABCDEF*			
	BFDCAE			
	CDEFBA			
	ECABFD			
	FEBADC			

Table 5.22 ANOVA Table for Latin-Square Design

Source	DF	SS	MS	$F_{(calc)}$
Treatments (tire brand)	$p - 1$	$\sum \frac{Y_{i..}^2}{p} - \frac{(Y...)^2}{p^2}$	$\frac{SS_{treat}}{DF_{treat}}$	$\frac{MS_{treat}}{MSE}$
Rows—block 1 (tire location)	$p - 1$	$\sum \frac{Y_{.j.}^2}{p} - \frac{(Y...)^2}{p^2}$	$\frac{SS_{row}}{DF_{row}}$	$\frac{MS_{row}}{MSE}$
Columns—block 2 (car)	$p - 1$	$\sum \frac{Y_{..k}^2}{p} - \frac{(Y...)^2}{p^2}$	$\frac{SS_{col}}{DF_{col}}$	$\frac{MS_{col}}{MSE}$
Error	$p^2 - 1 - 3(p - 1)$	by subtraction	$\frac{SSE}{DF_{error}} = MSE$	
Total	$p^2 - 1$	$\sum \sum \sum Y_{ijk}^2 - \frac{(Y...)^2}{p^2}$		

[a] p = number of treatments and also number of blocks for each constraint.

Latin squares designs only study one treatment, which, in this example, is tire brands. The Latin-square design imposes two types of experimental constraints (blocks), which can then be separated from the overall error term. This allows the experimenter to understand the importance of the blocking effects as well as make the experiment more sensitive to real treatment effects.

5.6 FACTORIAL EXPERIMENTS

Although Section 5.5 dealt with a number of different randomized designs, there was one common limitation: only one factor could be studied in each case. Factorial experiments, on the other hand, allow the experimenter to study two or more factors and their individual effects on the measured response. In turn, each factor can be studied at two or more levels. The experiment involves data collection at every possible combination of factors and levels, which allows an estimation of effects for the main factors and the interactions between factors. Main effects can also be studied in more detail, depending on the number of levels chosen for each factor. When two or more levels are used, a linear effect is measured. As the number of levels increases, quadratic (three or more levels) and cubic (four or more levels) main effects can be estimated in order to understand the curvature of the response.

Although a great deal of information is available from a well designed factorial experiment, there is one other important characteristic of factorial designs that cannot be overlooked. *Factorial designs require, for most cases, a relatively large number of experimental runs*. For example, if an experimenter wanted to study the effects of 3 oven temperatures and 4 kinds of cake mix on the baking time, 12 experimental runs (3×4) would be needed, before any replicates. Inclusion of replicates to verify the experimental results would double the number of runs needed.

When designing an experiment to improve a process output, keep in mind that when using a full factorial design, each process variable (factor) that is added increases the number of experimental runs exponentially. For example, if a manufacturing team identified 7 factors they felt might affect the process output, it would require 128 runs if each of the factors was only studied at 2 levels (2^7). Unless unlimited resources are available for the experimentation (which is not a likely case in most industries), be careful not to let the experiment get too large. Sections 5.7 (Fractional Factorial Designs) and 5.8 (Screening Experiments) discuss techniques for studying many factors with many fewer runs than with a full factorial. These alternate techniques can identify which factors have an effect on the response so that the important factors can be included later in an additional experimentation using a full factorial design.

There are two commonly used approaches to factorial designs that will be discussed in this section. The first deals with situations that involve the study of only two factors at any number of levels for each factor. The other common approach is referred to as the 2^f design where f number of factors are studied at two levels. The use of 3^f designs (f factors studied at three levels each) will be discussed briefly.

5.6.1 Two-Factor Full Factorial Designs

Two-factor factorial designs are fairly simple to analyze because it is easy to visualize the experiment using data collection tables and graphs. The data analysis is relatively straightforward and easy to understand.

As with any experiment, the first step in a two-factor factorial, is to choose the factors to be studied and determine the levels of each factor to be included in the experiment. With that information, an experimental matrix can be constructed that can be used for data collection as well as data analysis. One factor is assigned to the rows and one factor is assigned to the columns (refer to Table 5.23).

Table 5.23 Experimental Matrix for Two-Factor Factorial Designs

Factor *B* (Rows)	Factor *A* (Columns)				
	Level 1	Level 2	Level 3	···	Level *a*
Level 1					
Level 2					
Level 3					
⋮					
Level 6					

To demonstrate the techniques for a two-factor full factorial design, an example for a facial tissue manufacturer will be used. The manufacturer needs to find a way to make their tissues stronger (less apt to rip when it is being removed from the dispenser). The technical group theorizes, based on their experience making tissues, that there are two factors that affect the strength of the paper: the concentration of fiber (pulp) in the original slurry and the pressure applied by the forming rolls during the manufacture of the paper. They *think* that these factors affect the strength, but they are not sure. They do not know the desirable levels for the factors with their current equipment nor do they know if any interaction effects exist between the fiber content and the roll pressure.

To study the fiber content and the roll pressure, the group decided to run a designed experiment. Because there are limits dictated by the paper machine construction, the highest and lowest values for roll pressure possible were 300 and 100 lb/in.^{2}, respectively. It was decided to run at these levels plus the midpoint (200 lb/in.^{2}) in order to get an idea of the curvature of the response. For fiber content, four levels were selected to be included in the experiment. The group put together the experimental table shown in Table 5.24.

Because fiber content (columns) has 4 levels and roll pressure (rows) has 3 levels, there are a total of 12 (4×3) experimental runs. The group randomized the order of the runs and collected the data holding everything else in the process constant. Five paper samples were collected for

Table 5.24 Experimental Matrix for Fiber Content–Roll Pressure Factorial Design

Roll Pressure (lb/in.^{2})	Fiber Content (%)			
	5	7	9	11
300				
200				
100				

Table 5.25 Data from Experimentation of Fiber Content–Roll Pressure Design

Roll Pressure	Fiber Content (%)								Row
(lb/in.^{2})	5		7		9		11		$(Y_{.j})$
	60		60		62		62		
	61		56		60		58		
300	58	294	56	283	59	308	58	293	1178
	56		56		64		57		
	59		56		63		58		
	55		59		58		50		
	52		58		62		48		
200	51	268	59	286	57	289	49	251	1094
	55		55		56		52		
	55		55		56		52		
	62		65		69		81		
	58		68		71		85		
100	65	321	66	323	73	360	91	436	1440
	65		62		73		90		
	71		62		74		89		
Column $(Y_{i.})$	883		892		957		980		$Y_{..} = 3712$

each "cell" or treatment combination, and the samples were measured for tear strength. The results were measured in pounds per square inch and the sum of the five results for each cell was also calculated. The completed data table is shown in Table 5.25.

To analyze the data, there are a number of steps involved:

1. Perform an analysis of variance to test the factorial model hypothesis.
2. If factor or interaction effects are present, perform tests on means to determine which have a significant effect on the response.
3. Graph the responses to visually show the factor effects and interactions.

The basic ANOVA table for a two-factor multilevel factorial experiment is given in Table 5.26. The basic ANOVA table can be expanded to include the curvature of each component. Depending on the degrees of freedom for the treatments (factors and interactions), the linear, quadratic, or cubic nature of the responses can be determined.

For the treatments (factors), 1 degree of freedom is needed to study the linear effects, another is needed to study the quadratic effects, and a third

Table 5.26 ANOVA Table for a Two-Factor Multilevel Factorial Design[a]

Source	DF	Sum of Squares	Mean Square	F Value
Treatments	$ab - 1$	$\sum\sum \frac{Y_{ij\cdot}^2}{n} - \frac{Y_{\cdots}^2}{N} = SS_{treat}$	$\frac{SS_{treat}}{DF_{treat}} = MS_{treat}$	$\frac{MS_{treat}}{MSE}$
Factor A	$a - 1$	$\sum \frac{Y_{i\cdot\cdot}^2}{nb} - \frac{Y_{\cdots}^2}{N} = SS_A$	$\frac{SS_A}{DF_A} = MS_A$	$\frac{MS_A}{MSE}$
Factor B	$b - 1$	$\sum \frac{Y_{\cdot j\cdot}^2}{na} - \frac{Y_{\cdots}^2}{N} = SS_B$	$\frac{SS_B}{DF_B} = MS_B$	$\frac{MS_B}{MSE}$
Interactions				
$A \times B$	$(a - 1)(b - 1)$	$SS_{treat} - SS_A - SS_B = SS_{A\times B}$	$\frac{SS_{A\times B}}{DF_{A\times B}} = MS_{A\times B}$	$\frac{MS_{A\times B}}{MSE}$
Error	$ab(n - 1)$	$SS_{tot} - SS_{treat} = SS_E$	$\frac{SS_E}{DF_E} = MS_E$	
Total	$abn - 1$	$\sum\sum\sum Y_{ijk}^2 - \frac{Y_{\cdots}^2}{N} = SS_{tot}$		

[a]Levels of $A = a$; levels of $B = b$; sample size per cell $= n$; $N = a \times b \times n$

is needed to study cubic effects. For interactions of two factors, there are a number of combinations that can be studied. All combinations of linear, quadratic, and cubic effects for both factors can be included if there are enough degrees of freedom:

$$\begin{array}{lll} A_L \times B_L & A_L \times B_Q & A_L \times B_C \\ A_Q \times B_L & A_Q \times B_Q & A_Q \times B_C \\ A_C \times B_L & A_C \times B_Q & A_C \times B_C \end{array}$$

where the subscripts L = linear effect
Q = quadratic effect
C = cubic effect

One degree of freedom is needed for each interaction combination to be studied.

For the roll pressure–fiber content example, there are three levels of roll pressure so there are 2 degrees of freedom. There are four levels of fiber content meaning there are 3 degrees of freedom:

Roll pressure, DF = 2	Fiber content, DF = 3
1 DF linear	1 DF = linear
1 DF quadratic	1 DF = quadratic
	1 DF = cubic

For the interactions between roll pressure and fiber content (RP × FC), there are 6 degrees of freedom:

$$1\ \text{DF} = \text{RP}_L \times \text{FC}_L \quad 1\ \text{DF} = \text{RP}_L \times \text{FC}_Q \quad 1\ \text{DF} = \text{RP}_L \times \text{FC}_C$$
$$1\ \text{DF} = \text{RP}_Q \times \text{FC}_L \quad 1\ \text{DF} = \text{RP}_Q \times \text{FC}_Q \quad 1\ \text{DF} = \text{RP}_Q \times \text{FC}_C$$

The total degrees of freedom for the treatments is 11 (2 + 3 + 6). The first column of the ANOVA table is shown in Table 5.27.

To determine the main factor effects and interaction effects, the following calculations are performed:

$$\sum\sum\sum Y_{ijk}^2 = (60)^2 + (61)^2 + (58)^2 + (56)^2 + (59)^2 + \cdots \text{ all individual points}$$
$$= 235{,}224$$

$$\sum \frac{Y_{i..}^2}{nb} = \frac{(883)^2 + (892)^2 + (957)^2 + (980)^2}{(5)(3)} = 230{,}106.8$$

$$\sum \frac{Y_{.j.}^2}{na} = \frac{(1178)^2 + (1094)^2 + (1440)^2}{(5)(4)} = 232{,}906$$

$$\sum\sum \frac{Y_{ij}^2}{n} = \frac{(294)^2 + (268)^2 + (321)^2 + (283)^2 + (286)^2 + \cdots + (436)^2}{5}$$
$$= 234{,}889.2$$

$$\frac{Y_{...}^2}{N} = \frac{(3712)^2}{60} = 229{,}649.1$$

The results of these calculations are used to complete column 2 in the ANOVA table. The results from columns 1 and 2 are used to calculate column 3. The results from column 3 are used to calculate the values for F. The F values are the ratio of the variance due to the factor in that row (between groups) to the variance due to random error (within groups). The completed ANOVA table is shown in Table 5.28.

Next, the linear, quadratic, and cubic effects can be interpreted. Appendix D.6 gives the linear combinations for each case. The completed table for roll pressure and the calculations for SS_{lin} and SS_{quad} are shown

Table 5.27 Partial ANOVA Table Showing Degrees of Freedom for Two-Factor Factorials[a]

Source	DF
Treatments	11
Roll pressure	2
Linear	1
Quadratic	1
Fiber content	3
Linear	1
Quadratic	1
Cubic	1
Roll pressure × fiber content interaction	6
$RP_L \times FC_L$	1
$RP_L \times FC_Q$	1
$RP_L \times FC_C$	1
$RP_Q \times FC_L$	1
$RP_Q \times FC_Q$	1
$RP_Q \times FC_C$	1
Error	48
Total	59

[a] a = roll pressure levels = 3; b = fiber content levels = 4; n = sample size per cell = 5; N = total number of samples = 3 × 4 × 5 = 60.

in Table 5.29. The equation used for calculating the sum of squares is

$$\frac{[\Sigma(\text{coeff}_i)(\text{cell total}_i)]^2}{(\Sigma\epsilon_j)^2(\text{sample size})}$$

Similarly, the completed table for fiber content and the calculations for SS_{lin}, SS_{quad}, and SS_{cubic} are shown in Table 5.30. The interaction sums of

Table 5.28 Completed ANOVA Table for Roll Pressure–Fiber Content Main Effects and Interaction Effects

Source	DF	Sum of Squares	Mean Square	*F* Value
Treatments	11	5240.1	476.4	68.06
Roll Pressure	2	3256.9	1628.5	232.64
Fiber Content	3	457.7	152.6	21.80
RP × FC Interaction	6	1525.5	254.3	36.33
Error	48	334.8	7.0	
Total	59	5574.9		

Table 5.29 Linear and Quadratic Sum of Squares Calculation for Roll Pressure

$Y_{\cdot j}$	1178	1094	1440	$\Sigma\xi_j^2$	λ
Linear	−1	0	1	2	1
Quadratic	1	−2	1	6	3

$$SS_{\text{lin}} = \frac{(-1(1178) + 0(1094) + 1(1440))^2}{2 \times 20} = 1716.1$$

$$SS_{\text{quad}} = \frac{(1178 - 2(1094) + 1440)^2}{6 \times 20} = 1540.8$$

squares are calculated using the same coefficients in matrix form. For the $RP_L \times FC_L$ interaction, the matrix shown in Table 5.31 is used.

The coefficients in the boxes are found by multiplying the intersections of rows and columns. The coefficients are then multiplied by the corresponding totals for each cell and the interaction effect is found by dividing the square of the total by the sample size n and the product of the two $\Sigma\xi_j^2$ values.

For $RP_L \times FC_L$, the calculation is

$$\frac{[3(294) + (283) - (308) - 3(293) - 3(221) - (323) + (360) + 3(436)]^2}{(5)(20)(2)}$$

$$SS_{L \times L} = 648$$

Similarly, for the $RP_L \times FC_Q$ interaction, the matrix is shown in Table 5.32.

Table 5.30 Linear, Quadratic, and Cubic Sum of Squares Calculation for Fiber Content

$Y_{i\cdot}$	883	892	957	980	$\Sigma\xi_j^2$	λ
Linear	−3	−1	1	3	20	2
Quadratic	1	−1	−1	1	4	1
Cubic	−1	3	−3	1	20	10/3

$$SS_{\text{lin}} = \frac{[-3(883) - 1(892) + 1(957) + 3(980)]^2}{20 \times 15} = 422.5$$

$$SS_{\text{quad}} = \frac{[1(883) - 1(892) - 1(957) + 1(980)]^2}{4 \times 15} = 3.2$$

$$SS_{\text{cubic}} = \frac{[-1(883) + 3(892) - 3(957) + 1(980)]^2}{20 \times 15} = 32.0$$

Table 5.31 Interaction Coefficient Matrix for Roll Pressure (Linear) and Fiber Content (Linear)

	FC_L			
RP_L	−3	−1	1	3
−1	3	1	−1	−3
0	0	0	0	0
1	−3	−1	1	3

For $RP_L \times FC_Q$, the calculation is

$$\frac{[-(294) + (283) + (308) - (293) + (321) - (323) - (360) + (436)]^2}{(5)(4)(2)}$$

$$SS_{L \times Q} = 152.1$$

The other interaction effects are found using the same technique and can be added to the ANOVA table (Table 5.33).

With the entire ANOVA complete, the next step is to determine if any of the factor or interaction effects are significant using the calculated F values. The model being tested in this example is

$$Y_{ijk} = \mu + RP_i + FC_j + RP \times FC_k + \epsilon_{ijk}$$

where Y_{ijk} = the response
μ = mean effect for all treatments
RP_i = the effect due to the roll pressure
FC_j = the effect due to the fiber content
$RP \times FC_k$ = the effect due to thee roll pressure–fiber content interaction
ϵ_{ijk} = random error

The degrees of freedom used to look up the F table values are those involved in the mean square calculations.

Table 5.32 Interaction Coefficient Matrix for Roll Pressure (Linear) and Fiber Content (Quadratic)

	1	−1	−1	1
−1	−1	1	1	−1
0	0	0	0	0
1	1	−1	−1	1

Table 5.33 Completed ANOVA Table Showing Linear, Quadratic, and Cubic Effects

Source	DF	Sum of Squares	Mean Square	F Value
Treatments	11	5240.1	476.4	68.06
Roll Pressure	2	3256.9	1628.5	232.64
Linear	1	1716.1	1716.1	245.16
Quadratic	1	1540.8	1540.8	220.11
Fiber Content	3	457.7	152.6	21.80
Linear	1	422.5	422.5	60.36
Quadratic	1	3.2	3.2	0.46
Cubic	1	32.0	32.0	4.57
Interactions RP × FC	6	1525.5	254.3	36.33
$RP_L \times FC_L$	1	648.0	648.0	92.57
$RP_L \times FC_Q$	1	152.1	152.1	21.73
$RP_L \times FC_C$	1	32.0	32.0	4.57
$RP_Q \times FC_L$	1	416.7	416.7	59.53
$RP_Q \times FC_Q$	1	276.0	276.0	39.43
$RP_Q \times FC_C$	1	0.7	0.7	0.10
Error	48	334.8	7.0	
Total	59	5574.9		

The F_{calc} and F_{table} values are listed in Table 5.34. As can be seen from the table, the calculations for F (F_{calc}) exceeded the F_{table} value in many instances. In these instances, the null hypotheses are rejected, indicating that a significant effect exist.

The results indicate that both roll pressure and fiber content have significant effects on paper strength. This is easier to see when graphed. The graph for the responses for these factors are shown in Figure 5.25. An alternate graph is shown in Figure 5.26.

It is understandable now why there were so many significant effects shown in Table 5.34. The plots indicate a very complicated set of responses. Changing roll pressure and fiber content does affect the paper strength. Both sets of results exhibit linear and quadratic behavior. There are interactions present that are shown by the response lines not being parallel.

To accomplish the goal of manufacturing paper with the highest strength, the experimental results indicate that the process should be run at a low roll pressure (100 lb/in.^{2}) and high fiber content (11%).

Whenever results from a designed experiment are calculated and graphed, they should be carefully reviewed for other information outside the original experimental objective. For example, in this experiment, the paper strength was almost constant when the roll pressure was run at 300 lb/in.^{2}, regardless of the fiber content of the slurry. If a paper strength of

Table 5.34 Significant Effects for Fiber Content–Roll Pressure Factorial Designed Experiment

	F_{calc} [a]	F_{table} ($\alpha = 0.01$)
Treatments	68.06*	2.6
Roll Pressure	232.64*	5.0
Linear	245.16*	7.2
Quadratic	220.11*	7.2
Fiber Content	21.80*	4.2
Linear	60.36*	7.2
Quadratic	0.46	7.2
Cubic	4.57	7.2
Interactions	36.33*	3.2
$L \times L$	92.57*	7.2
$L \times Q$	21.73*	7.2
$L \times C$	4.57	7.2
$Q \times L$	59.53*	7.2
$Q \times Q$	39.43*	7.2
$Q \times C$	0.10	7.2

250 lb/in.^{2} were acceptable, manufacturing might want to run at a roll pressure of 300 lb/in.^{2} so that the fiber content could vary somewhat without changing the strength. Under such conditions, the process may be more robust. This means that it is less sensitive to minor fluctuations in processing conditions. This is desirable when trying to maintain a consistent product output.

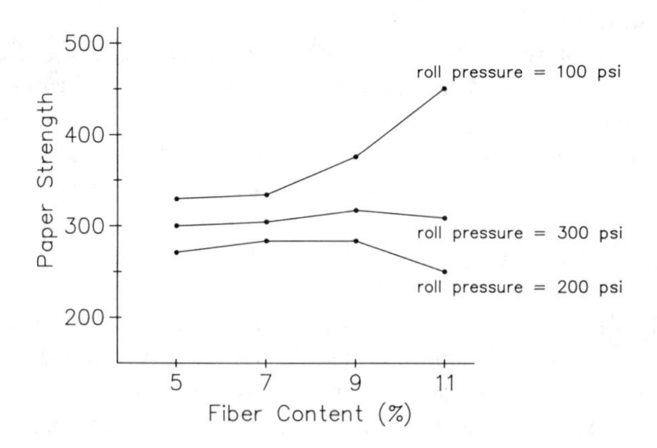

FIGURE 5.25. Paper strength response graph for fiber content at three levels of roll pressure.

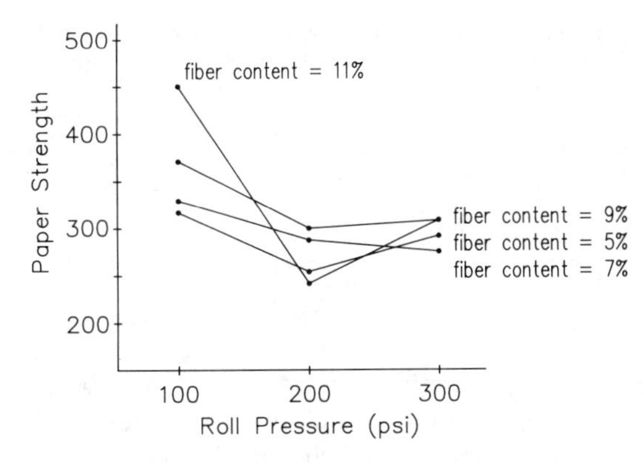

FIGURE 5.26. Paper strength response graph for roll pressure at four levels of fiber content.

5.6.2 2^f Factorial Designs

A form of full factorial design used often in industry is the 2^f factorial design, which has f factors studied at two levels. As with a two-factor experiment, every combination of factors and levels can be studied. Interaction effects can also be studied with 2^f designs, but unlike two-factor designs, linear, quadratic, or cubic contributions cannot be interpreted.

The number of experimental runs that have to be performed for a 2^f design is simply 2^f. For example, for three factors at two levels, 8 (2^3) runs are needed. Fór five factors at two levels, 32 (2^5) runs are needed. Because each factor is run at two levels, it is traditional to assign a $-$ to the low level and a $+$ to the high level. This makes the notation for the experiment and the calculations easier to use.

The simplest case of a 2^f factorial design is a 2^2 experiment: two factors studied at two levels resulting in four experimental combinations as shown in Figure 5.27.

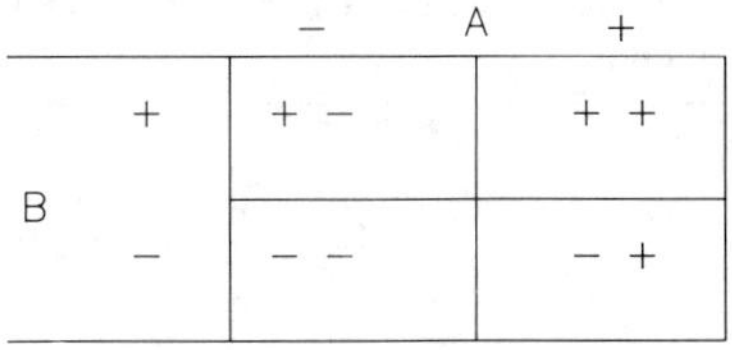

		A −	A +
B	+	+ −	+ +
B	−	− −	− +

FIGURE 5.27. Combination for two factors at two levels.

Table 5.35 Combinations of Two Factors at Two Levels

A	B
−	−
+	−
−	+
+	+

With factorial designs, interaction effects can also be evaluated, so it is useful to list the experimental combinations by the effects being measured: The factors and interactions are listed horizontally and the notation denoting the effect being measured is listed vertically. For the 2^2 example, start with the main factors A and B, and list all of the possible combinations (Table 5.35).

To find the interaction contribution to each combination $A \times B$, the levels of each factor A and B are multiplied together to form a third column. The results are given in Table 5.36, which is an orthogonal array. Each row is referred to as a *treatment combination* and represents the combinations of factors and levels that will be used for an experimental run. To designate the rows, another common notation for factorial designs is used. Every − is interpreted as a 1 and every + is interpreted as the lowercase letter of the factor. The first row in Table 5.36 would be designated as 1 (1×1) and the second row would be represented by a ($a \times 1$). The entire table is shown in Table 5.37. This method gives a notation corresponding to each treatment combination. This notation can be applied to a graphical representation of a 2^2 design as in Figure 5.28.

Once the experiment is run and the data captured, the effects of each factor and interaction can be separated and calculated. To calculate the effect of factor A, the average of the responses for A when factor B is at a high level and when B is at a low level are calculated. The effect of A

Table 5.36 Combinations of Two Factors at Two Levels and Their Interaction

A	B	AB
−	−	+
+	−	−
−	+	−
+	+	+

Table 5.37 Row Notation for Two-Factor Two-Level Design

	A	B	AB
(1)	−	−	+
a	+	−	−
b	−	+	−
ab	+	+	+

when B is low is

$$a - (1) = \bar{Y}_2 - \bar{Y}_1$$

The effect of A when B is high is

$$ab - b = \bar{Y}_4 - \bar{Y}_3$$

where $\bar{Y}_1$ = average response at (1)
$\bar{Y}_2$ = average response at a
$\bar{Y}_3$ = average response at b
$\bar{Y}_4$ = average response at ab

The effect is calculated from the average of these two equations:

$$\text{Effect of } A = \frac{(a - (1)) + (ab - b)}{2} = \frac{(\bar{Y}_2 - \bar{Y}_1) + (\bar{Y}_4 - \bar{Y}_3)}{2}$$

Similarly, using the matrix in Table 5.37, the effects of B and AB can be

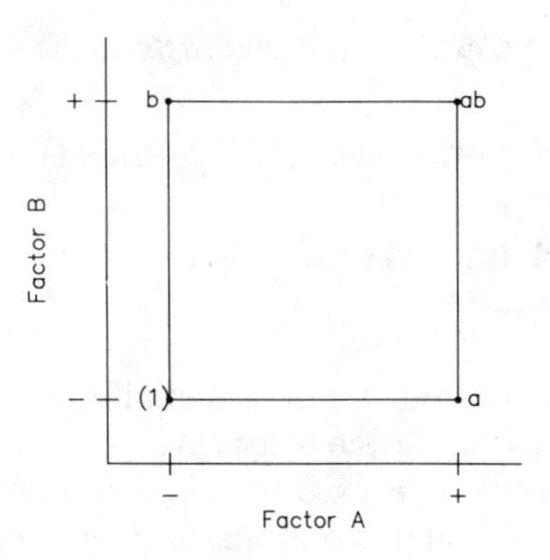

FIGURE 5.28. Graphical representation of a 2^2 design.

Table 5.38 Response Matrix for 2^2 Design

	A	B	AB
(1)	$-\bar{Y}_1$	$-\bar{Y}_1$	$+\bar{Y}_1$
a	$+\bar{Y}_2$	$-\bar{Y}_2$	$-\bar{Y}_2$
b	$-\bar{Y}_3$	$+\bar{Y}_3$	$-\bar{Y}_3$
ab	$+\bar{Y}_4$	$+\bar{Y}_4$	$+\bar{Y}_4$

found:

$$\text{Effect of } B = \frac{-(1) - a + b + ab}{2} = \frac{-\bar{Y}_1 - \bar{Y}_2 + \bar{Y}_3 + \bar{Y}_4}{2}$$

$$\text{Effect of } AB = \frac{(1) - a - b + ab}{2} = \frac{\bar{Y}_1 - \bar{Y}_2 - \bar{Y}_3 + \bar{Y}_4}{2}$$

A simpler method to get the equations is to plug the responses into the matrix (still maintaining the + and − signs). The matrix would then appear as in Table 5.38. To calculate the effects for each column, sum the values in the column and divide by the number of + signs in the column (the number of + signs will always equal the number of treatment combinations divided by 2). Using this method,

$$\text{Effect of } A = \frac{-\bar{Y}_1 + \bar{Y}_2 - \bar{Y}_3 + \bar{Y}_4}{2}$$

The effects are used to calculate the *contrasts* using the equation

$$\text{Contrast} = (n)(2^{f-1})(\text{effect})$$

where n = sample size per run
f = number of factors

The major purpose of the contrast is to simplify the mathematics for the ANOVA table calculations. The contrasts are used to calculate the sum of squares (SS), which in turn are used to calculate the mean square values and the F values. The calculated F value is compared with a table value for F to determine if the effects on the response are significant. To calculate the sum of squares:

$$SS = \frac{(\text{contrast})^2}{n2^f}$$

where n = sample size per run
f = number of factors

For factor A,

$$\text{Contrast } A = (n)(2^{f-1})\left(\frac{-\bar{Y}_1 + \bar{Y}_2 - \bar{Y}_3 + \bar{Y}_4}{2}\right)$$

$$SS_A = \frac{(\text{contrast } A)^2}{n2^f}$$

Similarly, for factor B and the interaction of $A \times B$,

$$\text{Contrast } B = (n2^{f-1})\frac{\left(-\bar{Y}_1 - \bar{Y}_2 + \bar{Y}_3 + \bar{Y}_4\right)}{2}$$

$$SS_B = \frac{(\text{contrast } B)^2}{n2^f}$$

$$\text{Contrast } AB = (n)(2^{f-1})\frac{\left(\bar{Y}_1 - \bar{Y}_2 - \bar{Y}_3 + \bar{Y}_4\right)}{2}$$

$$SS_{AB} = \frac{(\text{contrast } AB)^2}{n2^f}$$

The use of these equations will be made clearer with an example of a 2^3 design.

2^3 Factorial Example

A company wanted to determine the effect of molding conditions on the bursting strength of the finished part. They selected three factors to be included in the study: mold temperature, mold pressure, and process time. They planned a 2^3 factorial design and therefore had to select two levels

Table 5.39 Combinations of Main Factors—2^3 Design

A	B	C
−	−	−
+	−	−
−	+	−
+	+	−
−	−	+
+	−	+
−	+	+
+	+	+

for each factor:

		Low Level	High Level
A	Mold temperature (°F)	300	400
B	Mold pressure (lb/in.^{2})	2500	3500
C	Process time (min)	5	7

The next step in a 2^3 factorial is to list the treatment combinations/contrasts using the proper notation. (Appendix D.4 gives the matrices for designs of up to five factors.) Both main effects and interactions can be measured using full factorials. With three factors in a full factorial, there will be eight (2^3) combinations.

The team started with all of the possible combinations of A, B, and C as shown in Table 5.39. To find the interactions (AB, AC, BC, and ABC), they multiplied the coefficients for the corresponding columns as shown in Table 5.40. Finally, they added the row notation (1), a, b, ab, and so on as shown in Table 5.41.

Table 5.40 Combinations of Interactions—2^3 Design

AB	AC	BC	ABC
+	+	+	−
−	−	+	+
−	+	−	+
+	−	−	−
+	−	−	+
−	+	−	−
−	−	+	−
+	+	+	+

Table 5.41 Complete Matrix of Combinations for 2^3 Design

	A	B	AB	C	AC	BC	ABC
(1)	−	−	+	−	+	+	−
a	+	−	−	−	−	+	+
b	−	+	−	−	+	−	+
ab	+	+	+	−	−	−	−
c	−	−	+	+	−	−	+
ac	+	−	−	+	+	−	−
bc	−	+	−	+	−	+	−
abc	+	+	+	+	+	+	+

Note that the proper order for listing the factors and the interactions is to complete all combinations of A and B (main factors and interactions), then add C with all of its combinations. (If a fourth factor were added, it would be added to the right of column ABC).

The company collected the data for the experiment and, because the parts were relatively inexpensive to produce, they collected three data points for each set of conditions. The data are given in Figure 5.29. The response was the bursting strength of the parts measured in pounds per square inch. The average for each cell was calculated and is circled.

To calculate the factor effects, the average of each factor is found at the high levels and low levels of the other factors. For factor A, this means the average response when A is at a high level minus when A is at a low level:

$$\text{Effect}_A = \frac{(a + ab + ac + abc) - ((1) + b + c + bc)}{4}$$

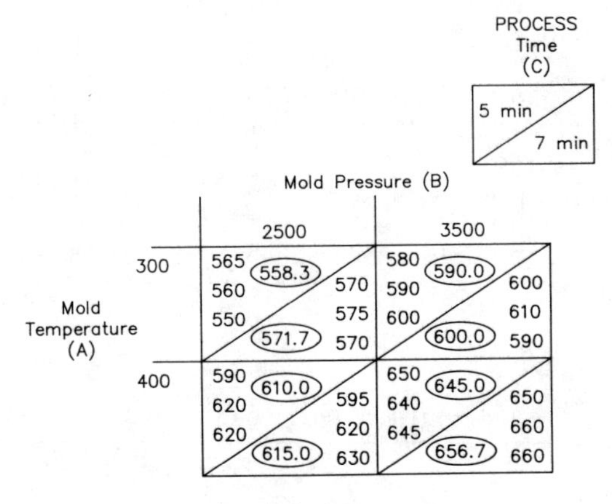

FIGURE 5.29. Experimental results for 2^3 molding experiment.

Similarly:

$$\text{Effect}_B = \frac{(-1) - a + b + ab - c - ac + bc + abc}{4}$$

$$\text{Effect}_C = \frac{(-1) + a - b - ab + c + ac + bc + abc}{4}$$

The interactions can be found using the same method. The values can be calculated by plugging in the average responses for each cell.

The experimental results, the average responses, are shown in Table 5.42. These responses are used to calculate the effects:

$$\text{Effect}_A = \frac{-558 + 610 - 590 + 645 - 572 + 615 - 600 + 657}{4}$$

$$= 51.8$$

$$\text{Effect}_B = \frac{-558 - 610 + 590 + 645 - 572 - 615 + 600 + 657}{4}$$

$$= 34.2$$

$$\text{Effect}_{AB} = \frac{558 - 610 - 590 + 645 + 572 - 615 - 600 + 657}{4}$$

$$= 4.2$$

$$\text{Effect}_C = \frac{-558 - 610 - 590 - 645 + 572 + 615 + 600 + 657}{4}$$

$$= 10.0$$

$$\text{Effect}_{AC} = \frac{558 - 610 + 590 - 645 - 572 + 615 - 600 + 657}{4}$$

$$= -1.7$$

$$\text{Effect}_{BC} = \frac{558 + 610 - 590 - 645 - 572 - 615 + 600 + 657}{4}$$

$$= 0.8$$

$$\text{Effect}_{ABC} = \frac{-558 + 610 + 590 - 645 + 572 - 615 - 600 + 657}{4}$$

$$= 2.5$$

Table 5.42 Complete Matrix of Combinations for 2^3 Design with Experimental Results

	A	B	AB	C	AC	BC	ABC	$\bar{Y}$
(1)	−	−	+	−	+	+	−	558
a	+	−	−	−	−	+	+	610
b	−	+	−	−	+	−	+	590
ab	+	+	+	−	−	−	−	645
c	−	−	+	+	−	−	+	572
ac	+	−	−	+	+	−	−	615
bc	−	+	−	+	−	+	−	600
abc	+	+	+	+	+	+	+	657

To calculate the contrasts, the equation is:

$$\text{Contrast} = (n)(2^{f-1})(\text{effect})$$

where n = sample size = 3 for this example
f = number of factors = 3

$$\text{Contrast}_A = (3)(2^{3-1})(\text{effect } A) = (12)(51.8) = 621.6$$

Similarly

$$\text{Contrast } B = 410.4$$

$$\text{Contrast } AB = 250.4$$

$$\text{Contrast } C = 120.0$$

$$\text{Contrast } AC = -20.4$$

$$\text{Contrast } BC = 9.6$$

$$\text{Contrast } ABC = 30.0$$

To calculate the sum of squares, the following equation is used:

$$SS = \frac{(\text{contrast})^2}{n2^f}$$

The sum of squares for factor A is

$$SS_A = \frac{(\text{contrast } A)^2}{3(2^3)} = \frac{(621.6)^2}{24} = 16099$$

The other sum of squares values are calculated similarly with the results:

$$SS_B = 7018$$

$$SS_{AB} = 106$$

$$SS_C = 600$$

$$SS_{AC} = 17.3$$

$$SS_{BC} = 3.8$$

$$SS_{ABC} = 37.5$$

Because only two levels are investigated in a 2^n factorial, there is only 1 degree of freedom for each of the factor effects. The ANOVA table with the degrees of freedom and the sum of squares columns completed is shown in Table 5.43. The total degrees of freedom is equal to the sample size ($n = 3$ in this case) times the number of cells, or contrasts (in this case, 8) minus 1:

$$DF_{tot} = (n \times i) - 1$$

$$= (3 \times 8) - 1 = 23$$

The degrees of freedom for the area can be found by adding up the degrees of freedom for the main factors and their interactions and then subtracting this quantity from DF_{tot}. In this case, $DF_{error} = 23 - 7 = 16$. Note that because this experiment was not replicated, the error term is confounded with the interaction effects.

Table 5.43 Partial ANOVA Table Showing Degrees of Freedom

Source	DF	SS
Main Factors		
A (mold temperature)	1	16,099
B (mold pressure)	1	7,018
C (process time)	1	600
Interactions		
$A \times B$	1	106
$A \times C$	1	17.3
$B \times C$	1	3.8
$A \times B \times C$	1	37.5
Error	16 (by subtraction)	
Total	23	

In order to find SS total, the same equation as in Section 5.6.1 is used:

$$SS_{tot} = \sum \sum \sum Y_{ijk}^2 - \frac{(Y...)^2}{N}$$

Using the individual response for the three samples of the eight treatment combinations, two components of the equation can be calculated:

$$\sum \sum \sum Y_{ijk}^2 = (565)^2 + 560^2 + 550^2 + 570^2 + 575^2 + \cdots$$

$$= 8{,}834{,}500$$

$$\frac{(Y...)^2}{N} = \frac{(14540)^2}{24} = 8808816.7$$

These can be used to calculate the total sum of squares SS_{tot}:

$$SS_{tot} = 8834500 - 8808816.7 = 25683.3$$

The sum of squares for the error term can be calculated by subtracting the sum of squares for the main factors and the interactions from the total sum of squares. The mean squares (MS) values are found by dividing the sum of squares values by the degrees of freedom. Finally, the F values are calculated by dividing the MS value for each factor and interaction by the mean square value of the error. The completed ANOVA table is shown in Table 5.44.

The calculated F values are then compared to the table values for 1 DF/16 DF (see Appendix B.6). For $\alpha = 0.01$ (99% confidence), $F_{table} = 8.53$. This indicates with a very high degree of confidence that temperature

Table 5.44 Completed ANOVA Table for Three-Factor Two-Level Molding Experiment

Source	DF	SS	MS	F Value
A	1	16,099	16,099	142
B	1	7,018	7,018	62
C	1	600	600	5.3
$A \times B$	1	106	106	0.94
$A \times C$	1	7.3	7.3	0.06
$B \times C$	1	3.8	3.8	0.03
$A \times B \times C$	1	37.5	37.5	0.33
Error	16	1811	113	
Total	23	25,683		

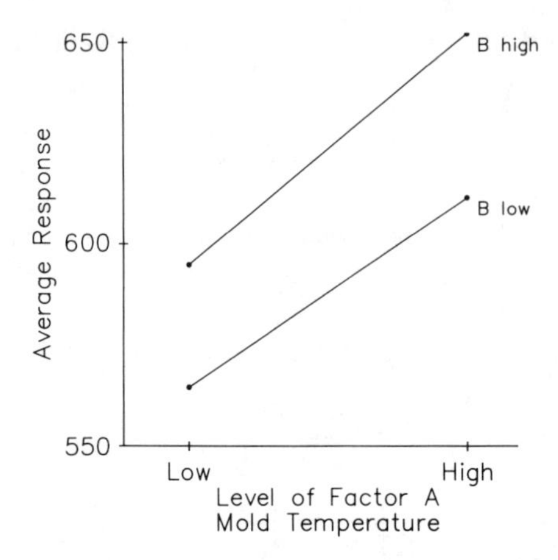

FIGURE 5.30. Graphical results for molding experiment with process time at a high level.

A and pressure *B* have significant effects on the part strength. This can be confirmed by plotting the results. Figure 5.30 shows the plot for part strength versus the levels of the temperature and the mold pressure.

The fact that the lines in Figure 5.30 are parallel indicates that there is no interaction between *A* and *B*, which confirms the calculated experimental results. It also indicates that high levels of *A* and *B* factors produce the strongest part. Similar graphs can be constructed for the other interaction effects.

The desirable levels for each factor can also be interpreted from the factor effects calculated earlier. The sign on the calculated effect indicates the direction of the effect. If the sign is positive, then the response will increase as the process is adjusted from the low level to the high level for that factor. If the sign is negative, then the response will decrease as the process is adjusted from the low level to the high level for that factor.

In the molding example, the main factor effects were all positive. This means that as each of the factors was increased from its low level to its high level, the response (strength) increased. Because the objective of the experiment was to increase the strength, the process should be set at the high levels for temperature, pressure, and time and data should be collected to verify the results.

The last point to discuss about 2^f factorial designs is the case where all of the runs cannot be performed under homogeneous conditions (due to time constraints or some other external factor). In these cases, it may be desirable to block the experiment into smaller portions. Unfortunately,

Table 5.45 Combinations of Three Factors at Two Levels and Their Interactions

A	*B*	*C*	*ABC*
−	−	−	−
+	−	−	+
−	+	−	+
+	+	−	−
−	−	+	+
+	−	+	−
−	+	+	−
+	+	+	+

blocking causes a certain level of confounding (confusing the effect of the block with another effect), but if the blocking is done correctly, the confounding can be limited to the higher-order interactions, which usually are not important.

The earlier example had three factors at two levels (2^3). Suppose the design were split in half and run with four of the treatment combinations on one day and four on another (day = block in this case). It would be desirable to limit the confounding to the *ABC* interaction. In order to come up with the two blocks, the *ABC* contrast could be used as the block generator.

To use the *ABC* interaction as the block generator, first list the combinations of *A*, *B*, and *C*. Then add the *ABC* contrast as shown in Table 5.45. Now, every *A*, *B*, and *C* combination corresponding to a + in the *ABC* contrast is in block 1 and every *A*, *B*, and *C* corresponding to a − in the *ABC* contrast is in block 2 as shown in Table 5.46.

The blocks can now be performed on separate days (or pieces of equipment or whatever the chosen blocking factor was) and then the data analyzed as usual. The only effect that cannot be measured is the *ABC* interaction because it will be confounded with the blocking effect. Any interaction can be used as block generators. For example, if there was the

Table 5.46 Combinations of 2^3 Design with *ABC* Blocking

Block 1			Block 2		
A	*B*	*C*	*A*	*B*	*C*
+	−	−	−	−	−
−	+	−	+	+	−
−	−	+	+	−	+
+	+	+	−	+	+

Table 5.47 2^3 Design Showing Blocking Using Second Order Interactions

			Block Generators	
A	*B*	*C*	*AB*	*BC*
−	−	−	+	+
+	−	−	−	+
−	+	−	−	−
+	+	−	+	−
−	−	+	+	−
+	−	+	−	−
−	+	+	−	+
+	+	+	+	+

Block 1 (+ +)			Block 2 (+ −)			Block 3 (− +)			Block 4 (− −)		
A	*B*	*C*	*A*	*B*	*C*	*A*	*B*	*C*	*A*	*B*	*C*
−	−	−	+	+	−	+	−	−	−	+	−
+	+	+	−	−	+	−	+	+	+	−	+

need to block a second time to run the experiment over four days, the experimenter would have to use some second-order interactions to further confound the experiment. This is shown in Table 5.47. In this case, all of the two-way interactions would be confounded with blocking effects, but the main effects could still be determined as usual.

More complete tables of blocking possibilities are available in *Statistics for Experimenters* by Box, Hunter, and Hunter (1978) published by John Wiley & Sons, Inc.

5.6.3 3^f Full Factorial Designs

Another type of factorial design, probably less commonly used than two-factor designs or 2^f designs, is the 3^f factorial design. In this case, f represents the number of factors being studied, but this time each factor is studied at three levels. The added level also adds a degree of freedom for each factor, which allows evaluation of linear and nonlinear (quadratic) effects (assuming the levels are equally spaced within the experimental region).

The simplest example is a 3^2 design: two factors each studied at three levels. To demonstrate the analysis, an example for an adhesive manufacturer will be worked through. An adhesive manufacturer wanted to understand the effects of temperature and humidity on the speed of the cure of a new adhesive product. They anticipated that the effects may not be linear, so they chose their factor levels so that they were equally spaced

Table 5.48 Data Table for Two-Factor Three-Level Experiment on Adhesive Cure Speed

Humidity	Temperature A (°F)		
B (%)	65	75	85
90	5 (11) 6	8 (16) 8	9 (19) 10
70	6 (12) 6	7 (13) 6	7 (15) 8
50	5 (11) 6	4 (8) 4	2 (5) 3

over the experimental region:

A	Temperature (°F)	65	75	85
B	Humidity (%)	50	70	90

The researchers went out and collected the data. They settled on two data points per cell, because the adhesive was expensive, so a total of 18 runs were performed. The data are shown in Table 5.48 along with the summed value for each cell.

The procedure for constructing the ANOVA table follows. Each factor has 2 degrees of freedom because of the fact that three levels are being tested. This means that both linear and quadratic effects of the main factors and the interactions can be evaluated. As before, the main factors are analyzed first:

$$SS_{tot} = \sum \sum Y_{ij}^2 - \frac{(Y..)^2}{N}$$

$$= 5^2 + 6^2 + 6^2 + 6^2 + 5^2 + 6^2 + \cdots - \frac{(110)^2}{18} = 73.8$$

$$SS_A = \frac{(11 + 12 + 11)^2 + (16 + 13 + 8)^2 + (19 + 15 + 5)^2}{6} - \frac{(110)^2}{18}$$

$$= 2.1$$

$$SS_B = \frac{(11 + 16 + 19)^2 + (12 + 13 + 15)^2 + (11 + 8 + 5)^2}{6} - \frac{(110)^2}{18}$$

$$= 43.1$$

Table 5.49 Partial ANOVA Table for Main Effects

Source	DF	Sum of Squares	Mean Square	F Value
A (temperature)	2	2.1	1.1	
A_L	1			
A_Q	1			
B (humidity)	2	43.1	21.6	
B_L	1			
B_Q	1			
$A \times B$ Interaction	4			
$A_L \times B_L$	1			
$A_L \times B_Q$	1			
$A_Q \times B_L$	1			
$A_Q \times B_Q$	1			
Error	9			
Total	17	73.8		

The ANOVA table (Table 5.49) can be started with the degrees of freedom and the results of the sum of squares calculations. To evaluate the interaction effects and to separate the main effects into linear and quadratic components, refer to Appendix D.6 to find the appropriate coefficients. This can be made easier to understand by labelling each of the cells in the experiments using subscripts as in Table 5.50.

Now, using the table in Appendix D.6, the coefficients for a linear equation are found:

$$-1 \quad 0 \quad +1$$

Similarly, the coefficients for a quadratic equation are found:

$$+1 \quad -2 \quad +1$$

With this information, contrasts (linear combinations) can be generated that can be used to calculate the sums of squares for the main effects. For the linear effect of A, we want to compare the results with A at a low level to the results when A is at the middle level and also to the results when A is at a high level:

$$\underbrace{a_0b_0 + a_0b_1 + a_0b_2}_{A \text{ low level}} \; \underbrace{a_1b_0 + a_1b_1 + a_1b_2}_{A \text{ middle level}} \; \underbrace{a_2b_0 + a_2b_1 + a_2b_2}_{A \text{ high level}}$$

Table 5.50 Cell Matrix Notation

Humidity B (%)	Temperature A (°F)		
	65	75	85
90	a_0b_2	a_1b_2	a_2b_2
70	a_0b_1	a_1b_1	a_2b_1
50	a_0b_0	a_1b_0	a_2b_0

Therefore, the linear contrast would be

$$A_L: -a_0b_0 - a_0b_1 - a_0b_2 + (a_1b_0)(0)$$
$$+(a_1b_1)(0) + (a_1b_2)(0) + a_2b_0 + a_2b_1 + a_2b_2$$

The quadratic contrast would be

$$A_Q: a_0b_0 + a_0b_1 + a_0b_2 - 2(a_1b_0) - 2(a_1b_1)$$
$$- 2(a_1b_2) + a_2b_0 + a_2b_1 + a_2b_2$$

Similarly,

$$B_L: - a_0b_0 - a_1b_0 - a_2b_0 + (0)(a_0b_1)$$
$$+(0)(a_1b_1) + (0)(a_2b_1) + a_0b_2 + a_1b_2 + a_2b_2$$
$$B_Q: a_0b_0 + a_1b_0 + a_2b_0 - 2(a_0b_1) - 2(a_1b_1)$$
$$-2(a_2b_1) + a_0b_2 + a_1b_2 + a_2b_2$$

To get the coefficients for the interaction effects, the appropriate coefficients for the main effects are multiplied together. For example, the $A_L \times B_L$ interaction effect is

$$a_0b_0 + (0)(a_0b_1) - (a_0b_2) + (0)(a_1b_0) + (0)(a_1b_1)$$
$$+(0)(a_1b_2) - a_2b_0 + (0)(a_2b_1) + a_2b_2$$

To make it easier to do the calculations, all the calculations can be put into table form (Table 5.51). The contrast for each treatment combination

Table 5.51 Matrix of Treatment Combinations for Calculating Contrast Values

	Treatment Combination								
Label	a_0b_0	a_0b_1	a_0b_2	a_1b_0	a_1b_1	a_1b_2	a_2b_0	a_2b_1	a_2b_2
Experimental Value	11	12	11	8	13	16	5	15	19
A_L	−1	−1	−1	0	0	0	+1	+1	+1
A_Q	+1	+1	+1	−2	−2	−2	+1	+1	+1
B_L	−1	0	+1	−1	0	+1	−1	0	+1
B_Q	+1	−2	+1	+1	−2	+1	+1	−2	+1
A_LB_L	+1	0	−1	0	0	0	−1	0	+1
A_LB_Q	−1	+2	−1	0	0	0	+1	−2	+1
A_QB_L	−1	0	+1	+2	0	−2	−1	0	+1
A_QB_Q	+1	−2	+1	−2	+4	−2	+1	−2	+1

can be calculated and the sums of squares are given by

$$SS_i = \frac{(\text{contrast}_i)^2}{n\Sigma\ \text{coefficients}_i^2}$$

where n = sample size
$\Sigma\ (\text{coefficients})_i^2$ = sum of the squares of the nine coefficients

$$\text{Contrast } A_L = -11 - 12 - 11 + 5 + 15 + 19 = 5$$

$$SS_{AL} = \frac{(5)^2}{2 \times 6} = 2.1$$

$$\text{Contrast } A_Q = 11 + 12 + 11 - 2(8) - 2(13) - 2(16) + 5 + 15 + 19 = -1$$

$$SS_{AQ} = \frac{(-1)^2}{2 \times 18} = 0.03$$

Similarly, the contrasts and the sum of squares for the other main factors and the interactions can be calculated. These can be added to the ANOVA table as shown in Table 5.52.

Based on the F table values, the significant effects at 99% confidence level are marked in the table with an asterisk. The results indicate that factor B (humidity) has a significant effect on the cure speed and it is largely a linear effect. Factor A (temperature) is not significant at 99%.

There is a significant linear-by-linear interaction between temperature and humidity present. This can be confirmed graphically as shown in

Table 5.52 Complete ANOVA Table for Adhesive Cure Speed Designed Experiment

Source	DF	Sum of Squares	Mean Squares	F Value[a]
A_{temp}	2	2.1	1.1	3.6
A_L	1	2.1	2.1	7.0
A_Q	1	0.03	0.03	0.1
$B_{humidity}$	2	43.1	21.6	72.0*
B_L	1	40.3	40.3	134.3*
B_Q	1	2.8	2.8	9.3
Interactions	4	25.6	6.4	21.3*
$A_L \times B_L$	1	24.5	24.5	81.7*
$A_L \times B_Q$	1	0.7	0.7	2.3
$A_Q \times B_L$	1	0.2	0.2	0.7
$A_Q \times B_Q$	1	0.2	0.2	0.7
Error	9	3.0	0.3	
Total	17	73.8		

[a]Significant effects at 99% confidence level are marked with an asterisk.

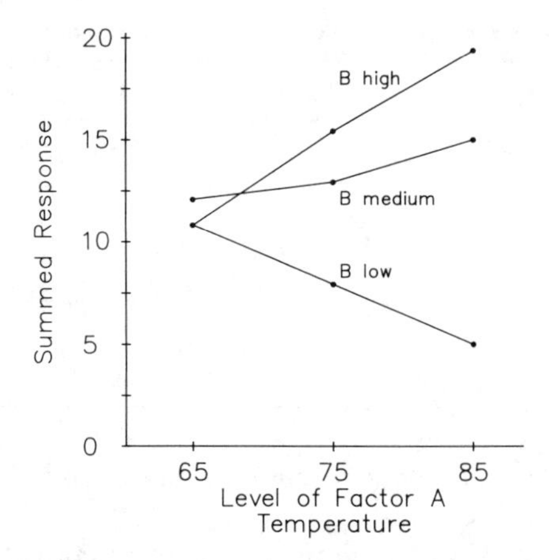

FIGURE 5.31. Effects of humidity and temperature on cure speed.

Figure 5.31. The shape of the plotted lines indicates that the results are largely linear. The lines cross, confirming that there are strong interactions present.

A 3^2 design is fairly easy to manage. A 3^3 is designed and analyzed using the same methods, but is more difficult to manage because there are 27 runs instead of 9. Both linear and quadratic effects and interactions can be calculated for a 3^3 design but not as easily as for a 3^2 design.

5.7 FRACTIONAL FACTORIAL DESIGNS

Section 5.6 discussed a number of different factorial designs for use when it is desirable or necessary to evaluate all main effects and all levels of interaction. Although these designs are useful, there comes a point in many experiments where there are too many factors to be studied and the size of a full factorial make it too expensive or impractical. In these cases, it makes more sense to run a fractional (or partial) factorial, which still allows the researcher to evaluate the main factor effects and the key interaction effects but requires fewer runs. This is accomplished by confounding the higher-order interactions (three-way and higher), which are often insignificant anyway, with the main effects and two-way interactions.

In Section 5.6.2 (2^f Full Factorial Designs), there was a brief discussion about the use of blocking to expand the versatility of the factorial design. The technique allows the experimenter to add a blocking variable (performing the experiment over a couple of days, or with operators on different shifts) without losing much of the power of the full factorial design. Fractional factorials utilize this same approach to reduce the number of

Table 5.53 Combinations of Three Factors at Two Levels and Their Interactions

	A	*B*	*AB*	*C*	*AC*	*BC*	*ABC*
(1)	−	−	+	−	+	+	−
a	+	−	−	−	−	+	+
b	−	+	−	−	+	−	+
ab	+	+	+	−	−	−	−
c	−	−	+	+	−	−	+
ac	+	−	−	+	+	−	−
bc	−	+	−	+	−	+	−
abc	+	+	+	+	+	+	+

runs needed in the experiment without losing the ability to understand the main factor effects.

It is easier to visualize the derivation of a fractional factorial through the use of an example of a 2^3 factorial transformed into a fractional factorial. The treatment combinations (orthogonal array) for a 2^3 full factorial are given in Table 5.53. (See Section 5.6.2 for more details on developing the array.)

Because most high-order (three or more factor) interactions are not significant, they are likely choices to be confounded with something else. Therefore, the higher-order interactions can be set as the defining relation. In this example, the eight-run experiment is separated into two four-run experiments that separate treatment combinations according to the sign convention of *ABC* as shown in Table 5.54. By doing this, a situation is created where there are "aliases"—more than one factor with the same treatment combination. After looking more closely at the upper

Table 5.54 Separation of Eight-Run Experiment into Two Four-Run Experiments Based on the Value of Interaction *ABC*

	ABC (+)						
	A	*B*	*AB*	*C*	*AC*	*BC*	*ABC*
a	+	−	−	−	−	+	+
b	−	+	−	−	+	−	+
c	−	−	+	+	−	−	+
abc	+	+	+	+	+	+	+

	ABC (−)						
	A	*B*	*AB*	*C*	*AC*	*BC*	*ABC*
(1)	−	−	+	−	+	+	−
ab	+	+	+	−	−	−	−
ac	+	−	−	+	+	−	−
bc	−	+	−	+	−	+	−

portion of Table 5.54, it can be clearly seen that

$$AC = B$$
$$A = BC$$
$$C = AB$$

These relationships are called pre-aliases. Additionally, in the $ABC(-)$ matrix in the lower portion of Table 5.54,

$$A = -BC$$
$$B = -AC$$
$$C = -AB$$

By running the experiment defined by the upper portion of Table 5.54 only (four runs instead of eight), the main effects are confounded with the interactions. However, *by assuming that the interactions are unimportant*, the main effects can still be estimated. This may not seem like an important shortcut with a 2^3 design that only has 8 runs to begin with, but when dealing with many factors, it can be very advantageous.

If it is necessary to reduce the runs even further (increase the confounding), simply choose another "defining relation." By further increasing confounding, the number of runs can be reduced further. Once again, this sacrifices the ability to estimate some more of the interaction effects.

Another method for developing a fractional factorial is the addition of a new factor in place of a high-order interaction. For example, suppose there are four factors to be investigated but a 2^4 (16-run) experiment is too costly and time consuming. Because again it is safe to assume that third-order interactions are negligible, treatment ABC can be renamed for a new factor D as in Table 5.55. This creates an eight-run experiment that investigates four factors instead of the 16 run 2^4 experiment. This is just another way of approaching the use of "*defining relations*" *to create confounding*. In this case, all the main effects are confounded with the

Table 5.55 Substitution of a Fourth Factor for the Third-Order Interaction *ABC*

	A	B	C	$ABC = D$
(1)	−	−	−	−
a	+	−	−	+
b	−	+	−	+
ab	+	−	−	−
c	−	−	+	+
ac	+	−	+	−
bc	−	+	+	−
abc	+	+	+	+

three-way interaction, and each two-way interaction is confounded with another two-way interaction.

The data analysis is done using the same methods as were used for full factorial experiments. When interpreting the results, the experimenter must keep in mind the confounding scheme before drawing too many conclusions about the main factor effects and the two-way interaction effects.

There is another method for handling fractional factorial data analysis. Here is an example:

Company XYZ has discovered that, on start-up with a new polymer product, they are suffering from poor product yields, which is costly because the raw materials are very expensive. The engineering manager decided to run a designed experiment to determine which processing factors affect product yield. After brainstorming and voting and ranking, the project team identified five process parameters to be included in the study:

A Temperature of compounder
B Raw material mix time
C Grinding speed
D Grinder screen size
E Resin viscosity

With the factors identified, the design could be selected. The first design recommended by the team was a full factorial using two levels for each factor (2^5). This experiment would require 32 runs. Because of the expense of carrying out each run, the group decided to explore fractional factorial designs. After further discussion about the process and the factors chosen, the decision was made to run a "half factorial"—an experiment requiring only 16 runs. In order to come to this conclusion, the group assumed that:

1. Third- and higher-order interactions would not be significant and could be confounded with the main and two-way interactions. This would still allow them to measure the main factor and two-way interaction effects.
2. In order to eliminate the need for replication to find an estimate of experimental error, the group agreed that most two-way interactions would probably be insignificant. They selected a few of those columns to be used to estimate the experimental error.

The group went ahead and designed their experimental matrix. They started with an orthogonal array, which is the same array used for a 2^4 full factorial.

Table 5.56 Five-Factor Fractional Factorial Design

Run	A	B	C	D	E
1	−	−	−	−	+
2	+	−	−	−	−
3	−	+	−	−	−
4	+	+	−	−	+
5	−	−	+	−	−
6	+	−	+	−	+
7	−	+	+	−	+
8	+	+	+	−	−
9	−	−	−	+	−
10	+	−	−	+	+
11	−	+	−	+	+
12	+	+	−	+	−
13	−	−	+	+	+
14	+	−	+	+	−
15	−	+	+	+	−
16	+	+	+	+	+

To add another factor without adding more runs, they renamed the four-way *ABCD* interaction factor *E*, which generated the fifth column in the experiment (Table 5.56). This resulting design matrix describes the factor levels for the 16 experimental runs. They were then able to fill in the column describing each run using the standard lowercase notation (Table 5.57).

Table 5.57 Combinations for a Five-Factor Fractional Factorial Design

Run		A	B	C	D	E
1	(−1)	−	−	−	−	+
2	*a*	+	−	−	−	−
3	*b*	−	+	−	−	−
4	*ab*	+	+	−	−	+
5	*c*	−	−	+	−	−
6	*ac*	+	−	+	−	+
7	*bc*	−	+	+	−	+
8	−*de*	+	+	+	−	−
9	*d*	−	−	−	+	−
10	*ad*	+	−	−	+	+
11	*bd*	−	+	−	+	+
12	−*ce*	+	+	−	+	−
13	*cd*	−	−	+	+	+
14	−*be*	+	−	+	+	−
15	−*ae*	−	+	+	+	−
16	−*e*	+	+	+	+	+

Table 5.58 Fractional Factorial Example of Yield Results with Yates Algorithm Calculations

Run		Yield Results	Column 1	Column 2	Column 3	Column 4
1	(1)	26.3	51.3	231.3	486.6	892.7
2	*a*	25.0	180.0	255.3	406.1	416.1
3	*b*	38.0	59.3	136.9	227.4	434.7
4	*ab*	142.0	196.0	269.2	188.7	326.1
5	*c*	20.3	58.1	102.7	265.4	156.3
6	*ac*	39.0	78.8	124.7	169.3	148.1
7	*bc*	45.0	60.3	31.3	192.6	135.9
8	−*de*	151.0	208.9	157.4	133.5	104.5
9	*d*	22.6	−1.3	128.7	24.0	−80.5
10	*ad*	35.5	104.0	136.7	132.3	−38.7
11	*bd*	30.2	18.7	20.7	22.0	−96.1
12	−*ce*	48.6	106.0	148.6	126.1	−59.1
13	*cd*	22.8	12.9	105.3	8.0	108.3
14	−*be*	37.5	18.4	87.3	127.9	104.1
15	−*ae*	33.1	14.7	5.5	−18.0	119.9
16	−*e*	175.8	142.7	128.0	122.5	140.5

Remember the major assumption that was made by the group. All three- or higher-order interactions were probably insignificant and could be "aliased" with main effects. The other "half factorial," that the group decided not to use, can be found by using the opposite signs in the E column. After the experimental matrix was finalized by the group, the experiment was run and the data were collected for each of the 16 runs. The yield results were recorded in units of percent (Table 5.58).

The Yates algorithm was used to calculate the sum of squares and the mean square effects for each run. The steps after putting the runs in the standard notation order are:

- Column 1 The first item is the sum of the first pair of observations $((1) + a)$. The second item is the sum of the second pair of observations $(b + ab)$. Continue until all pairs are used. Then use the differences between pairs for the remaining items (i.e. value for item 9 is $(1) - a$).
- Column 2 Use the same procedure but with the column 1 values.
- Column 3 Use the same procedure but with the column 2 values.
- Column 4 Same procedure using column 3 values.

These calculated values are also included in Table 5.58.

The mean square effects are found by dividing column 4 by 8 (which is equal to one-half of the total number of runs). The sum of squares values are found by squaring the values in column 4 and dividing by 16:

$$\text{MS} = \frac{\text{column 4}}{8}$$

$$\text{SS} = \frac{(\text{column 4})^2}{16}$$

This is similar to what was done with full factorials in Section 5.6.2. Column 4 is analogous to the contrasts calculated for full factorial experiments. Using these equations, the team added the MS and SS values to their data (Table 5.59). Because the first row (1) is not measuring an effect, the values were not calculated.

In order to get an estimate of the experimental error, the group decided to choose some insignificant two-way interactions. They agreed unanimously that factor D, grinder screen size, would not have any significant two-way interactions. Therefore, they averaged the mean square

Table 5.59 Complete ANOVA Table for Polymer Yield Fractional Factorial[a]

Run		Column 4	MS	SS	F Test
1	(1)	892.7			
2	a	416.1	52.0	10,821	12.48*
3	b	434.7	54.3	11,810	13.24*
4	ab	326.1	40.8	6,646	9.95*
5	c	156.3	19.5	1,527	4.76
6	ac	148.1	18.5	1,376	4.51
7	bc	135.9	17.0	1,154	4.15
8	$-de$	104.5	13.1	683	3.20
9	d	−80.5	−10.1	405	2.46
10	ad	−38.7	−4.8	94	1.17
11	bd	−96.1	−12.0	577	2.93
12	$-ce$	−59.1	−7.4	218	1.80
13	cd	108.3	13.5	733	3.29
14	$-be$	104.1	13:0	677	3.17
15	$-ae$	119.9	15.0	899	3.66
16	$-e$	140.5	17.6	1,234	4.29

[a]The asterisk indicates significant factors. See text for additional description.

effects values for the D interactions to get an estimate of the error:

de	−13.1
ad	−4.8
bd	−12.0
cd	13.5
	−16.4

$$\text{MS error} = \frac{|-16.4|}{4} = 4.1$$

The F test was then performed on the remaining effects using the equation

$$F = \frac{\text{MS factor}}{\text{MS error}}$$

To determine significance, they chose an α level of 0.05 and looked up the F table value for 1 DF and 4 DF. (Remember, they had 4 degrees of freedom to calculate MS error.) The table value is 7.71. There were a few significant factors, which the team marked with an asterisk in Table 5.59. The team decided that they would do more experimentation to optimize the levels of the temperature of the compounder (A) and the raw material mix time (B) because they recognized that the effects of both of these factors and their interaction ($A \times B$) were all significant.

The team acquired information about all 5 factors in only 16 runs by utilizing the factional factorial approach and the Yates algorithm. This greatly reduced the cost of the experimentation and the amount of time necessary to get results.

When fractional factorials are expanded to study a large number of factors (six or more), they are typically called screening experiments. Section 5.8 discusses various techniques for designing and analyzing screening experiments.

5.8 SCREENING EXPERIMENTS

Although the use of more traditional experimental designs, such as factorial designs, provides the most complete picture of what is going on in an experimental region, they are not always the most practical. In many cases, it is very unclear what processing or material related factors affect the property being measured. Many factors must be investigated to determine which ones have significant effects on the response and which should be included in further experimentation. A screening design allows the investigation of many different factors in a minimum of experiments and is the most practical and cost effective to study many factors at once.

Screening designs are actually one form of fractional factorial design. There are a number of different methods for performing screening experiments, but clearly there are two methods that are recognized industry-wide: Plackett–Burman designs and Taguchi methods. Both are very powerful tools that use orthogonal arrays for designing and analyzing the experiments. Both will be covered in the following sections. Screening designs for mixture experiments (a very special case with special constraints) will be covered in Sections 5.9.3 and 5.9.4.

Screening experiments have been well defined by the originators, Dr. Plackett and Dr. Burman (1946), since the 1940s. Since then, the original techniques have been improved by simplifying the data analysis, thus making the designs easier to administer. Taguchi methods, a derivation of Plackett–Burman methods popularized by Dr. Genichi Taguchi, a Japanese quality control expert, have been in practice in Japan since the 1950s. In the last 5–10 years they have received a great deal of interest from American industry and have become a well-known experimental tool. Although both methods draw a certain level of criticism from traditional statisticians, they have proven in case after case that they work and are thus widely used.

In general, the concept behind screening designs is simple. A linear model of the form

$$y = b_0 + b_1 x_1 + b_2 x_2 + \cdots + b_i x_i$$

is used. The data are fitted to it to determine the coefficients (b_i). Each variable (x_i) is studied at two levels (usually designated by $+1$ and -1) and the absolute values of the coefficients are ranked to determine the relative effect that each factor had on the response. The sign of the coefficients determines the direction that the response moves as the factor is changed from the low level (-1) to the high level ($+1$). If the coefficient is positive, the response value will increase; if the coefficient is negative, the response value will decrease. Small coefficients lead to the assumption that the associated factor does not have a significant effect on the response over the tested range and can be ignored or left out of the next phase of testing.

Before screening experiments can be used, the variation in the process being studied must be understood. It is best if the process under study is a controlled (stable) process. Then factors affecting product quality and variation can be identified and the process capability can be improved. However, if the process is not stable or if it is in the design stage with the process stability unknown, screening designs can still be used; additional runs may be required to get a better estimate of the experimental error or background noise of the process.

The preferred factor levels identified during screening experiments are usually not the optimum levels, but they act as a good starting point for the next phase of experimentation whether it is a factorial-type experiment or EVOP. Surprisingly enough, significant improvements can be realized in many instances just by adjusting the factors to the levels indicated in the screening experiment.

Screening experiments, for the most part, use predetermined designs, which must then be randomized, repeated, and replicated as with any other experimental design. It must be remembered that the purpose of screening experiments is to determine the major effects. Information about interactions can be extracted, but most of that information is lost or incomplete.

5.8.1 Plackett–Burman Screening Designs

There are nine basic steps involved in running a Plackett–Burman screening design. Not all of these steps are unique to Plackett–Burman designs: Steps 1 through 3 and step 9 apply to all designs. Each step will be discussed individually.

Step 1: Lay the Groundwork

Determine what improvement is needed and how it will be measured.

Set Experiment Objectives

As with any experiment, there has to be an objective—something that is targeted for improvement. It may be a product characteristic or it may be a process output or yield. There may even be multiple characteristics that need to be monitored for change during the experiment. It is important to agree on and document the objectives of the experiment so that everyone involved understands them.

Define the Process

Next, the process to be used for the experiment has to be defined. Don't forget about the extended process (customers or suppliers). If factors outside the process have a potential effect on the output, then include them.

Select a Measurement System

The other fundamental step that has to be completed in the initial stages is the selection of a measurement system. The measurement system used should be stable reproducible, and repeatable. If possible, use variable data (numbers). If only attribute data are available (product appearance, for example), force them to be variable data by assigning number rankings. Remember, the measurement system used must be precise and accurate enough to be sensitive to small changes in the process output. The error associated with the measurement system must be as small as possible in order to keep the sensitivity of the experiment as high as possible.

Step 2: Brainstorm

Generate a list of potential factors. Get everyone involved with the process involved in the brainstorming—people from different departments and even different companies. Free flow of ideas should be encouraged: No idea is a bad idea. Generate a list of things that could affect, positively or negatively, the product or process characteristic that has been targeted for improvement. All ideas are accepted as potential factors. (More information about brainstorming can be found in Section 5.2.2.)

Step 3: Design Selection

Choose factors and setting levels.

Factor Selection

There are many techniques that can be used to narrow down the list of potential factors to those that will be used in the experiment. Voting and ranking is probably the simplest; it helps to eliminate personal bias and keep the level of objectivity high. There are some rules of thumb:

- Keep the factors practical, feasible, and cost effective.
- It is better to have too many factors than too few. Remember, this is a screening experiment and it is designed to look at many factors.

Choose Factor Levels

With the final list of factors selected, get the process experts together to choose the two levels of each factor that will be studied. Choose levels so that the process will be operating on the edges of the operating window but will still be able to make product that can be tested. Be BOLD—the levels should be far enough apart so that their effect on the process, if there is one, can be detected. If there is concern about some of the combinations (whether the process will make measurable product), run preliminary range-finding trials to establish the processing boundaries.

Matrix Selection

Plackett–Burman experiments use predetermined design matrices. They are classified by the number of experimental runs they require: 8-run,

Table 5.60 Plackett–Burman 8-Run Design Matrix

TC	*A*	*B*	*C*	*D*	*E*	*F*	*G*
1	+	+	+	−	+	−	−
2	−	+	+	+	−	+	−
3	−	−	+	+	+	−	+
4	+	−	−	+	+	+	−
5	−	+	−	−	+	+	+
6	+	−	+	−	−	+	+
7	+	+	−	+	−	−	+
8	−	−	−	−	−	−	−

12-run, 16-run, 20-run, 24-run, 28-run, and so forth. As discussed in Section 5.8, these screening designs handle f factors with $f + 1$ experimental runs. That means that the 8-run matrix can study 7 factors, the 12-run matrix can study 11 factors, and so on. Once the number of factors that will be studied has been determined, select the next largest matrix that will accommodate all of the factors. The 8-run matrix is shown in Table 5.60.

If you have	*Use*
up to 7 factors	8-run matrix
8–11 factors	12-run matrix
12–15 factors	16-run matrix
16–19 factors	20-run matrix

Don't eliminate factors only to be able to use a smaller matrix. As always, there is a trade-off; experimenting on more factors provides more information, but it costs more too. Any "extra columns" in the matrix that are not occupied by factors can be used as dummy variables, which will be

Table 5.61 2^3 Full-Factorial Design Matrix

	A	*B*	*AB*	*C*	*AC*	*BC*	*ABC*
(1)	−	−	+	−	+	+	−
a	+	−	−	−	−	+	+
b	−	+	−	−	+	−	+
ab	+	+	+	−	−	−	−
c	−	−	+	+	−	−	+
ac	+	−	−	+	+	−	−
bc	−	+	−	+	−	+	−
abc	+	+	+	+	+	+	+

Table 5.62 8-Run Plackett–Burman Matrix with the Treatment Combination Order Rearranged

TC	*A*	*B*	*C*	*D*	*E*	*F*	*G*
1	+	+	+	−	+	−	−
7	+	+	−	+	−	−	+
4	+	−	−	+	+	+	−
6	+	−	+	−	−	+	+
8	−	−	−	−	−	−	−
3	−	−	+	+	+	−	+
2	−	+	+	+	−	+	−
5	−	+	−	−	+	+	+

discussed shortly. (A number of design matrices are provided in Appendix E.)

Orthogonality
Section 5.3.1 discussed orthogonal arrays and used the orthogonal array shown in Table 5.61, (the matrix for a 2^3 factorial design) for the examples.

How does this orthogonal array relate to the Plackett–Burman 8-run matrix shown in Table 5.60? Following the rules of orthogonality, it can be proven that the Plackett–Burman matrix is orthogonal (even though this may not be as obvious as with the 2^3 matrix). Table 5.62 shows the Plackett–Burman matrix with the run order rearranged.

After comparing the rearranged Plackett–Burman matrix to the 2^3 array, some similarities can be seen. For instance, Plackett–Burman column *A* is the inverse of column *C* in the 2^3 matrix. Other similarities include:

Plackett–Burman column	*A*	*B*	*C*	*D*	*E*	*F*	*G*
2^3 column	$-C$	BC	$-ABC$	$-AB$	AC	B	A

It becomes easier to see that not only is the Plackett–Burman matrix an orthogonal array, but the columns are the same (or simply the inverse in a few cases) as the 2^3 matrix! This happens because by assuming that interactions are insignificant (which forms the basis for designing a screening experiment), the interactions columns are renamed as main effects. (Remember, this is the method for creating a fractional factorial. Screening designs are in the family of fractional factorial designs.) Orthogonal arrays form the foundation for many types of experimental designs.

Dummy Variables

Dummy variables can be used when there aren't enough factors to fill the Plackett–Burman matrix that has been chosen. For example, if there are 10 factors, a 12-run matrix would be selected. The 12 run matrix would handle 11 factors, which leaves 1 unassigned factor that can be used as a dummy variable. The effects of the dummy variables are analyzed in the same way that the factor effects are analyzed.

The effect measured for dummy variables is an estimate of the experimental error or normal variation in the process. (Remember, there won't be any deliberate changes in the process with that factor because there isn't a process variable assigned to it. The process is allowed to vary on its own.) The results for the dummy variables are very useful. If their effects are small and insignificant, there is a good chance that the experiment was sensitive enough to detect the real factor effects. If, on the other hand, the dummy variable effects are significant, then there is a good chance that the experiment was not sensitive to true factor effects and measures need to be taken to find a way to reduce the level of experimental error.

Step 4: Finalize the Design—Replicate, Reflect, Randomize

Once the design matrix is selected, there are three other steps that should be considered before the design is considered complete.

Replicate

Replication, repeating some or all of the experimental runs, should be used as much as possible in the experiment. Replication allows for the estimation of the amount of experimental error there is in the experiment. To replicate, repeat the experiment, randomizing the treatment combinations once again. There is always a cost penalty associated with performing more runs, so the benefits of additional data will have to be weighed against the cost penalty. If dummy variables are used, there is less need for replication because the effect of the dummy variable may provide an estimate of the experimental error.

Reflect

In addition to repeating runs, the experiment can be reflected. This means that the signs of all of the factor levels in the experiment are reversed. An example of an 8-run experiment that was replicated and reflected is given in Table 5.63.

Reflecting the experiment allows for reduction in the level of confounding that is taking place. Remember that in order to study so many factors with so few runs, the interaction effects were confounded with the main

Table 5.63 Plackett–Burman Eight-Run Design, Reflected and Replicated

	Run	Mean	A	B	C	D	E	F	G
	1	+	+	+	+	−	+	−	−
	2	+	−	+	+	+	−	+	−
	3	+	−	−	+	+	+	−	+
	4	+	+	−	−	+	+	+	−
	5	+	−	+	−	−	+	+	+
	6	+	+	−	+	−	−	+	+
	7	+	+	+	−	+	−	−	+
	8	+	−	−	−	−	−	−	−
Reflected	9	+	−	−	−	+	−	+	+
original	10	+	+	−	−	−	+	−	+
	11	+	+	+	−	−	−	+	−
	12	+	−	+	+	−	−	−	+
	13	+	+	−	+	+	−	−	−
	14	+	−	+	−	+	+	−	−
	15	+	−	−	+	−	+	+	−
	16	+	+	+	+	+	+	+	+
Replicated	$R1$	+	+	+	+	−	+	−	−
original	$R2$	+	−	+	+	+	−	+	−
	$R3$	+	−	−	+	+	+	−	+
	$R4$	+	+	−	−	+	+	+	−
	$R5$	+	−	+	−	−	+	+	+
	$R6$	+	+	−	+	−	−	+	+
	$R7$	+	+	+	−	+	−	−	+
	$R8$	+	−	−	−	−	−	−	−
Replicated	$R9$	+	−	−	−	+	−	+	+
reflected	$R10$	+	+	−	−	−	+	−	+
	$R11$	+	+	+	−	−	−	+	−
	$R12$	+	−	+	+	−	−	−	+
	$R13$	+	+	−	+	+	−	−	−
	$R14$	+	−	+	−	+	+	−	−
	$R15$	+	−	−	+	−	+	+	−
	$R16$	+	+	+	+	+	+	+	+

effects. Reflecting the experiment reduces the amount of confounding, which results in a higher level of confidence that the effects being measured are main factor effects. As with replication, reflection adds experimental runs, so the benefits must be weighed against the cost penalty.

Randomize

A *must-do* for all experiments is randomization. This means that the run order of the treatment combinations is randomized. Assignment of factors to columns should also be randomized although there may be experiments where a dummy factor is placed in a specific column in order to obtain information on a specific interaction. Randomization reduces the effect of external sources of variation, such as time of day, equipment warm-up or cool-down phenomena, tool wear, raw material variation, and so on. There are a number of different methods that can be used to truly randomize, such as random number generators or random number lists (Appendix B.7). Always randomize the experiment as much as possible. It does not add runs, so there is minimum penalty.

Block Randomization

In some cases, depending on the factors being used, it may not be practical to completely randomize the run order. For example, if one factor is process temperature, it may not be practical to adjust the temperature up and down every run. (That may be an added source of variation if the same set point is not hit every time, not to mention the added time needed between runs to heat or cool the equipment and reach steady state.) Again, it is important to weigh the pros and cons. Instead, it may be better to run all of the low temperature runs first and then all of the high temperature runs. This is called block randomization. The runs within each block (temperature) should still be randomized. Be careful, though, and only do this if there is much to gain in time, money, or process stability. This approach carries a risk of confounding effects. In this example, the temperature effect will be confounded with the time effect.

Step 5: Perform the Experiment

With the design matrix (replicated, reflected, and randomized) in hand, it is time to run the experiment and collect data. For each run, move across the corresponding row and set each factor at the indicated level (− or +). Run the process, allow it to reach steady state, and collect the sample or data point. Collect at least two data points and more if possible at each set of conditions to enable estimation of the experimental error. Make note of any unusual observations during the experiment. When all of the data are collected, calculate the mean (arithmetical average) and ranges (high value–low value) for each run.

Step 6: Determine the Factor Effects

The original design matrix is used to calculate the factor effects.

Plug Values into the Matrix

Use the mean values for each row. Starting with the first row, multiply the mean for that row by the sign in each column and replace each location in the row by that product. For example, in an 8-run experiment, the first row may look something like

	Mean	A	B	C	D	E	F	G	$\overline{Y}$	R
Run 1	+	−	+	+	+	−	+	−	$\overline{Y}_1$	R_1

After multiplying $\overline{Y}_1$ by all of the members of the row, it would look like

	Mean	A	B	C	D	E	F	G	$\overline{Y}$	R
Run 1	+	$-\overline{Y}_1$	$+\overline{Y}_1$	$+\overline{Y}_1$	$+\overline{Y}_1$	$-\overline{Y}_1$	$+\overline{Y}_1$	$-\overline{Y}_1$	$\overline{Y}_1$	R_1

Similarly, row 2 would become

	Mean	A	B	C	D	E	F	G	$\overline{Y}$	R
Run 2	$+\overline{Y}_2$	$+\overline{Y}_2$	$-\overline{Y}_2$	$-\overline{Y}_2$	$+\overline{Y}_2$	$+\overline{Y}_2$	$+\overline{Y}_2$	$-\overline{Y}_2$	$\overline{Y}_2$	R_2

Notice that the rules of multiplication apply:

$$(-) \times (-) = (+)$$

$$(-) \times (+) = (-)$$

$$(+) \times (+) = (+)$$

Continue this multiplication/substitution for *all the rows*.

Calculate Factor Effects

With the matrix substitution complete, it is time to focus on the columns. For each column, sum all of the values. (Remember the sign conventions. Some of the values are negative and some are positive.) Then divide that value by the number of pluses that were in that column in the original matrix. This should be the total number of runs divided by 2. *The number in the denominator will be the same for all columns*. The results of these calculations for each column are the net effects for each factor. (This is very similar to the techniques used to find the contrasts in a 2^f factorial.)

Interpreting the Effects

The size of the effect relates to the magnitude of change that can be expected in the process output due to changing the factor from one level to another. The sign of the effect is also important.

A positive effect means that the response will increase as the factor is changed from the low level to the high level.

A negative effect means that the response will decrease as the factor is changed from the low level to the high level.

Be careful not to draw any conclusions about the factors yet. These effects still have to be compared to the level of experimental error that was experienced in the experiment in order to determine if the effects are significant or could just be due to normal variation.

Step 7: Test the Effects for Significance

In order to determine whether the factor effects observed in the experiment were real or just normal variation in the process, an estimate of the level of variation present in the experiment (called experimental error) must first be calculated. Then the factor effects are compared to the experimental error. This is a form of hypothesis test that will determine whether or not the factor effect is outside the normal control limits of the process.

Dr. Charles Holland (1983) has worked to simplify the analysis techniques originally introduced by Plackett and Burman. He has reduced the analysis to a few simple steps.

Calculate the Experimental Error

The experimental error s_{ee} is found using the formula

$$s_{ee} = \frac{\overline{R}}{d_2}$$

where $\overline{R}$ = is the average range
d_2 = table value based on sample size

The average range $\overline{R}$ is simply

$$\overline{R} = \frac{\text{sum of all ranges } R}{\text{number of runs}}$$

The constant d_2 is found from Table 5.64.

Calculate the Standard Deviation for an Effect

The standard deviation for an effect, s_{eff}, is found using the equation

$$s_{eff} = \frac{2s_{ee}}{\sqrt{Tn}}$$

where s_{ee} = experimental error
T = total number of runs
n = sample size per run

Table 5.64 Table for d_2 Constant Depending on Sample Size n

n	d_2	n	d_2
		11	3.173
2	1.128	12	3.258
3	1.693	13	3.336
4	2.059	14	3.407
5	2.326	15	3.472
6	2.534	16	3.532
7	2.704	17	3.588
8	2.847	18	3.640
9	2.970	19	3.689
10	3.078	20	3.735

Calculate the Test Statistic
For Plackett–Burman design analysis, a form of a t test is used: A value is calculated based on the t-table value and is compared to the effects measured. To use the t table, an α level (confidence level) is selected and the degrees of freedom must be calculated. The typical choice for α for these designs is 0.05 (which means that 1 out of every 20 times an effect will be considered statistically significant when it isn't really). Degrees of freedom are calculated using the equation

$$\mathrm{DF} = T(n - 1)$$

where again T = number of runs
n = number of samples per run

The test statistic, called the two sigma effect (TSE), can then be found:

$$\mathrm{TSE} = (t_{\mathrm{table}}) \times (s_{\mathrm{eff}}) \quad \text{for } \alpha/2 = 0.025, \mathrm{DF}$$

This is the test statistic.

Test the Factor Effects for Significance
Compare the TSE to the absolute value of all of the factor effects:

If TSE is greater than the factor effect, then the factor effect is not significant.
If TSE is less than the factor effect, then the factor effect is significant.

If the dummy variable calculates to be significant, then the estimate of experimental error was too low or there is a significant interaction in that column confounding the dummy variable effect. This potential interaction should be studied further as part of future experimentation.

Step 8: Run a Confirmation Run

Once the significant factors and preferred levels (based on the sign of the effect) are identified, a confirmation run should be performed. Set up the process using the desired setting and collect data. The results should be nearly equal to or better than the best results achieved in the experiment. Why is it important to make a confirmation run? Remember that assumptions were made that interactions would not be important in order to justify doing a screening experiment in the first place. The confirmation run will give a good indication as to the correctness of those assumptions.

Also, at this point, compare the results to the initial experimental objectives. Was the level of improvement that was needed achieved? If not, look at factors not included in the original experimentation.

Step 9: Adjust the Process and Plan for Improvement

The process changes that led to improvement can be incorporated immediately to see immediate gains, but keep in mind that what has been learned has not approached optimization of the process. Screening designs are used to identify the important factors and are intended to be the first step in the journey of continual improvement. A factorial design or response surface experiment designed around those significant factors may be a logical next step.

Side Step: Interactions

Even though the interaction effects were confounded with the main effects in the initial design, there is still a way to graphically interpret the two-way interactions. It is a qualitative method but it can still provide useful information, particularly if the results of the confirmation run indicate the presence of a strong interaction.

Drawing the Graph

1. Choose the two factors to be studied. (We'll call them factor 1 and factor 2).
2. Construct the axes for the graph. The x-axis will have two hash marks that are labeled "low level" and "high level". The y-axis scale will depend on the response values.

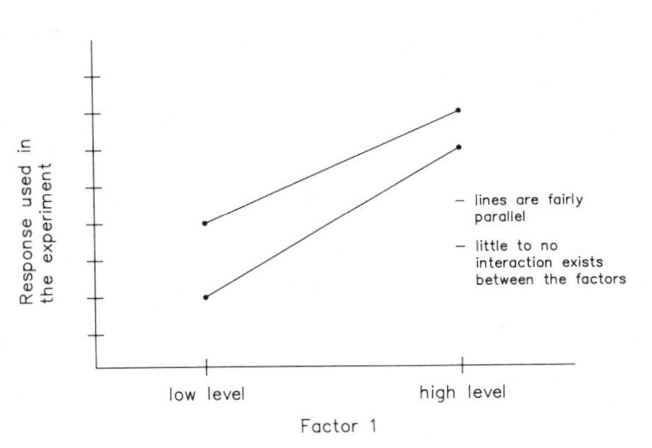

FIGURE 5.32. Response graph showing little or no interaction between factors.

3. Using the columns in the design matrix corresponding to these factors, calculate the following averages:
   When factor 2 is low:
   - the average of factor 1 when it is low (−);
   - the average of factor 1 when it is high (+).

   When factor 2 is high:
   - the average of factor 1 when it is low (−);
   - the average of factor 1 when it is high (+).
4. Plot two lines on the graph: one that corresponds to factor 2 when it is low and the other that corresponds to factor 2 when it is high.

Interpretation of Graphs

Parallel or nearly parallel lines indicate that little to no interactions exist. In Figure 5.32, the lines are fairly parallel, which means little to no interaction exists between the factors. Alternatively, nonparallelism indicates that an interaction is present. In Figure 5.33, the lines are clearly nonparallel, which means a strong interaction exists between the factors.

Interactions can be studied further in the follow-up experimentation.

Plackett–Burman Experiment Example

A screw-injection-molded part is not meeting dimensional tolerances. The part must measure 2.000 ± 0.003 in. in length. Excessive shrinkage is resulting in average part lengths of 1.990 in. A customer inspects all of the incoming parts and has threatened to take their business elsewhere if the problem isn't taken care of soon.

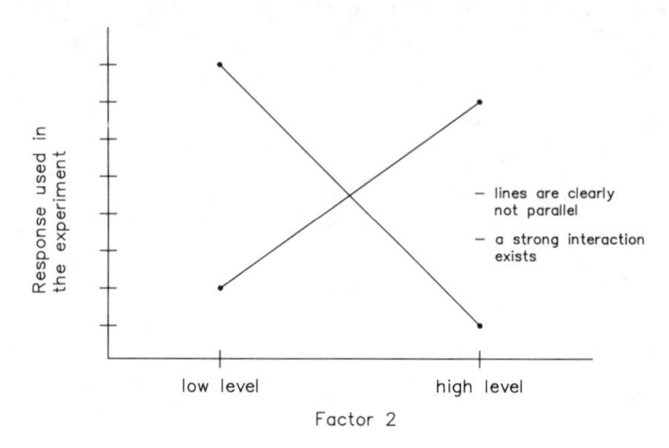

FIGURE 5.33. Response graph showing interaction between factors.

The objective: To identify the processing factors that affect part shrinkage and adjust the process to achieve a nominal length of 2.000.
The process: The molding machine and the material.
The measurement: Calipers.

All of the people involved with the job got together to brainstorm possible factors. They produced a list of factors, which was reduced to seven items by voting and ranking. The seven factors were:

Mold temperature
Cure time
Injection pressure
Back pressure
Injection time
Material viscosity
Hold Time

Then, the "experts" at running the process set bold levels for each factor:

Factor	High Level (+)	Low Level (−)
A Mold temperature (°F)	370	320
B Cure time (s)	60	30
C Injection pressure (lb/in.^{2})	2500	1500
D Back pressure (lb/in.^{2})	200	50
E Injection time (s)	12	6
F Material viscosity	Soft	Stiff
G Hold time (s)	30	15

Table 5.65 Plackett–Burman Design with Treatment Combination Order Randomized within Mold Temperature Blocks

Run	TC	*A*	*B*	*C*	*D*	*E*	*F*	*G*
1	1	+	+	+	−	+	−	−
2	4	+	−	−	+	+	+	−
3	7	+	+	−	+	−	−	+
4	6	+	−	+	−	−	+	+
5	5	−	+	−	−	+	+	+
6	3	−	−	+	+	+	−	+
7	8	−	−	−	−	−	−	−
8	2	−	+	+	+	−	+	−

Because there are seven factors, the logical matrix choice for this experiment is the 8-run matrix.

To finalize the design, the run order was randomized. Also, the decision was made to collect five data points for each run in order to get an estimate of the experimental error. The finalized design is shown in Table 5.65.

Block randomization was used for the factor "mold temperature" in order to keep the experiment cost effective and to avoid added experimental error due to constant adjustment of the mold temperature. Because of the length of time and the amount of material needed for each experimental run, the decision was made not to replicate or reflect the experiment.

To run the experiment, the press was set up for the first set of conditions and allowed to reach steady state. Five parts were collected and measured. Then the press was adjusted for the next set of conditions. This was repeated until samples were collected for all eight runs. The average value for each run (and the range) were calculated (Table 5.66). To calculate the factor effects, the average values for each run were first

Table 5.66 Molding Process Experimental Data

Run	$\bar{Y}$	R
1	1.992	0.008
2	1.990	0.008
3	1.998	0.009
4	1.995	0.005
5	1.998	0.009
6	2.000	0.007
7	1.991	0.006
8	2.001	0.005
		$\bar{R} = 0.0071$

Table 5.67 Molding Process Experimental Data Added to the Plackett–Burman Design Matrix

Run	A	B	C	D	E	F	G
1	+1.992	+1.992	+1.992	−1.992	+1.992	−1.992	−1.992
2	+1.990	−1.990	−1.990	+1.990	+1.990	+1.990	−1.990
3	+1.998	+1.998	−1.998	+1.998	−1.998	−1.998	+1.998
4	+1.995	−1.995	+1.995	−1.995	−1.995	+1.995	+1.995
5	−1.998	+1.998	−1.998	−1.998	+1.998	+1.998	+1.998
6	−2.000	−2.000	+2.000	+2.000	+2.000	−2.000	+2.000
7	−1.991	−1.991	−1.991	−1.991	−1.991	−1.991	−1.991
8	−2.001	+2.001	+2.001	+2.001	−2.001	+2.001	−2.001

plugged into the matrix shown in Table 5.67. For each column, the values are added (to determine the sum) and then divided by the number of pluses in each column (four) to calculate the factor effects:

	A	B	C	D	E	F	G
Sum	−0.015	+0.013	+0.011	+0.013	−0.005	+0.003	+0.017
Effect	−0.004	+0.003	+0.003	+0.003	−0.001	+0.001	+0.004

The effects indicate the magnitude and direction that each factor influences the part length. For instance, by increasing the mold temperature from its low level to its high level, the average part length decreases by 0.004 in. Similarly, by changing the hold time from low level to high level, the average part length increases by 0.004 in.

Now, the effects were tested for statistical significance. First, the estimate of experimental error, s_{ee}, was calculated:

$$s_{ee} = \frac{\overline{R}}{d_2}$$

where $\overline{R}$ = average range = 0.0071
d_2 = table value for $n = 5 = 2.326$

so

$$s_{ee} = \frac{0.0071}{2.326} = 0.0031$$

Next, the estimate of the standard deviation of an effect was calculated:

$$s_{\text{eff}} = \frac{2s_{\text{ee}}}{\sqrt{Tn}} = \frac{2(0.0031)}{\sqrt{8.5}} = 0.00098$$

where T = number of runs
n = sample size

To do the hypothesis test by calculating the TSE, the degrees of freedom must be found:

$$\text{DF} = T(n-1) = 8(5-1) = 32$$

With an α level of 0.05, the t-table value is 2.042. The TSE now could be found:

$$\begin{aligned}\text{TSE} &= (t_{1-\alpha/2,\,\text{DF}}) \times (s_{\text{eff}}) \\ &= (2.042)(0.00098) \\ &= 0.0021\end{aligned}$$

This value was then compared to the absolute values of the individual factor effects. The effects that were greater the TSE value were significant:

> 0.0020	< 0.0020
A	*E*
B	*F*
C	
D	
G	

Because the objective of the experiment was to increase the size of the parts to 2.000 (from 1.990), it is important to identify the significant factors and their levels corresponding to an increase in the response (part size).

A confirmation run would consist of the following factors and levels:

Factor	Level
A Mold temperature	Low
B Cure time	High
C Injection pressure	High
D Back pressure	High
G Hold time	High

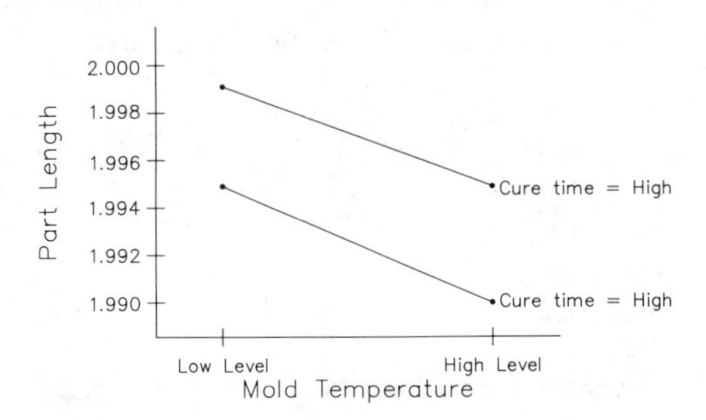

FIGURE 5.34. Response graph for mold temperature and cure time.

The results of the confirmation run should produce parts with an average length of at least 2.001 (the highest value in the experiment). If there is a large difference in the results from this value, it indicates that there may be interaction effects present or that a key factor was missed in the original experimentation.

Suppose there was a suspicion that there was a significant interaction between mold temperature and cure time that effected part length. A plot will confirm or deny this as shown in Figure 5.34.

When cure time is low (experimental runs 2, 4, 6, and 7):

Average when mold temperature is low (runs 6 and 7)

$$\frac{2.000 + 1.991}{2} = 1.9955$$

Average when mold temperature is high (runs 2 and 4)

$$\frac{1.990 + 1.991}{2} = 1.9905$$

When cure time is high (experimental runs 1, 3, 5, and 8):

Average when mold temperature is low (runs 5 and 8)

$$\frac{1.998 + 2.001}{2} = 1.9995$$

Average when mold temperature is high (runs 1 and 3)

$$\frac{1.992 + 1.998}{2} = 1.995$$

The two lines are nearly parallel, so there does not appear to be an interaction between mold temperature and cure time. Because the experiment indicated that low mold temperature is desirable, an interaction effect does not have to be considered before selecting a cure time. (The high level of cure time is desirable, as was indicated earlier.)

5.8.2 Taguchi Screening Experiments

The Taguchi methods are Plackett–Burman designs modified by Genichi Taguchi, a Japanese engineer and quality control expert. Dr. Taguchi's methods encompass the entire scope of design and improvement with an emphasis on improving quality without adding capital or material costs. These methods focus on adjusting process parameters to optimize process output and to reduce variation. Although Taguchi addresses many aspects of quality improvement, this chapter will only discuss the use of his screening experiments: their application, design, and analysis.

Taguchi Screening Experiments

Because the Taguchi methods are derived from Plackett–Burman screening designs, they too are meant to investigate a large number of factors and determine the major effects. The same approach is used (as with other methods) to identify the factors that will be included in the experiment. All of the independent variables (factors) and dependent variables must be measurable, and all of the conditions surrounding the experiment must be well defined. Also, a clear objective must be stated and understood by everyone involved in the experiment.

As with other screening experiments, Taguchi methods use orthogonal arrays—design matrices that are condensed forms of full factorial design matrices. This condensation is accomplished by assuming that there are no interactions present between three or more factors (or that they are small enough to be insignificant). Two-way interactions are considered insignificant relative to main factor effects. However, this is not always the case so Taguchi includes an interaction table to help design into the experiment a way to evaluate key two-way interactions. Taguchi stresses, however, that the first priority for any experiment is to understand the main effects, and then as a secondary priority, evaluate interaction effects.

The Taguchi L8 design matrix with its corresponding interaction table are shown in Table 5.68. The interaction table indicates to the experimenter where confounding exists in the matrix so that the experimenter can either design around the potential for confounding or assign a specific interaction effect to a specific column to study it in the experiment. This can be shown by looking at columns 3 and 5 in the design matrix. If we replace the 1s with the value of $+1$ and the 2s with the value of -1, then

Table 5.68 Taguchi L8 Design Matrix with Interactions

Taguchi L8 Matrix

TC	Column 1	2	3	4	5	6	7
1	1	1	1	1	1	1	1
2	1	1	1	2	2	2	2
3	1	2	2	1	1	2	2
4	1	2	2	2	2	1	1
5	2	1	2	1	2	1	2
6	2	1	2	2	1	2	1
7	2	2	1	1	2	2	1
8	2	2	1	2	1	1	2

L8 Interaction Table

Column	1	2	3	4	5	6	7
(1)	—	3	2	5	4	7	6
(2)		—	1	6	7	4	5
(3)			—	7	6	5	4
(4)				—	1	2	3
(5)					—	3	2
(6)						—	1
(7)							—

columns 3 and 5 look like:

Column 3	Column 5
+1	+1
+1	−1
−1	+1
−1	−1
−1	−1
−1	+1
+1	−1
+1	+1

If the +1 and −1 are multiplied together across each row, the following column is generated:

+1
−1
−1
+1
+1
−1
−1
+1

This corresponds with column 6 in the design matrix, which indicates that column 6 represents a main effect plus a two-way interaction between columns 3 and 5. This is called confounding: It is impossible in the design to separate the two effects.

Taguchi assumes that two-way interactions are not significant, but if the experimenter believes that there are some important interactions to be studied, they can be assigned to appropriate columns. The interaction table helps to find the column location of those interactions. To use Table 5.68, first we find the row that locates the lower column number of interest. This row is indicated by the column number in parentheses. In the preceding example, we looked at the 3 × 5 interaction. In this case, we find the row designated (3). Next, we locate the column number of the other factor of interest as it is listed across the top of the table. This is column 5 in the example. The point at which that column and the row we found earlier intersect is the column in the design matrix that represents the two-way interaction between factors 3 and 5; Table 5.69 demonstrates how to locate the 3 × 5 interaction.

Because the location of the interaction effects can be identified, the design can study them if desired. An L8 design can evaluate seven terms. These seven can be any combination of main effects and two-way interaction effects. There are some guidelines, however, that should be kept in mind when designing the experiment. Remember that most two-way interactions do not exist. It is usually better to test more independent variables (main effects) than to include a lot of interactions. For the same reasons, don't move up to the next larger design only to include more interactions. This will create more experiments, and, again, most interactions are not significant.

There are five basic steps for designing the experiment, based on the understanding of the interaction table and confounding effects.

Table 5.69 Use of the L8 Interaction Table to Identify Interaction Location

L8 Interaction Table

Column	1	2	3	4	5	6	7
(1)	—	3	2	5	4	7	6
(2)		—	1	6	7	4	5
(3)			—	7	6	5	4
(4)				—	1	2	3
(5)					—	3	2
(6)						—	1
(7)							—

indicates the 3 × 5 interaction is found in column 6

1. Determine the number of factors and interactions that will be included in the experiment. (Brainstorming, cause-and-effect diagrams, and voting and ranking are all useful for this step.) Select high and low levels for each factor that represent the experimental region of interest.
2. Select the Taguchi design matrix that will be able to investigate the number of factors that have been identified. Remember, don't create a lot of extra runs to evaluate interactions because they probably aren't significant enough to worry about. Choose the smallest Taguchi design for the number of factors identified.
3. Assign to columns the variables that will also be evaluated for interactions. Then, using the interaction table, identify the columns that need to be used for those interactions (label them).
4. Assign the remaining variables to the remaining columns.
5. Remove all unassigned columns and interaction columns from the matrix.

The matrix that is left is the actual experimental matrix that will be run. The high and low levels for each factor can be substituted for the 1 and 2 values in the matrix.

Another approach can be used in situations where it is desirable to experiment at more than two levels for a given factor. For example, suppose five factors were identified for an experiment, but one factor, perhaps because it is a costly raw material or because there is a reactivity involved, really has to be investigated at more than two levels. This can be accomplished by matrix multiplication similar to that used to understand interaction effects. Start with the L8 design matrix as shown in Table 5.70.

Columns 1, 2, and 3 can be combined to form a modified L8 design matrix where factor 1 is investigated at four levels (levels 1, 2, 3, and 4) as shown in Table 5.71.

Table 5.70 L8 Table

	Factors						
Run	1	2	3	4	5	6	7
1	1	1	1	1	1	1	1
2	1	1	1	2	2	2	2
3	1	2	2	1	1	2	2
4	1	2	2	2	2	1	1
5	2	1	2	1	2	1	2
6	2	1	2	2	1	2	1
7	2	2	1	1	2	2	1
8	2	2	1	2	1	1	2

Table 5.71 Modified Taguchi Table to Investigate One Factor at Four Levels

Run	Factors 1	4	5	6	7
1	1	1	1	1	1
2	1	2	2	2	2
3	2	1	1	2	2
4	2	2	2	1	1
5	3	1	2	1	2
6	3	2	1	2	1
7	4	1	2	2	1
8	4	2	1	1	2

With the design completed, the usual rules of randomization, replication, and reflection apply to establish the final experimental design before the data are collected. Once the data are collected, the matrix has to be analyzed to determine which factors significantly affect the response.

Data Analysis

With all of the experiments performed and the data collected, the results are analyzed in two steps. First, the mean effects are determined. Then, the variation analysis, which can be equally as important, is performed.

The analysis of the mean effects is very straightforward. The following steps are performed:

1. Calculate the average of all the responses measured $\bar{\bar{Y}}$.
2. Calculate the average response for all of the experiments run at the high level for the first factor $\bar{Y}_{1H}$ and then the average response for all of the experiments run at the low level for the first factor $\bar{Y}_{1L}$.
3. Repeat step 2 for the high and low levels for the remaining factors. The same thing is done for the interaction columns treating the 1 as the high level and 2 as the low level.
4. Calculate the effect of each factor at its high level and its low level.

$$E_{iH} = \bar{Y}_{iH} - \bar{\bar{Y}}$$

$$E_{iL} = \bar{Y}_{iL} - \bar{\bar{Y}}$$

where $\bar{\bar{Y}}$ = average of all responses
E_{iH} = effect of high level of factor i
E_{iL} = effect of low level of factor i

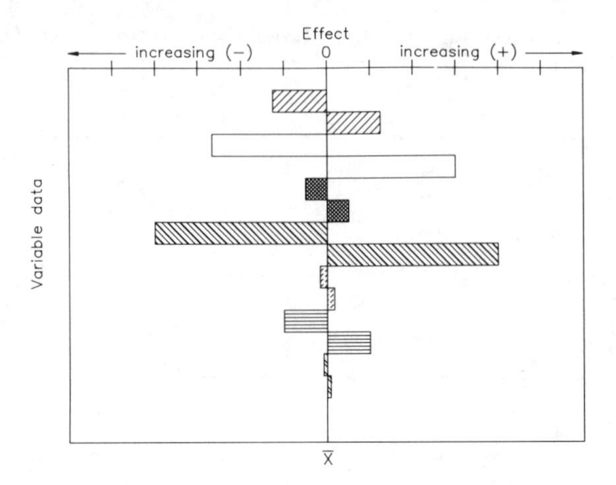

FIGURE 5.35. Bar graph of main factor effects showing relative magnitude and direction.

In order to visualize the effects, it is sometimes helpful to graph the results on a bar graph. The format is shown in Figure 5.35.

5. Calculate the overall effect E_i of each factor

$$E_i = E_{iH} - E_{iL}$$

The effects of each factor can then be compared to the objective of the experiment. If the goal was to increase the response (i.e., the yield or percent acceptable), it is desirable to make adjustments in the levels that increase the experimental output. Similarly, it may be desirable to lower the response (i.e., percent cracks, number of defects); therefore, the levels can be chosen that lower the response. The statistical significance of the effects can be determined using simple statistical methods (ANOVA) and utilizing an estimate of the experimental error. (Experimental error can be estimated by running replicates of one or more experimental runs and measuring the variation present, or by using a pooled standard deviation from all the runs.) Taguchi, however, in a break from Plackett–Burman philosophies, does not stress the use of statistical analysis. He states that if an adjustment leads to an improvement—make the adjustment.

Interactions can also be handled graphically if the magnitude of the effect suggests that the interaction is significant. In those cases, a simple line graph is used (see Figure 5.36). Taguchi recommends plotting the responses of all factors using line graphs. An example of an experimental output is graphed in Figure 5.37. Optimization recommendations can easily be recognized. For this case, the combination that would give the

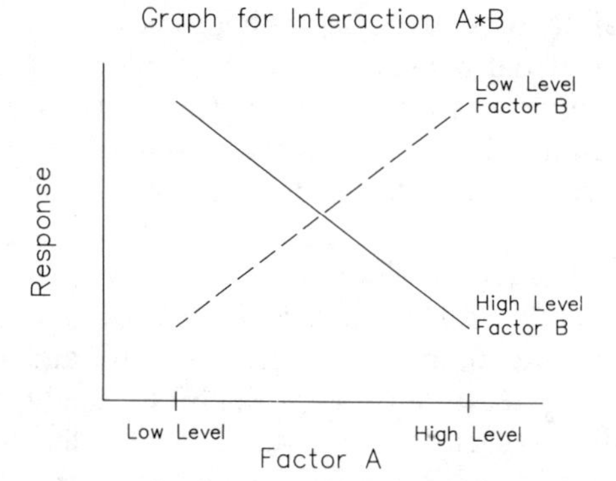

FIGURE 5.36. Line graph of responses showing $A * B$ interaction.

highest response would be chosen for a follow-up confirmation trial:

$$C_1 \quad B_2 \quad D_1 \quad A_1 \quad E_2$$

The purpose of doing a confirmation run is to check the reproducibility of the results from the initial experiment. The results of the confirmation run

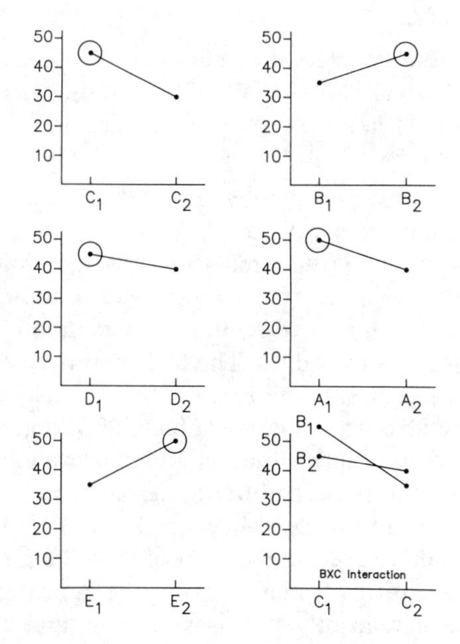

FIGURE 5.37. Graphs of responses for main factors.

should be equal to or better than the predication made based on the screening design. If the results are equal, then reproducibility is good. If the results are significantly different (better or worse), it indicates that there are interaction effects or unidentified main factors still affecting the process output. Noise (common cause) factors may also be affecting the process output.

A noise factor is a factor whose level either cannot or will not be set or maintained but could still affect the process output. Examples include humidity or other weather effects, time of day, and day of the week. Taguchi uses the signal-to-noise ratio (S/N) to handle those factors that can affect product variation but that may be difficult to control. The reason behind this is simple. Adjustments can be made to a process to improve the output, but if the variation is large, the results may still be bad. Taguchi uses the signal-to-noise (S/N) ratio to handle the issue of normal variation. There are three ratios commonly used:

1. Less is better (desire is to minimize a property)
2. More is better (desire is to maximize a property)
3. Target is best (adjust to a desired level)

There are three types of noise described by Taguchi:

1. Outer noise (environmental conditions)
2. Inner noise (deterioration of parts, materials, etc.)
3. Between-product noise (piece-to-piece variation)

The goal is to strive to continually reduce the level of noise in a process.

Why are these important? What if, for example, the objective of the experiment was to improve the strength of a welded joint. There are two ways to improve the values: make adjustments to improve the mean strength or make adjustments to reduce the variation in the process, thus eliminating many of the low values. The S/N approach takes both of these into account—not just raising the *mean value*. Taguchi suggests setting up a separate noise matrix to study the effects of various sources of noise, thus indicating ways to reduce variation and improve quality. Some factors may appear in both the noise and the design matrix.

An example of this approach is shown in Table 5.72. There is a separate matrix, called the outer array, to the right of the L8 matrix for studying the noise factors. The entire L8 matrix has to be repeated for each of the noise factors. This significantly increases the size (and cost) of the experiment, but sometimes the information can be useful.

Table 5.72 Taguchi Matrix with Seven Control and Three Noise Factors

	Design Matrix							Noise Matrix				
									i	ii	iii	iv
								M	1	2	1	2
	A	*B*	*C*	*D*	*E*	*F*	*G*	*N*	1	1	2	2
L8	1	2	3	4	5	6	7	*O*	2	1	1	2
1	1	1	1	1	1	1	1					
2	1	1	1	1	2	2	2					
3	1	2	2	1	1	2	2					
4	1	2	2	1	2	1	1					
5	2	1	2	1	2	1	2					
6	2	1	2	1	1	2	1					
7	2	2	1	1	2	2	1					
8	2	2	1	1	1	1	2					

To calculate S/N ratios, the following calculations are performed for each run:

	MSD	S/N
Less is better	$\dfrac{Y_1^2 + Y_2^2 + Y_3^2 + \cdots + Y_n^2}{n}$	$-10 \log \text{MSD}$
More is better	$\dfrac{\dfrac{1}{Y_1^2} + \dfrac{1}{Y_2^2} + \dfrac{1}{Y_3^2} + \cdots + \dfrac{1}{Y_n^2}}{n}$	$-10 \log \text{MSD}$
Target is best	—	$10 \log \dfrac{\bar{Y}_2}{s^2}$

where Y = response values for the run
n = sample size measured
$\bar{Y}$ = average response for the run
s = sample standard deviation for the run

The higher the S/N ratio, the better; it gives an indication of the magnitude of effect that the control factors have on the process output compared to process noise (normal variation). Taguchi recommends graphing the S/N responses in the same way the control factor responses are graphed; see Figure 5.38.

The experimenter should choose factor levels that reduce variation first (high S/N outputs) and then choose the control factors that deal the

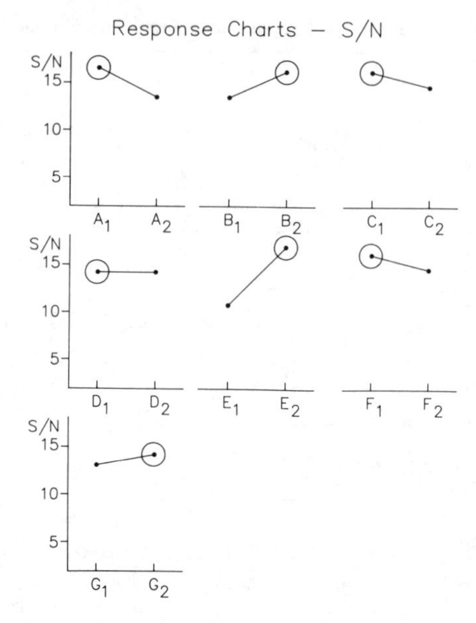

FIGURE 5.38. Graphs of Taguchi signal-to-noise responses.

process back to the target. That way the result is a process *on target* with *less variation*.

Most of the examples discussed here dealt with an L8 design. There are other Taguchi matrices that can be selected.

L12 Orthogonal Array

In an L12 array (Table 5.73), the interactions are distributed to all columns uniformly. *There is no interaction table* so interactions can't be

Table 5.73 Taguchi L12 Matrix

No.	1	2	3	4	5	6	7	8	9	10	11
1	1	1	1	1	1	1	1	1	1	1	1
2	1	1	1	1	1	2	1	2	2	2	2
3	1	1	2	2	2	1	2	1	2	2	2
4	1	2	1	2	2	1	1	2	1	1	2
5	1	2	2	1	2	2	2	2	1	2	1
6	1	2	2	2	1	2	2	1	2	1	1
7	2	1	2	2	1	1	2	2	1	2	1
8	2	1	2	1	2	2	2	1	1	1	2
9	2	1	1	2	2	2	1	2	2	1	1
10	2	2	2	1	1	1	1	2	2	1	2
11	2	2	1	2	1	2	1	1	1	2	2
12	2	2	1	1	2	1	2	1	2	2	1

Table 5.74 Taguchi L16 Design Matrix and Interaction Table (Interactions between Two Columns)

	L16 Matrix														
No.	1	2	3	4	5	6	7	8	9	10	11	12	13	14	15
1	1	1	1	1	1	1	1	1	1	1	1	1	1	1	1
2	1	1	1	1	1	1	1	2	2	2	2	2	2	2	2
3	1	1	1	2	2	2	2	1	1	1	1	2	2	2	2
4	1	1	1	2	2	2	2	2	2	2	2	1	1	1	1
5	1	2	2	1	1	2	2	1	1	2	2	1	1	2	2
6	1	2	2	1	1	2	2	2	2	1	1	2	2	1	1
7	1	2	2	2	2	1	1	1	1	2	2	2	2	1	1
8	1	2	2	2	2	1	1	2	2	1	1	1	1	2	2
9	2	1	2	1	2	1	2	1	2	1	2	1	2	1	2
10	2	1	2	1	2	1	2	2	1	2	1	2	1	2	1
11	2	1	2	2	1	2	1	1	2	1	2	2	1	2	1
12	2	1	2	2	1	2	1	2	1	2	1	1	2	1	2
13	2	2	1	1	2	2	1	1	2	2	1	1	2	2	1
14	2	2	1	1	2	2	1	2	1	1	2	2	1	1	2
15	2	2	1	2	1	1	2	1	2	2	1	2	1	1	2
16	2	2	1	2	1	1	2	2	1	1	2	1	2	2	1

L16 Interaction Table

Column	1	2	3	4	5	6	7	8	9	10	11	12	13	14	15
(1)	—	3	2	5	4	7	6	9	8	11	10	13	12	15	14
(2)		—	1	6	7	4	5	10	11	8	9	14	15	12	13
(3)			—	7	6	5	4	11	10	9	8	15	14	13	12
(4)				—	1	2	3	12	13	14	15	8	9	10	11
(5)					—	3	2	13	12	15	14	9	8	11	10
(6)						—	1	14	15	12	13	10	11	8	9
(7)							—	15	4	13	12	11	10	9	8
(8)								—	1	2	3	4	5	6	7
(9)									—	3	2	5	4	7	6
(10)										—	1	6	7	4	5
(11)											—	7	6	5	4
(12)												—	1	2	3
(13)													—	3	2
(14)														—	1
(15)															—

studied. Of course, that also means that confounding isn't a problem and experimental results are more reproducible.

L16 Design Matrix

As with an L8 matrix, the L16 matrix (Table 5.74) has a corresponding interaction table.

Table 5.75 Taguchi L9 Matrix

	Factor			
Run	*A*	*B*	*C*	*D*
1	1	1	1	1
2	1	2	2	2
3	1	3	3	3
4	2	1	2	3
5	2	2	3	1
6	2	3	1	2
7	3	1	3	2
8	3	2	1	3
9	3	3	2	1

L9 Matrix

The L9 matrix (Table 5.75) studies four factors at three levels each with nine experimental runs. The L9 design offers an opportunity to determine if the main factor effects are linear or not because it studies the factors at three levels.

5.8.3 Comparison: Plackett–Burman and Taguchi

Both Plackett–Burman and Taguchi methods investigate large numbers of factors in a quick, cost-effective manner to achieve a desired response. Although there are some differences in their approaches, the methods are essentially the same.

Both methods use a fractional factorial approach (orthogonal arrays) to reduce the number of experiments by assuming that third- and higher-order interactions are nonexistent. Both methods recommend that the factors themselves should be chosen to avoid a known interaction; both concentrate on two levels of experimentation for each variable involved; both require some degree of replication to establish statistical comparisons; and both can be analyzed using simple mathematics that does not require the use of computers or complicated programs.

There are also four basic differences between the two. Some of these differences are real and some are perceptions:

1. *Orthogonal arrays.* Both methods use orthogonal arrays, but the arrays appear to be different. Plackett and Burman use an array chosen to purposely spread the interactions so that they are less likely to interfere with the separation of the main effects. The Taguchi array is a transformation of the Plackett–Burman array. It

makes no effort to spread the interactions. The array appears more logical (easier to relate to a 2^3 factorial) and mathematically elegant. This allows Taguchi methods to look at interactions more easily than Plackett–Burman techniques. Interactions can be assigned to their correct columns in the matrix and can be interpreted. However, interactions can also be evaluated using the Plackett–Burman approach as shown in Section 5.8.1.

2. *Data analysis.* The philosophies and methods for data analysis are different for the two designs. The Plackett–Burman approach is more cautious or conservative; it involves a modified t test, which is used to determine if the effects that are measured can be considered significant (based on the estimate of the level of normal variation present). The error in the interpretation of the results is that a change will not be made in a factor level when it should be made. Taguchi, on the other hand, says that the process should be adjusted in the direction that the factor effect suggests. There is no test of significance performed and the error in the interpretation of the data, when it occurs, is that a change will be made in a factor level when it shouldn't be made.
3. *Means versus variance.* The Plackett–Burman method concentrates solely on the comparison of means and the effects each of the factors has on the mean. Taguchi looks at the means as a method for improvement, but also looks even further at the level of variation present in the process. Taguchi places much greater emphasis on the reduction of the variation in the process. The use of a separate noise matrix in combination with a design matrix is a breakthrough in experimental design offered by Taguchi. Before Taguchi introduced the signal-to-noise ratio, screening designs were used by most experimenters to look only at the effects on the mean value. The introduction of the S/N ratio encouraged experimenters to look at effects on variation as well as effects on the mean. Today, many experimenters prefer to separate the analyses of the effects on the mean and the effects on variation. The effects on the mean are analyzed as outlined earlier using $\bar{Y}$. The effects on variation are analyzed by substituting $\log s$ or $-\log s$ for $\bar{Y}$. These analyses can be performed either with Taguchi methods or Plackett–Burman techniques. Many experimenters use the Plackett–Burman techniques for their use of statistics to determine the significance of an effect. We recommend the separation of the effects analysis and the use of the Plackett–Burman analysis techniques in most cases.
4. *Beyond the Basic DOE.* The Plackett–Burman technique was developed as a mathematical model (probably because at the time there

was a lack of computing power) and, as developed, did not have the accessory methods used by Taguchi. Accessory methods, such as the use of response graphs, selection of control factors, and classified attribute analysis, can be applied to Plackett–Burman designs as easily as to Taguchi designs. One area that Taguchi methods are more powerful is the investigation of more than two levels of a factor.

In selecting a design method, the following guidelines are recommended. Either technique can be used in most cases. Plackett–Burman designs should be used where it is desirable to statistically determine the significance of an effect. If multiple (> 2) factor levels need to be incorporated, then Taguchi methods are the best choice.

5.9 MIXTURE DESIGN TECHNIQUES

There are many experimental techniques available for handling process variables that can be varied independently of each other. Examples of such variables include processing temperature, cycle time, or material type. However, frequently the process industries deal with product composition issues that add other constraints to the experiment. The traditional high- and low-level selection method will not work in these cases because ultimately the sum of the fractions of the constituents must equal 1.0. Because of this additional constraint, special experimental methods designed to handle mixtures must be employed.

As with more traditional experimental techniques, there are different methods for handling the different test situations that arise. Response surface techniques exist for studying situations involving a small number of constituents. They provide the greatest amount of information about the effects of each constituent within the entire test. There are also screening experiments for determining the main effects of many constituents in a quick and efficient manner. These designs are helpful for determining which constituents should be included in a second experiment—most likely a response surface experiment. Finally, there are special techniques for handling mixtures where the levels of the constituents are constrained. (For example, in an experiment to optimize the levels of flour, sugar, and butter in a cake, it is not practical to vary all of the constituents from 0 to 100%; it is already known that *some* flour is needed to make a cake and that other ingredients are needed as well.)

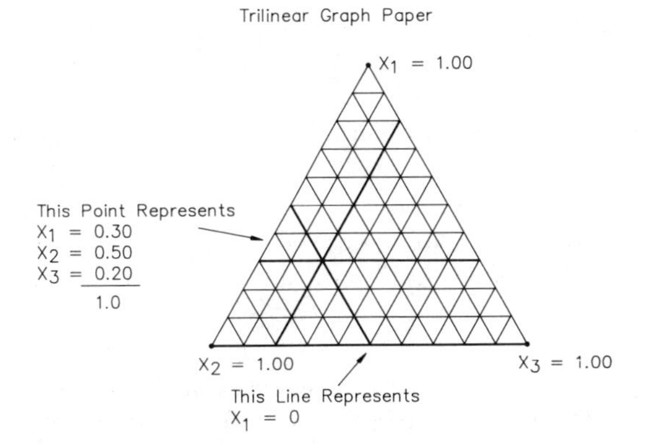

FIGURE 5.39. Triangular grid for three-component design.

The main constraint ruling mixture designs is the fact that the proportions of each of the components must satisfy the following constraints:

$$0 \leq x_i \leq 1.0$$

$$x_1 + x_2 + \cdots + x_k = 1.0$$

These constraints define an experimental region that is simplex in nature (triangular for three components, tetrahedral for four components, etc.) versus the cuboidal region used in standard factorial type designs.

Graphical Inspection

In order to view the experimental region graphically, trilinear graph paper is routinely used. Figure 5.39 illustrates the use of the trilinear coordinates for a three-component design. The region is a triangle with a component assigned to each of the vertices; each vertex has a corresponding base, located directly opposite it. The proportion of each component can be varied from zero, corresponding to the baseline, to one, corresponding to the vertex. The area in the interior of the triangle, therefore, represents combinations of all three components at varying proportions.

Trilinear graph paper is very useful for visualizing the experimental region of a three component design and it can also be used to display the region of a four component design. This is accomplished by displaying three of the components at fixed levels of the fourth component. Each level of the fourth component requires a separate trilinear graph. Then, the graphs are manually stacked to develop a three-dimensional experimental region. If it is decided to plot components x_1, x_2, and x_3 at

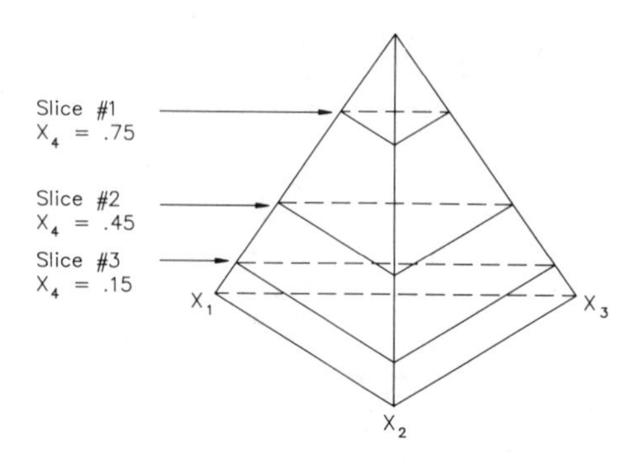

FIGURE 5.40. Plots of the experimental regions for four-component design.

various fixed levels of x_4, a new set of coordinates would have to be developed to maintain the graphical constraint that the sum of the proportions must equal 1.0. The new coordinates (x_1', x_2', and x_3') are easily calculated using the following formulas:

$$x_1' = x_1 + \frac{x_4}{3}$$

$$x_2' = x_2 + \frac{x_4}{3}$$

$$x_3' = x_3 + \frac{x_4}{3}$$

The plots generated at various levels of x_4 can then be "stacked" to visualize the experimental region of the four components. Figure 5.40 illustrates this technique.

Mixture Design Selection

As with any experiment, the ultimate goal is to obtain the maximum amount of accurate and useful results with a minimum expenditure of resources. There are a number of considerations that should be taken into account when developing the experiment strategy. The number of components to be studied is the most important consideration. If there are only a few components of interest (between two and five), a manageable experiment can be designed that will allow for a fairly complete understanding of the entire region. Response surface designs are used for nonconstrained situations (Section 5.9.1) whereas extreme vertices designs (Section 5.9.2) are the choice for constrained variables. If there are many components to be studied (greater than five) a screening experiment is more practical and should be used first to identify the key components to be included in a second, more comprehensive experiment. There are two forms of screening experiments: one for unconstrained situations (Section 5.9.3) and one for constrained regions (Section 5.9.4). The use of constraints is one of the most important aspects of the experiment that must be determined before an experimental strategy can be selected.

There are a few specialized cases that can't be overlooked when choosing an experimental strategy. In cases where there are known inert components in a formula (by inert we mean that they do not affect the properties being measured), the best strategy is to hold the level of the inert constituents constant. In doing so, the experimental region can be restricted to the active components and the total number of experiments will be minimized. In some cases, mixture designs may not be necessary when dealing with component proportions where the majority of the mixture is made up of one component and the rest of the components are only present in small amounts (less than 5%). Here, is usually sufficient to use a factorial design and vary the levels of the trace components independently. Any differences in the trace component proportions are adjusted by changing the level of the major component (also called the "slack" variable).

Once the experimenter has identified the components to be tested, has determined if there are any constraints on any of the proportions, and has reviewed the possibility of the existence of any of the special situations cited in the preceding discussion, a decision can be made about the experimental strategy. There are a number of variations of each design type involving different numbers of experimental runs, different levels of repetition, and varying degrees of theoretical model check points. One design option from each design type will be discussed in some depth in subsequent sections. This will be followed by a brief overview of some of the other design options.

The reader should keep in mind that much of the material presented here describes how to design an experiment to collect data needed to develop a mathematical model. The actual solution of the model, for the most part, often requires and is greatly aided by the use of computer programs. There are many good commercial software packages available on the market to handle these designs.

5.9.1 Mixture Response Surface Designs

The most commonly used strategy for experimenting with mixtures is the response surface design. This approach is used in situations where there are fewer than six components to be studied and there are no constraints on the component proportions other than the two universal constraints that are common to all mixture designs:

$$0 \le x_i \le 1.0$$

$$x_1 + x_2 + \cdots + x_k = 1.0$$

The use of the response surface design allows the experimenter to develop a mathematical model that can predict the responses in the entire experimental region. This approach is no different from the use of models with other experimental methods except that the mixture constraints previously noted simplify the traditional mathematical models, allowing for the use of fewer coefficients. For example, in the simplest case, a linear model with two variables (constituents), the analysis starts with the equation for independent variables:

$$Y = b_0 + b_1 x_1 + b_2 x_2$$

By imposing the mixture constraint $x_1 + x_2 = 1$, the equation becomes

$$Y = b_0(x_1 + x_2) + b_1 x_1 + b_2 x_2$$

$$Y = b_0 x_1 + b_0 x_2 + b_1 x_1 + b_2 x_2$$

$$Y = b_1' x_1 + b_2' x_2$$

where $b_1' = b_0 + b_1$
$b_2' = b_0 + b_2$

Similar reductions can be made to the quadratic and cubic models commonly used with these designs. The models for the two- and three-component mixture models are given in Table 5.76. Another important aspect of the mathematical model is that the quadratic and special cubic terms do not represent interaction effects. Instead, they correspond to blending effects. In other words, the significant coefficients of the quadratic terms indicate that there is a quadratic deviation from the linear response surface. Similarly, significant coefficients in the special cubic equation indicate a cubic deviation from the quadratic and linear response surfaces.

With an understanding of the mathematical models involved, the experiment can be designed. For a linear model, only the pure components

Table 5.76 Models for Two- and Three-Component Mixture Experiments

Two components	
Linear	$y = b_1x_1 + b_2x_2$
Quadratic	$y = b_1x_1 + b_2x_2 + b_{12}x_1x_2$
Special cubic	N/A
Three components	
Linear	$y = b_1x_1 + b_2x_2 + b_3x_3$
Quadratic	$y = b_1x_1 + b_2x_2 + b_3x_3 + b_{12}x_1x_2 + b_{13}x_1x_3 + b_{23}x_2x_3$
Special cubic	$y = b_1x_1 + b_2x_2 + b_3x_3 + b_{12}x_1x_2 + b_{13}x_1x_3 + b_{23}x_2x_3 + b_{123}x_1x_2x_3$

need to be included in the experiment. Normally, a quadratic or a cubic model should be used to accurately describe the total response surface. Therefore, an experiment should be designed to allow for the calculation of at least the quadratic coefficients and, preferably, calculation of the special cubic coefficients. To describe this in more detail, a three-component simplex will be used as an example. All of the steps taken with the three-component design can also be taken with the four- and five-component designs. (Those designs are given in Appendix F.)

In order to satisfy the requirements of good experimentation and to generate enough information to calculate the necessary coefficients, the experimental points need to cover the entire experimental region, should be replicated/repeated sufficiently to understand the experimental error and sensitivity of the experiment, and should offer a way to check the accuracy of the proposed mathematical model. In order to achieve all of these objectives, design points can be selected as shown in Table 5.77.

Table 5.77 Designated Points to Cover the Entire Experimental Region

	x_1	x_2	x_3	Y
1	1	0	0	Y_1
2	0	1	0	Y_2
3	0	0	1	Y_3
4	1/2	1/2	0	Y_{12}
5	1/2	0	1/2	Y_{13}
6	0	1/2	1/2	Y_{23}
7	1/3	1/3	1/3	Y_{123}
8	2/3	1/6	1/6	Y_{1123}
9	1/6	2/3	1/6	Y_{1223}
10	1/6	1/6	2/3	Y_{1233}

In Table 5.77, blends 1, 2, and 3 represent the pure components. Blends 4, 5, and 6 represent the combinations of two components at a time. Blend 7, also called the centroid, is often repeated two or three times to get a better estimate of experimental error. Blends 8, 9, and 10 are called check blends and are used to check the accuracy of the model.

To solve for the coefficients of the model, the following equations are used:

$$b_1 = Y_1$$

$$b_2 = Y_2$$

$$b_3 = Y_3$$

$$b_{12} = 4Y_{12} - 2(Y_1 + Y_2)$$

$$b_{13} = 4Y_{13} - 2(Y_1 + Y_3)$$

$$b_{23} = 4Y_{23} - 2(Y_2 + Y_3)$$

$$b_{123} = 27Y_{123} - 12(Y_{12} + Y_{13} + Y_{23}) + 3(Y_1 + Y_2 + Y_3)$$

Once the model has been determined, the check blends (8, 9, and 10) are used to determine how well the model fits the data. A simple F test is performed comparing the variance of the lack of fit s_{LOF}^2 to the error variance of the average response (experimental error). The lack of fit variance is given by

$$s_{\text{LOF}}^2 = \frac{1}{k}\left[\left(\bar{Y}_{m1} - Y_{p1}\right)^2 + \left(\bar{Y}_{m2} - Y_{p2}\right)^2 + \cdots + \left(\bar{Y}_{mi} - Y_{pi}\right)^2\right]$$

where k = number of check blends (three in this case)
$\bar{Y}_{mi}$ = measured response of check blend i
Y_{pi} = response of point i as predicted by the model

The value for s_{LOF}^2 is compared to s_{error}^2, which can be calculated using the equation

$$s_{\text{error}}^2 = \left[\frac{(n_1 - 1)s_1^2 + (n_2 - 1)s_2^2 + \cdots + (n_k - 1)s_k^2}{(n_1 - 1) + (n_2 - 1) + \cdots + (n_k - 1)}\right] \Big/ k$$

where n_k = observations made for every check blend
k = number of check blends
s_k = standard deviation of sample for each check blend

If the F ratio ($F = s^2_{\text{LOF}}/s^2_{\text{error}}$) does not exceed the tabulated F value for k and υ degrees of freedom (where k number of blend checks and $\upsilon = \Sigma(n_k - 1)$, then the model can be used with a high degree of confidence that the conclusions drawn from the results are correct. (See Section 5.4.5 for a more detailed discussion about F tests.)

Pseudocomponents

In some cases, the components used are actually mixtures of several other components. In these cases, it may be useful to manipulate the data in order to present them in terms of pure components. The pure components can be found by using the simple formula

$$z_j = a_j + [1 - (a_1 + a_2 + \cdots + a_n)]x_j$$

where z_j = pure component proportion
x_j = pseudocomponent proportion
a_j = lower bound of j component
a_n = lower bounds of components through n (this includes a_j)

Assume that the components x_1, x_2, and x_3 were in fact made up of the combinations

$$x_1 = 0.20z_1 + 0.40z_2 + 0.40z_3$$

$$x_2 = 0.10z_1 + 0.70z_2 + 0.20z_3$$

$$x_3 = 0.40z_1 + 0.10z_2 + 0.50z_3$$

Through experimentation, we found that the optimal levels of components x_1, x_2, and x_3 were:

$$x_1 = 0.20$$

$$x_2 = 0.60$$

$$x_3 = 0.20$$

The lower bounds of the pure components (a_j) were:

$$a_1 = 0.10$$

$$a_2 = 0.10$$

$$a_3 = 0.20$$

Knowing the optimal values of x_j and the lower bounds a_j, we can solve for z_j:

$$z_1 = 0.10 + (1 - 0.40)(0.20) = 0.22$$

$$z_2 = 0.10 + (1 - 0.40)(0.60) = 0.46$$

$$z_3 = 0.20 + (1 - 0.40)(0.20) = 0.32$$

So, to produce the optimum blend using pure components, the combination $z_1 = 0.22$, $z_2 = 0.46$, and $z_3 = 0.32$ would be used.

This transformation can also be carried over to the mixture model. The coefficients of the equation can either be expressed in terms of pure components or pseudocomponents. First, solve the transformation formula for the pseudocomponents:

$$x_{ij} = \frac{z_{ij} - a_j}{1 - (a_1 + a_2 + \cdots + a_n)}$$

For this case the equations are

$$x_1 = \frac{z_1 - 0.10}{(1 - 0.4)}$$

$$x_2 = \frac{z_2 - 0.10}{(1 - 0.4)}$$

$$x_3 = \frac{z_3 - 0.20}{(1 - 0.4)}$$

These values for x_1, x_2, and x_3 can be substituted into the model equation.

It is easier to understand the mixture response surface methodology by working through an example with three components.

In an example, a local pub is trying to develop a new house drink in an attempt to draw in new business. The owner, who owns stock in a liquor company, has established that the drink will only contain three liquors:

A	vermouth
B	gin
C	bourbon

The head bartender is given the responsibility to come up with a mixture of these three liquors that will taste the best. The bartender, who happens

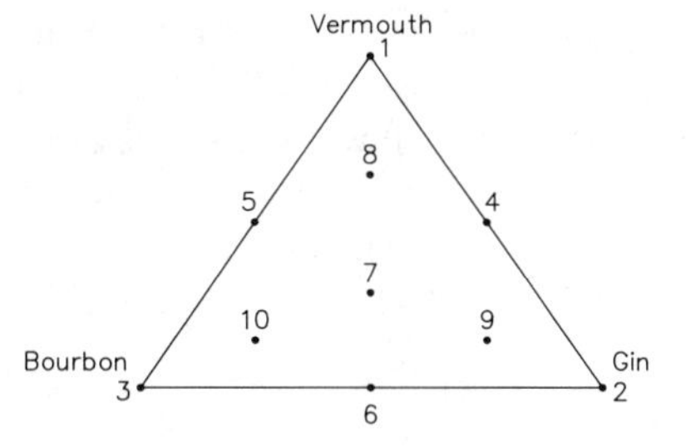

FIGURE 5.41. Mixture design for mixed drink experiment.

to have a college degree in statistics, decides to use a mixture experiment to determine which combination tastes best using the fewest trials. Some of the regular customers are asked to rate the taste of each drink on a scale of 1 (worst) to 100 (best), and the bartender will average their results to come up with the responses.

Table 5.78 Drink Design Proportions

Drink	Vermouth	Gin	Bourbon
1	1	0	0
2	0	1	0
3	0	0	1
4	1/2	1/2	0
5	1/2	0	1/2
6	0	1/2	1/2
7	1/3	1/3	1/3
8	2/3	1/6	1/6
9	1/6	2/3	1/6
10	1/6	1/6	2/3

Table 5.79 Drink Design Proportions with Responses

Drink	Vermouth	Gin	Bourbon	Y_a	Y_b	Y_c	$\bar{Y}$
1	1	0	0	10	7	11	9.3
2	0	1	0	23	25	20	22.7
3	0	0	1	31	28	26	28.3
4	1/2	1/2	0	18	23	23	21.3
5	1/2	0	1/2	31	30	26	29.0
6	0	1/2	1/2	55	60	45	53.3
7	1/3	1/3	1/3	72	75	73	73.3
8	2/3	1/6	1/6	18	25	23	22.0
9	1/6	2/3	1/6	43	38	42	41.0
10	1/6	1/6	2/3	90	88	86	88.0

The bartender decides to use the design in Figure 5.41. Ten drinks are mixed as shown in Table 5.78. Portions of the 10 drinks are given to each of three judges and they write down their scores. The bartender then consolidates the scores in one table and calculates the averages (Table 5.79). Now, the coefficient for the special cubic model can be found using the calculations:

$$b_1 = \bar{Y}_1 = 9.3$$

$$b_2 = \bar{Y}_2 = 22.7$$

$$b_3 = \bar{Y}_3 = 28.3$$

$$b_{12} = 4(\bar{Y}_{12}) - 2(\bar{Y}_1 + \bar{Y}_2) = 4(21.3) - 2(9.3 + 22.7) = 21.2$$

$$b_{13} = 4(\bar{Y}_{13}) - 2(\bar{Y}_1 + \bar{Y}_3) = 4(29) - 2(9.3 + 28.3) = 40.8$$

$$b_{23} = 4(\bar{Y}_{23}) - 2(\bar{Y}_2 + \bar{Y}_3) = 4(53.3) - 2(22.7 + 28.3) = 111.2$$

$$\begin{aligned} b_{123} &= 27(\bar{Y}_{123}) - 12(\bar{Y}_{12} + \bar{Y}_{13} + \bar{Y}_{23}) + 3(\bar{Y}_1 + \bar{Y}_2 + \bar{Y}_3) \\ &= 27(73.3) - 12(21.3 + 29.0 + 53.3) + 3(9.3 + 22.7 + 28.3) \\ &= 916.8 \end{aligned}$$

The model is, therefore,

$$\begin{aligned} Y = {} & 9.3x_1 + 22.7x_2 + 28.3x_3 + 21.2x_1x_2 + 40.8x_1x_3 \\ & + 111.2x_2x_3 + 916.8x_1x_2x_3 \end{aligned}$$

Using the check blends (drinks 8, 9, and 10), the bartender tests the model.

For drink 8:

$$x_1 = 0.667$$

$$x_2 = 0.167$$

$$x_3 = 0.167$$

$$\begin{aligned} Y_{\text{model}} = {} & (9.3)(0.667) + (22.7)(0.167) + (28.3)(0.167) \\ & + (21.2)(0.667)(0.167) + (40.8)(0.667)(0.167) \\ & + (111.2)(0.167)(0.167) \\ & + (916.8)(0.667)(0.167)(0.167) \end{aligned}$$

$$Y_{\text{model}} = 41.8$$

For drink 9:

$$x_1 = 0.167$$

$$x_2 = 0.667$$

$$x_3 = 0.167$$

$$Y_{\text{model}} = 54.4$$

For drink 10:

$$x_1 = 0.167$$

$$x_2 = 0.167$$

$$x_3 = 0.667$$

$$Y_{\text{model}} = 58.8$$

To check the fit of the model, the bartender calculates the variance of the measurements to the model and compared that to the error variance. Variance due to lack of fit:

$$s_{\text{LOF}}^2 = \frac{1}{k}\left[\left(\bar{Y}_{m_8} - Y_{p_8}\right)^2 + \left(\bar{Y}_{m_9} - Y_{p_9}\right)^2 + \left(\bar{Y}_{m_{10}} - Y_{p_{10}}\right)^2\right]$$

$$= \frac{1}{3}\left[(22 - 41.8)^2 + (41 - 54.4)^2 + (88 - 58.8)^2\right]$$

$$= \frac{1}{3}(392 + 179.6 + 852.6)$$

$$s_{\text{LOF}}^2 = 474.7$$

where k = number of check blends
$\bar{Y}_m$ = measured response
Y_p = predicted response

Variance due to error:

$$s_{\text{error}}^2 = \left[\frac{(n_8 - 1)s_8^2 + (n_9 - 1)s_9^2 + (n_{10} - 1)s_{10}^2}{(n_8 - 1) + (n_9 - 1) + (n_{10} - 1)} \right] \Big/ k$$

$$= \left[\frac{2(3.6)^2 + 2(2.65)^2 + (2)(2)^2}{2 + 2 + 2} \right] \Big/ 3$$

$$s_{\text{error}}^2 = 2.66$$

where k = number of check blends
n_k = number of samples per check blend
s_k = standard deviation of data for each check blend

Then an F test is performed:

$$F_{\text{calc}} = \frac{s_{\text{LOF}}^2}{s_{\text{error}}^2} = \frac{474.7}{2.66} = 178.5$$

The error due to lack of fit is significant, so the model is *not* adequate for predicting the experimental region. What can the bartender do now? There are two choices.

1. Choose the best drink out of the experimental drinks (drink 10).
2. Try to fit the data to a different model such as quadratic, linear, or cubic.

The bartender decides to try the quadratic model:

$$Y = b_1 x_1 + b_2 x_2 + b_3 x_3 + b_{12} x_1 x_2 + b_{13} x_1 x_3 + b_{23} x_2 x_3$$

Again the blends are checked to determine if the model works.
Quadratic model:

s	$\bar{Y}_m$	Y_p
Drink 8	22	24.8
Drink 9	41	37.3
Drink 10	88	41.7

$$s_{\text{LOF}}^2 = \tfrac{1}{3}\left[(22 - 24.8)^2 + (41 - 37.3)^2 + (88 - 41.7)^2\right]$$

$$= \tfrac{1}{3}(7.84 + 13.69 + 2143)$$

$$= 721.5$$

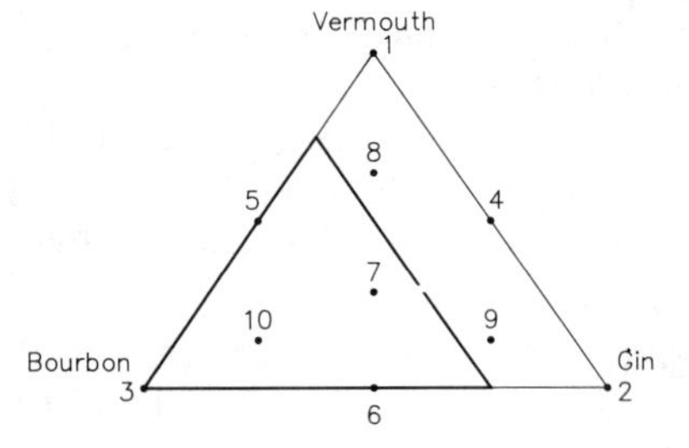

FIGURE 5.42. Region for future experimentation.

Still, the F test indicates a lack of fit:

$$\frac{s^2_{\text{LOF}}}{s^2_{\text{error}}} = \frac{721.5}{2.66} = 271.2$$

The model seems to do well in most of the region (it predicts drinks 8 and 9 fairly well), but check drink 10 is nowhere near the predicted value. At this point, the bartender chooses drink 10, because it did represent the best combination in the experiment.

What else could have been done?

1. A few more drink combinations could have been added to the experiment.
2. Linear or cubic models could have been tried for fit.
2. An experiment could have been run in a smaller region to focus on the area around drink 10, which, because of its high response values, seems interesting (Figure 5.42).

It's probably better that the bartender wait, however, to see what the reaction of more customers is to drink 10 before conducting further experiments.

5.9.2 Extreme Vertices Designs

In many instances when dealing with mixtures, one or more of the components has both an upper and lower constraint. This causes the experimental region to be different from the simplex configuration. (Note: If only the lower constraints exist, the region can still be described using a simplex and the techniques described in Section 5.9.1 should be used.) Establishing the experimental strategy and identifying the proper blends to describe the region becomes much more complicated when an upper constraint is introduced as well.

The first thing to do is to understand the component upper and lower bounds and use them to define the experimental region. To show this, a three-component example will be discussed first, followed by a discussion on a four-component experiment.

In the three component example, the goal of the experiment is to optimize a product that is comprised of three components with the following constraints:

$$0.3 \leq x_1 \leq 0.9$$

$$0.1 \leq x_2 \leq 0.4$$

$$0.2 \leq x_3 \leq 0.8$$

It is best to construct a triangle diagram to visualize this (see Figure 5.43). The heavy black line is the outline of the experimental region that satisfies all of the constraints. With three components, there is a total of 12 "constraint plane intersections" calculated from the equation:

$$\text{Constraint plane intersections} = n * 2^{n-1}$$

where n is the number of components.

This means there are 12 possible combinations of the three components, however, not all 12 will fall into the allowable experimental region as determined by the component constraints. It is possible to see that there will only be four extreme vertices because there are only four vertices in the experimental region. To find the blends describing the vertices, the following algorithm, described by McLean and Anderson

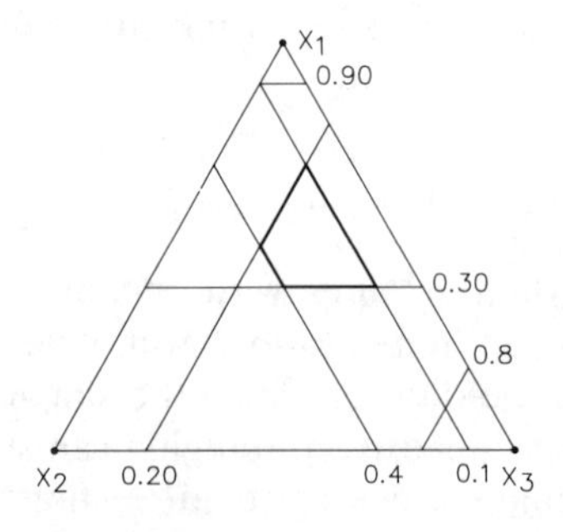

FIGURE 5.43. Extreme vertices design plot.

(1966), is used:

1. List all possible factorial combinations using lower and upper bounds for each component except the last one. This will generate 2^{n-1} points (n = number of components). Repeat this leaving x_{n-1} blank, x_{n-2} blank, and so on until you have $n * 2^{n-1}$ combinations.
2. Fill in the level of the component that was left blank so that the total of each line is 1.0.
3. Check the levels of all components versus the constraints to determine which points satisfy the experimental region.
4. Eliminate duplicates at this point. Replicate blends can be chosen later.

For the three-component example given in Figure 5.43 the first step is to list all the possible factorial combinations. As stated earlier, there are 12 intersections, so there will be 12 combinations as shown in Table 5.80. Then, using the lower constraints (−) and upper constraints (+), plug in the component levels for each line. At this point, one component is still left blank as shown in Table 5.81.

Next, fill in the blank component so that each line will add up to 1.0. Table 5.82 shows the completed component levels.

Finally, look at each case and determine if all points satisfy all constraints. Obviously, each combination containing a negative value is invalid. Also, eliminate any duplicates. As was seen in Figure 5.43, there are

Table 5.80 List of Possible Factorial Combinations for Three-Component Design

	Factorial Combinations			Component Levels[a]		
	x_1	x_2	x_3	x_1	x_2	x_3
1	−	−				
2	+	−				
3	−	+				
4	+	+				
5	−		−			
6	+		−			
7	−		+			
8	+		+			
9		−	−			
10		+	−			
11		−	+			
12		+	+			

[a] $0.3 \leq x_1 \leq 0.9$; $0.1 \leq x_2 \leq 0.4$; $0.2 \leq x_3 \leq 0.8$.

Table 5.81 Addition of Upper and Lower Constraint Values

	Factorial Combinations			Component Levels		
	x_1	x_2	x_3	x_1	x_2	x_3
1	−	−		0.3	0.1	
2	+	−		0.9	0.1	
3	−	+		0.3	0.4	
4	+	+		0.9	0.4	
5	−		−	0.3		0.2
6	+		−	0.9		0.2
7	−		+	0.3		0.8
8	+		+	0.9		0.8
9		−	−		0.1	0.2
10		+	−		0.4	0.2
11		−	+		0.1	0.8
12		+	+		0.4	0.8

Table 5.82 Complete Design for Three-Component Three-Level Mixture Experiment[a]

	Factorial Combinations			Component Levels			
	x_1	x_2	x_3	x_1	x_2	x_3	
1	−	−		0.3	0.1	0.6*	(3)
2	+	−		0.9	0.1	0.0	
3	−	+		0.3	0.4	0.3*	(2)
4	+	+		0.9	0.4	−0.3	
5	−		−	0.3	0.5	0.2	
6	+		−	0.9	−0.1	0.2	
7	−		+	0.3	−0.1	0.8	
8	+		+	0.9	−0.7	0.8	
9		−	−	0.7	0.1	0.2*	(4)
10		+	−	0.4	0.4	0.2*	(1)
11		−	+	0.1	0.1	0.8	
12		+	+	−0.2	0.4	0.8	

[a]The four vertices are indicated by asterisks.

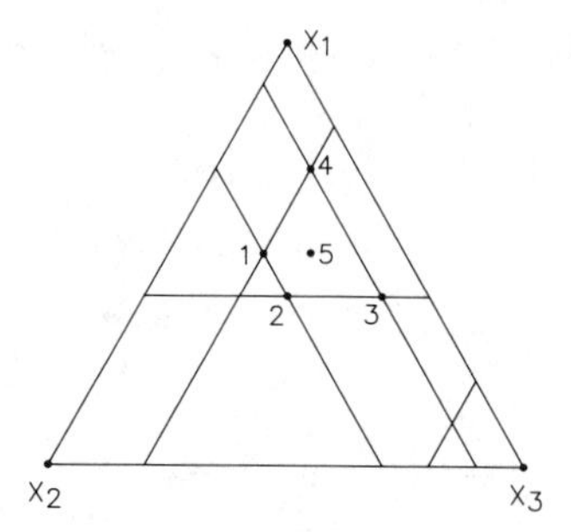

FIGURE 5.44. Centroid added to extreme vertices design plot.

only four vertices remaining: they are identified in the table with an asterisk. These blends should be included in the experiment and, as with other mixture designs, the centroid should also be included in the experiment. The centroid is found by taking the average of all of the vertices:

$$x_{1C} = \frac{0.3 + 0.3 + 0.7 + 0.4}{4} = 0.425$$

$$x_{2C} = \frac{0.1 + 0.4 + 0.1 + 0.4}{4} = 0.25$$

$$x_{3C} = \frac{0.6 + 0.3 + 0.2 + 0.2}{4} = 0.325$$

The centroid can be added to the diagram (Figure 5.44).

Even with the extreme vertices and the centroid calculated, the design is not complete until we review it to assure that it is practical and feasible. In some cases, two vertices will be so close together that the most practical and economic approach may be to use an average of the two and only test one blend. This is the case in this three-component example. The five bends that have been identified so far have been numbered in Figure 5.44. Vertices 1 and 2 are close together and could conceivably be replaced by a single blend (the averages of 1 and 2). The new blend would be

$$x_1 = \frac{0.4 + 0.3}{2} = 0.35$$

$$x_2 = \frac{0.4 + 0.4}{2} = 0.40$$

$$x_3 = \frac{0.3 + 0.2}{2} = 0.25$$

This is designated by an $\times$ on Figure 5.45.

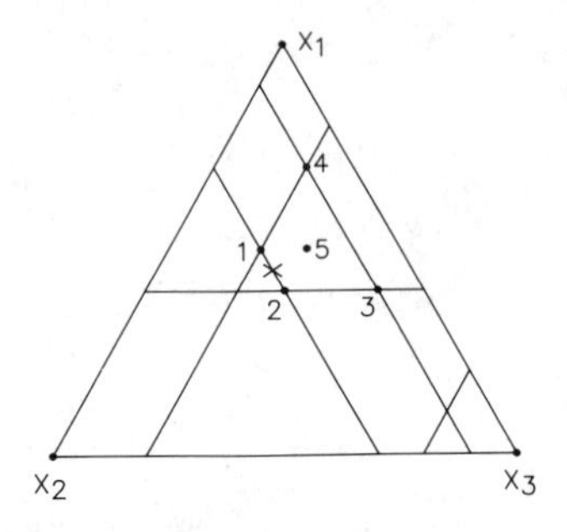

FIGURE 5.45. Averaging close blends to form a single blend for experimentation.

Another consideration is the distance between each of the vertices. It is not desirable to have the points spread too far apart as this would leave large areas of the experimental region unrepresented by the trial blends.

With three component designs, the long edges (the long sides of the polygon defining the region) can easily be seen by plotting the vertices on tilinear paper. Midpoints of the long edges are found by taking the averages of the surrounding vertices. For example, the experimenter may decide to add a midpoint to the line between vertices 3 and 4 on Figure 5.46. The resulting point (represented by a box on the diagram) is

$$x_1 = \frac{0.3 + 0.7}{2} = 0.5$$

$$x_2 = \frac{0.1 + 0.1}{2} = 0.1$$

$$x_3 = \frac{0.6 + 0.2}{2} = 0.4$$

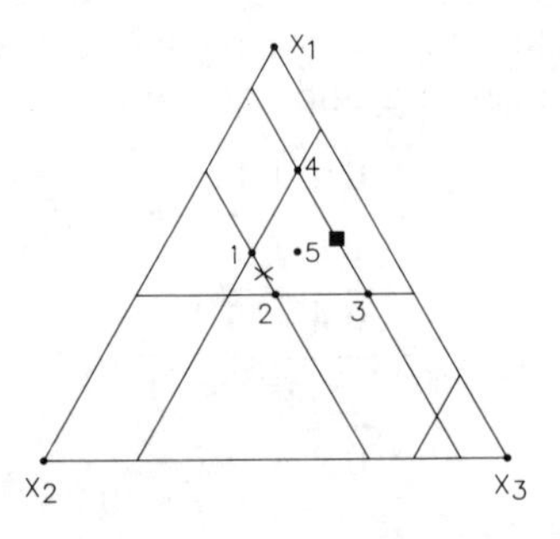

FIGURE 5.46. Addition of a midpoint between vertices 3 and 4.

Unfortunately, in experiments with more than three components, it is not possible to visually determine where there are long edges. In those cases, the lengths of the edge can be determined mathematically using the formula

$$d_{ij}^2 = \left[\frac{x_{1i} - x_{1j}}{b_1 - a_1}\right]^2 + \left[\frac{x_{2i} - x_{2j}}{b_2 - a_2}\right]^2 + \cdots + \left[\frac{x_{ni} - x_{nj}}{b_n - a_n}\right]^2$$

where a_i = minimum level of x_i
b_i = maximum level of x_i
d_{ij} = distance between vertices i and j

For this three-component example, the distance between vertices 3 and 4 can be calculated:

$$d_{34}^2 = \left[\frac{0.3 - 0.7}{0.9 - 0.3}\right]^2 + \left[\frac{0.1 - 0.1}{0.4 - 0.1}\right]^2 + \left[\frac{0.6 - 0.2}{0.8 - 0.2}\right]^2$$

$$= 0.444 + 0 + 0.444 = 0.888$$

In a complicated design, all edges should be calculated and those that are considerably longer (50% longer than any others) than the others should be supplemented with a midpoint.

Another issue that has to be addressed to set up an experiment with four or more components is the inclusion of all of the constraint plane centroids. In the three-component example, there was only one constraint plane formed by the vertices 1, 2, 3, and 4. The centroid was the average of those four points. With four or more components, the experimental region has more than one constraint plane. The planes must be identified according to the vertices and there will be a maximum of $2n$ constraint planes. Each centroid can be calculated using the average of the vertices forming each plane. The overall centroid is then found by taking the average of all of the vertices.

Analysis of Results

Because of the complexity of the extreme vertices design, it is very difficult to calculate results by hand, determine the model, and test for fit. As mentioned in Section 5.9, computer programs are available that will do this.

5.9.3 Mixture Screening Experiments

In many instances, it is not easy to identify the three or four components in a mixture that affect the critical properties. A long list of components may be generated. Response surface experiments are powerful tools when dealing with five or fewer components, but are not well suited for six or more components because many experiments would be required. In those cases, a screening experiment is performed that allows the determination of which components have major and significant effects on the properties of the mixture. With that information (and after limited experimentation), the identified components can be included in a response surface experiment to optimize the region. Those components not found to have a significant effect on the critical properties can be ignored in further experimentation.

The reason that more components can be studied in a screening experiment with a minimum amount of experimentation is that a linear model, rather than a quadratic or special cubic model, is used to "screen" the major component effects. A graphical depiction for a three-component case is given in Figure 5.47.

The linear behavior of each component is determined and then the coefficients are determined. The first-degree (linear) model is

$$Y = b_1x_1 + b_2x_2 + b_3x_3 + \cdots + b_nx_n$$

The steps in setting up, executing, and analyzing a mixture screening design are straightforward:

1. Identify the components (factors) to be tested ($n \geq 6$).
2. Decide whether the addition of even a small amount of the components may affect the properties of the mixture.

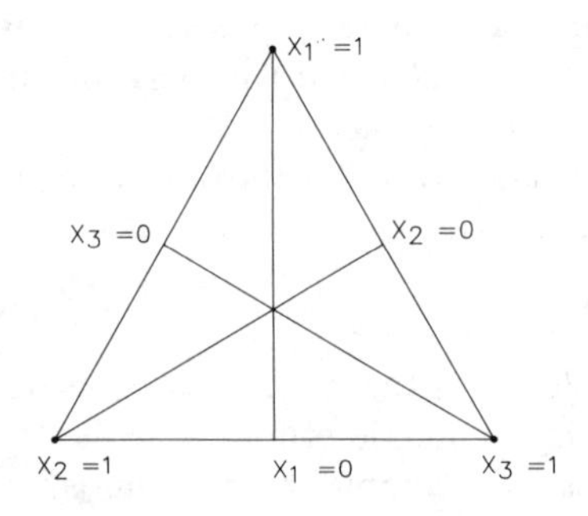

FIGURE 5.47. Three-component mixture screening experiment graph.

3. Choose a screening design.
   - VIC design (a design that includes vertex and interior points and the centroid) for no end effects.
   - VICE (a VIC design plus a study of the end points) to include end effects.
4. Determine test blends needed for experimentation.
   - $2n + 1$ for VIC.
   - $3n + 1$ for VICE.
5. Perform the experiment.
6. Plot the results.
7. Calculate regression coefficients and component effects.
8. Identify components with major effects to use in a response surface experiment.

A major decision point is step 2. When dealing with a case where even a very small amount of the components will affect the measured property, it is advised to include the points on the baseline (or the end points) for the component. Otherwise, only the vertices, interior points, and overall centroid (a VIC design) need to be included (see Figure 5.48).

VICE designs often aren't needed when studying many components. As the number of components increases, the overall centroid will move closer to the baseline, thus eliminating the need to include the end effects. The design points themselves are simple to determine for any number of components (see Table 5.83). With the design blends determined, their order should be randomized, and then the responses can be measured.

With the response data in hand, the next step is to graph the results. The response are plotted on the y axis and the component fractions are plotted along the x axis. If all four points (VICE) are used, there will be four points plotted for each component along the x axis. One point (the

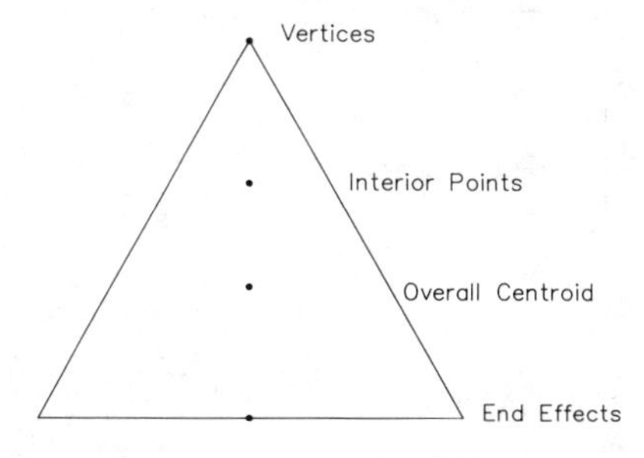

FIGURE 5.48. Design points where small amounts of component will cause a response.

Table 5.83 Mixture Screening Experiment Design Points

Number	Composition				
	x_1	x_2	x_3	...	x_n
1	1	0	0		0
2	0	1	0		0
3	0	0	1		0
⋮					
n	0	0	0		1
1	$(n+1)/2n$	$1/2n$	$1/2n$		$1/2n$
2	$1/2n$	$(n+1)/2n$	$1/2n$		$1/2n$
⋮					
n	$1/2n$	$1/2n$	$1/2n$		$(n+1)/2n$
1	$1/n$	$1/n$	$1/n$		$1/n$
1	0	$1/(n-1)$	$1/(n-1)$		$1/(n-1)$
2	$1/(n-1)$	0	$1/(n-1)$		$1/(n-1)$
⋮					
n	$1/(n-1)$	$1/(n-1)$	1/b		0

centroid) is shared by all components. Based on the slopes or the curvature of the lines, conclusions can be drawn about the effect of each component on the measured response. For example, Figure 5.49 shows a sample plot for a 10-factor screening design.

The results indicate that components 3 and 6 have a strong positive effect, factors 1, 2, 7, 5, and 9 have marginal effects (negligible), and factors 4, 8, 10 have strong negative effects on the response. With this information, the experimenter has enough information to make a decision

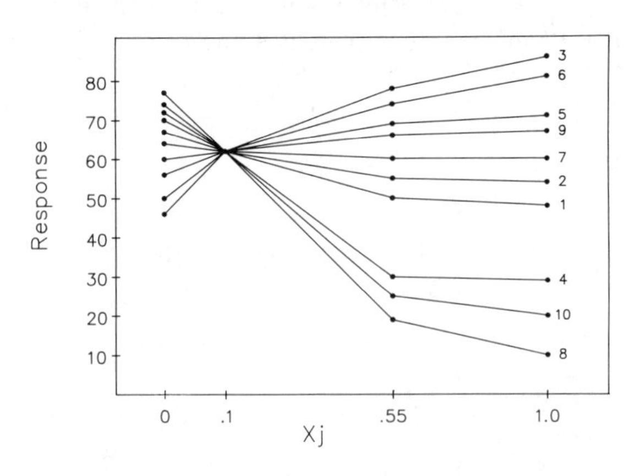

FIGURE 5.49. Response plot for a 10-factor screening experiment.

about further response surface experimentation design. In order to maximize the response, it would make sense to leave out or minimize components 4, 8, 10 because they have a negative effect on the response. Components 1, 2, and 7 may be held constant because they had negligible negative effects. Components 3, 6, 5, and 9 could be incorporated into a four-component response surface experiment.

Mathematical confirmation of the graphical results is recommended. The coefficients of the linear blending model are determined using multiple regression, usually with the help of a computer. Once the coefficients b_i are found, the factor effects E_i can be calculated:

$$E_i = (n - 1)^{-1}[nb_i(b_1 + b_2 + \cdots + b_n)]$$

where n = number of components and
i = component number

Using t values, the confidence limits of the effects can be calculated, and those results can be used to determine which components to include in the next phase of experimentation.

The equation for the confidence limits of an effect is

$$E_i \pm ts\sqrt{\frac{n}{(n-1)\left(\frac{5}{4} + (n-1)^{-2}\right)}}$$

where t = t table factor for $\alpha = 0.05$
s = experimental error standard deviation
n = number of components

Using this equation, confidence limits can be established for all of the components. The confidence limits represent the range that includes the true effect of each factor. If any of the confidence limit ranges include zero, it is safe to assume that the corresponding factors have no effect.

5.9.4 Extreme Vertices Screening Designs

As with unconstrained situations, it is not practical to run a full response surface design when there are many constrained components (six or more). In these cases, a screening design is recommended to investigate large numbers of components and determine which have a significant effect in order to justify their inclusion in a smaller response surface design. In extreme vertices screening designs, as with unconstrained designs, the ability to investigate a large number of components with a practical number of experimental blends is possible because only a linear blending model is used. Also, when doing an extreme vertices screening design, only certain vertices need to be tested to collect enough data to fit the linear model.

In Section 5.9.2, where constrained response surface designs are discussed, a method for calculating the vertices of the experimental region, developed by McLean and Anderson (1966), was presented. The same method can be used to determine the experimental blends for screening design. However, an easier method that requires fewer calculations is available: the XVERT algorithm. There are other algorithms (CADEX, Wynn-Exchange), but for the purpose of this section only, the XVERT algorithm will be reviewed.

There are a number of different types of computer software available for both designing these experiments and for analyzing the data, and the use of computers is recommended to perform the linear regression calculations. This chapter concentrates on fitting linear models in extreme vertices screening designs. The same methods are used to design experiments to fit quadratic models. The only difference with quadratic models is that more experimental blends must be tested from the list of possible blends generated by the algorithm.

A simple three-component example will be used to demonstrate the XVERT algorithm. The constraints are shown in Table 5.84.

The main steps in the XVERT algorithm are:

1. Rank the components by their range values from lowest to highest:

Component	Range
1 (A)	0.4
3 (B)	0.4
2 (C)	0.6

The components are relabeled with letters starting with A by their range order.

Table 5.84 Constraints for Three-Component Design Example

Component	Minimum	Maximum	Range
1	0.1	0.5	0.4
2	0.3	0.9	0.6
3	0.2	0.6	0.4

Table 5.85 Combinations of Two Factors at Two Levels

	A	
B	+	−
+	+ +	− +
−	+ −	− −

From *Technometrics*, Volume 18, Screening concepts and designs for experiments with mixtures, by R: D. Snee and D. W. Marquardt.

Table 5.86 Matrix for Level of Factors A and B

Blend	A	B
1	0.5	0.6
2	0.5	0.2
3	0.1	0.6
4	0.1	0.2

2. Develop a two-level factorial design using the first $n - 1$ components. Plackett–Burman matrices can also be used for large screening designs. Plug in the high and low values as indicated by the matrix.

   For this example, where $n = 3$ components, a two-level factorial must be developed for two $(n - 1)$ components.

   There are four blends of components A and B in a 2^2 experiment. These blends are shown in Table 5.85. Plug in the high and low levels of each component corresponding to the design as shown in Table 5.86.
3. Calculate the values for the nth component so that the sum of all components in each blend is 1.0. These values for component C are shown in Table 5.87.

Table 5.87 Mixture Screeing Experiment Design Showing Calculated Value for Factor C[a]

Blend	A	B	C (Calculated)
1	0.5	0.6	−0.1
2	0.5	0.2	0.3*
3	0.1	0.6	0.3*
4	0.1	0.2	0.7*

[a]Asterisks indicate the core blends that meet all constraints.

Table 5.88 Mixture Screening Experiment Design—Calculated Subsets

Blend	A	B	C	
1	0.5	0.6	−0.1	Lower limit of C 0.3 correction is +0.4
1a	0.1	0.6	0.3	Corrected A by −0.4
1b	0.5	0.2	0.3	Corrected B by −0.4

4. Identify the core points (those that satisfy all of the constraints for each component). The points that satisfy the constraint for C ($0.3 \leq C \leq 0.9$) are considered the core points. These are the points for blends 2, 3, and 4 and are indicated with asterisks in Table 5.87.
5. For those blends where the calculated value for component X_n is outside its constraint levels, set the value of X_n to either its upper or lower limit (whichever is closest) and correct the other components *one at a time* by the same amount that X_n was changed as shown in Table 5.88. This forms a subset.
6. For each subset, determine which points are vertices by comparing the component values to their constraints. All bends that meet the constraints are valid. If only one blend is found in a subset, add it to the list of core points.

   1a meets all constraints
   1b meets all constraints

   Both could be included in the experiment, but because they are repeats of blends 2 and 3, they can be discarded. In large designs with many subsets, one blend is chosen from each subset to keep the experiment manageable.
7. Calculate the centroid by taking the average for each component for all the blends to be included in the experiment (Table 5.89). The XVERT algorithm identified four blends for the three-component example (see Figure 5.50).

Table 5.89 Finalized Design for Mixture Screening Experiment with Centroid

	A	B	C
1	0.5	0.2	0.3
2	0.1	0.6	0.3
3	0.1	0.2	0.7
Centroid	0.23	0.33	0.44

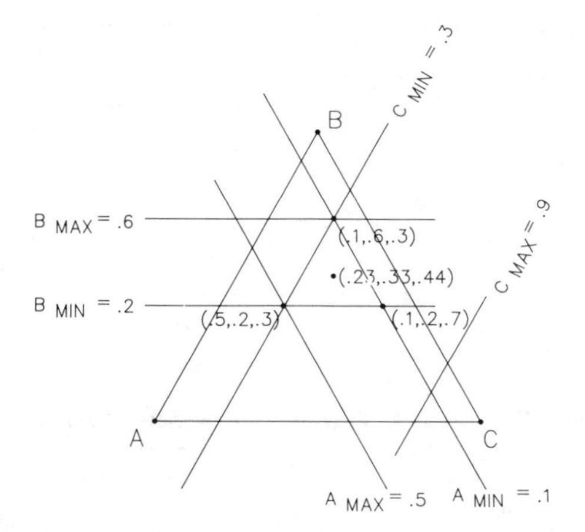

FIGURE 5.50. Identification of blends for a three-component design using the XVERT algorithm.

For more complicated examples (five or more components), it is easier to use the Plackett–Burman matrices in place of the 2^{n-1} two-level factorial design. (Plackett–Burman matrices are given in Appendix E.)

It is recommended to run between 5 and 10 blends more than there are components to be studied. Some replicates should also be run. For example, in a case where seven components are involved, the number of desired blends may be:

One for each component	7
Addition of 5 to 10	5
Centroid	1
Three replicates of centroid	3
Total	16

If a factorial design was used, with $n = 12$ (seven component blends plus five additional blends), it would require 2^{11} combinations. Instead, the 16-run Plackett–Burman design will be used. The constraints for the seven components to be studied are given in Table 5.90.

Again, the components are ranked by range from lowest to highest as shown in Table 5.91.

Now, the high and low values are plugged into the first six columns of the Placket–Burman matrix and the value for component G is calculated. The results are shown in Table 5.92 with asterisks indicating the core blends which meet all constraints.

Table 5.90 Constraints for Seven-Component Design Example

D	$0.10 \leq C_1 \leq 0.25$	$R = 0.15$
B	$0 \leq C_2 \leq 0.10$	$R = 0.10$
A	$0 \leq C_3 \leq 0.05$	$R = 0.05$
F	$0.20 \leq C_4 \leq 0.40$	$R = 0.20$
E	$0.05 \leq C_5 \leq 0.20$	$R = 0.15$
C	$0 \leq C_6 \leq 0.10$	$R = 0.10$
G	$0.10 \leq C_7 \leq 0.30$	$R = 0.20$

Table 5.91 Ranking the Ranges of the Seven Components

$A\ (C_3)$	$R = 0.05$
$B\ (C_2)$	$R = 0.10$
$C\ (C_6)$	$R = 0.10$
$D\ (C_1)$	$R = 0.15$
$E\ (C_5)$	$R = 0.15$
$F\ (C_4)$	$R = 0.20$
$G\ (C_7)$	$R = 0.20$

Table 5.92 Mixture Screening Design Showing Calculated Values for Factor G[a]

Blend	A	B	C	D	E	F	G (Calculated)	Correction
1	0.05	0.10	0.10	0.25	0.05	0.40	0.05	−0.05
2	0.05	0.10	0.10	0.10	0.20	0.20	0.25*	
3	0.05	0.10	0	0.25	0.05	0.40	0.15*	
4	0.05	0	0.10	0.10	0.20	0.40	0.15*	
5	0	0.10	0	0.25	0.20	0.20	0.25*	
6	0.05	0	0.10	0.25	0.05	0.20	0.35	+0.05
7	0	0.10	0.10	0.10	0.05	0.40	0.25*	
8	0.05	0.10	0	0.10	0.20	0.20	0.35	+0.05
9	0.05	0	0	0.25	0.05	0.20	0.45	+0.15
10	0	0	0.10	0.10	0.05	0.20	0.55	+0.25
11	0	0.10	0	0.10	0.05	0.40	0.35	+0.05
12	0.05	0	0	0.10	0.20	0.40	0.25*	
13	0	0	0	0.25	0.20	0.40	0.15*	
14	0	0	0.10	0.25	0.20	0.40	0.05	−0.05
15	0	0.10	0.10	0.25	0.20	0.20	0.15*	
16	0	0	0	0.10	0.05	0.20	0.65	+0.35

[a]Asterisks indicate the core blends that meet all constraints.

Table 5.93 Calculated Subsets for Seven-Component Design[a]

	Blend	*A*	*B*	*C*	*D*	*E*	*F*	*G*	
1	1a	0	0.10	0.10	0.25	0.05	0.40	0.10	✓*
	1b	0.05	0.05	0.10	0.25	0.05	0.40	0.10	✓
	1c	0.05	0.1	0.05	0.25	0.05	0.40	0.10	✓
	1d	0.05	0.10	0.10	0.20	0.05	0.40	0.10	✓
	1e	0.05	0.10	0.10	0.25	0	0.40	0.10	No
	1f	0.05	0.10	0.10	0.25	0.05	0.35	0.10	
6	6a	0.10	0	0.10	0.25	0.05	0.20	0.30	No
	6b	0.05	0.05	0.10	0.25	0.05	0.20	0.30	✓*
	6c	0.05	0	0.15	0.25	0.05	0.20	0.30	No
	6d	0.05	0	0.10	0.30	0.05	0.20	0.30	No
	6e	0.05	0	0.10	0.25	0.10	0.20	0.30	✓
	6f	0.05	0	0.10	0.25	0.05	0.25	0.30	✓
8	8a	0.10	0.10	0	0.10	0.20	0.20	0.30	No
	8b	0.05	0.15	0	0.10	0.20	0.20	0.30	No
	8c	0.05	0.10	0.05	0.10	0.20	0.20	0.30	✓*
	8d	0.05	0.10	0	0.15	0.20	0.20	0.30	✓
	8e	0.05	0.10	0	0.10	0.25	0.20	0.30	No
	8f	0.05	0.10	0	0.10	0.20	0.25	0.30	✓
9	9a	0.20	0	0	0.25	0.05	0.20	0.30	No
	9b	0.05	0.15	0	0.25	0.05	0.20	0.30	No
	9c	0.05	0	0.15	0.25	0.05	0.20	0.30	No
	9d	0.05	0	0	0.40	0.05	0.20	0.30	No
	9e	0.05	0	0	0.25	0.20	0.20	0.30	✓*
	9f	0.05	0	0	0.25	0.05	0.35	0.30	✓
10	10a	0.25	0	0.10	0.10	0.05	0.20	0.30	No
	10b	0	−0.25	0.10	0.10	0.05	0.20	0.30	No
	10c	0	0	0.35	0.10	0.05	0.20	0.30	No
	10d	0	0	0.10	0.35	0.05	0.20	0.30	No
	10e	0	0	0.10	0.10	0.30	0.20	0.30	No
	10f	0	0	0.10	0.10	0.05	0.45	0.30	No
11	11a	−0.05	−0.10	0	0.10	−0.05	0.40	0.30	✓
	11b	0	0.15	0	0.10	0.05	0.40	0.30	No
	11c	0	0.10	0.05	0.10	0.05	0.40	0.30	✓
	11d	0	0.10	0	0.15	−0.05	0.40	0.30	✓
	11e	0	0.10	0	0.10	0.10	0.40	0.30	✓
	11f	0	0.10	0	0.10	0.05	0.45	0.30	No

Table 5.93 Continued

	Blend	A	B	C	D	E	F	G	
14	14a	−0.05	0	0.10	0.25	0.20	0.40	0.10	No
	14b	−0	−0.05	0.10	0.25	−0.20	0.40	0.10	No
	14c	−0	0	0.05	0.25	0.20	0.40	0.10	✓
	14d	0	0	0.10	0.20	0.20	0.40	0.10	✓
	14e	0	0	0.10	0.25	−0.15	0.40	0.10	✓
	14f	0	0	0.10	0.25	0.20	0.35	0.10	✓
16	16a	0.35	0	0	0.10	0.05	0.20	0.30	No
	16b	0	0.35	0	0.10	0.05	0.20	0.30	No
	16c	0	0	0.35	0.10	0.05	0.20	0.30	No
	16d	0	0	0	0.45	0.05	0.20	0.30	No
	16e	0	0	0	0.10	0.40	0.20	0.30	No
	16f	0	0	0	0.10	0.05	0.55	0.30	No

[a]Asterisks indicate the subset blends that will be used in the experimentation.

The next step is to calculate the subsets for those blends that do not meet the constraints. These are shown in Table 5.9.3.

In order to make up the predetermined 12 blends, the 8 core blends and 4 blends from the subsets are used. Usually, the first blend from each subset is selected as a starting point. With this criteria (and making sure there are no duplicates), subsets 1a, 6b, 8c, and 9e are chosen.

Table 5.94 Complete Design for Seven-Component Screening Experiment

	A	B	C	D	E	F	G	
1	0.05	0.10	0.10	0.10	0.20	0.20	0.25	
2	0.05	0.10	0	0.25	0.05	0.40	0.15	
3	0.05	0	0.10	0.10	0.20	0.40	0.15	
4	0	0.10	0	0.25	0.20	0.20	0.25	
5	0	0.10	0.10	0.10	0.05	0.40	0.25	
6	0.05	0	0	0.10	0.20	0.40	0.25	
7	0	0	0	0.25	0.20	0.40	0.15	
8	0	0.10	0.10	0.25	0.20	0.20	0.15	
9	0	0.10	0.10	0.25	0.05	0.40	0.10	
10	0.05	0.05	0.10	0.25	0.05	0.20	0.30	
11	0.05	0.10	0.05	0.10	0.20	0.20	0.30	
12	0.05	0	0	0.25	0.20	0.20	0.30	
13	0.029	0.063	0.054	0.183	0.150	0.300	0.217	Replicate 3 times

The overall centroid is calculated by averaging the component values of the 12 blends. The result of the calculation of the overall centroid is

A	*B*	*C*	*D*	*E*	*F*	*G*
0.029	0.063	0.054	0.183	0.150	0.300	0.217

The final list of blends for this example is given in Table 5.94.

Similar XVERT algorithms can be used for quadratic models, but calculations are usually too involved to do by hand.

Once the data are collected, the results are fitted to the linear model using linear regression (usually with the help of a computer):

$$y = b_1x_1 + b_2x_2 + \cdots + b_nx_n$$

Once the coefficients (b_i) are calculated, conclusions can be drawn depending on their relative sizes.

5.10 RESPONSE SURFACE ANALYSIS

In most cases when designed experiments are being used, there are normally opportunities for process optimization (or continuing improvement) remaining after the significant factors are identified. This is where response surface analysis techniques fit. Response surface analysis uses a series of experiments with the factors (independent variables) that are already known to affect the response and allows the experimenter to generate mathematical equations that describe how the response behaves in the experimental regions. These equations from the experiments lead to an optimal point or, rather, optimal levels of the factors at which to run the process. Response surface analysis typically only looks at two factors.

The steps for performing a response surface analysis are:

1. Identify two critical factors affecting the response (from the results of a screening or a fractional factorial experiment).
2. Run a 2^2 full factorial experiment with the levels set based on the initial experimentation (Figure 5.51). Alternately, the results from a portion of a larger designed experiment could be used as the starting point. For example, results (responses) from the highlighted area in Figure 5.52 could be used in place of running a full factorial design in the region shown.
3. Use the experimental results to determine an equation that defines how the response behaves in the experimental region. Initially, a

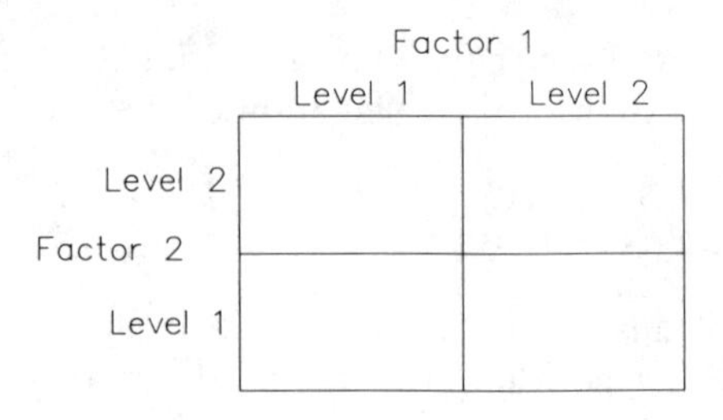

FIGURE 5.51. Typical 2 full factorial design used for initial response surface analysis.

first-order equation is used in the form of

$$Y = \beta_0 + \beta_1 X_1 + \beta_2 X_2 + \epsilon$$

where Y = contrast of the response
X_1 = factor 1
X_2 = factor 2
$\beta_0, \beta_1, \beta_2$ = constants calculated from the experiment
ϵ = error

4. Graph the response in the experimental region (Figure 5.53) by assigning values to the response and calculating contour lines based on X_1 and X_2.
5. Calculate the path of steepest ascent. The line representing this path is perpendicular to the contour lines.
6. Move up the path of steepest ascent (normally to the edge of the experimental region) and run another designed experiment (Figure 5.54). This experiment will not only have runs at the four corner points, but also at the center point.

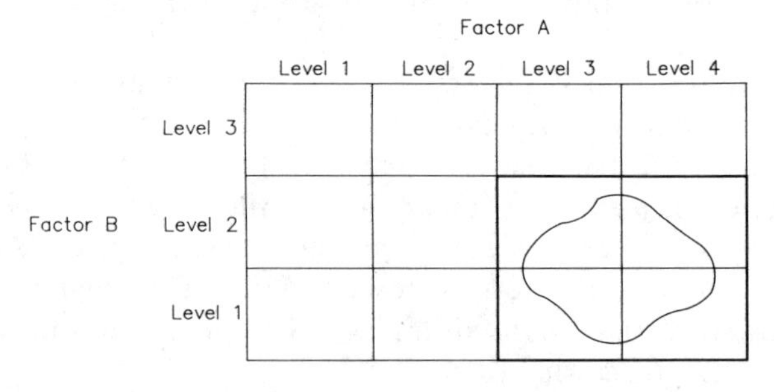

FIGURE 5.52. Using the results from a 2-factor, multi-level experiment for response surface analysis.

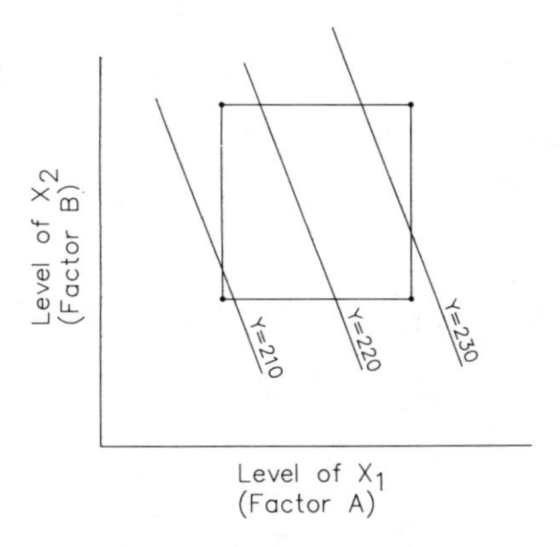

FIGURE 5.53. Response plotted in the experimental region.

7. Repeat steps 3 through 6 until the experimental results show the response is dropping off (Figure 5.55). Once this occurs, back up to the previous experiment and run a central composite design (CCD) (Figure 5.56). The CCD is used to determine if the contour lines in the region fit a higher-order (nonlinear) equation. The results of a CCD are complex to calculate. The calculations are best done by computer with a program such as SAS. For the brave, formulas for calculating the coefficients of a second-order design can be found in Appendix B of *How to Apply Response Surface Methodology* by John A. Cornell (published by the American Society for Quality Control). Complex formulas such as those used by Cornell are beyond the scope of this work.

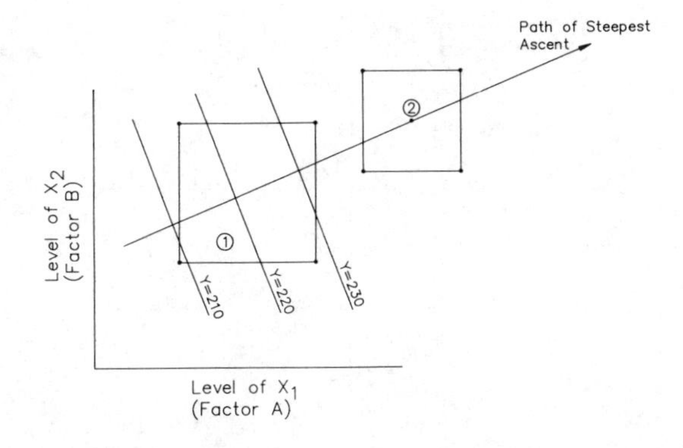

FIGURE 5.54. Secondary experimentation centered on the path of steepest ascent.

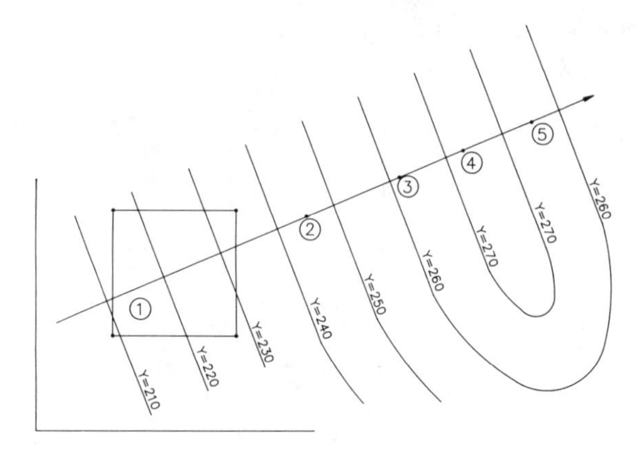

FIGURE 5.55. Response surface mapping using multiple experimentation central composite design.

In Chapters 5.9.1 and 5.9.2 (Mixture Response Surface Designs and Extreme Vertices Designs), response surface techniques were used to describe the experimental region defined by mixture components. The appropriate linear, quadratic, or cubic mathematical model (whichever fits the data best) was developed to predict the remaining responses. The same type of response surface techniques can be used for nonmixture examples. The following is an example of this technique:

In Chapter 5.6.3 (3^f Factorial Designs), an example was shown on investigating the effect of temperature and humidity on the cure time of a new adhesive. It was found that humidity had a strong significant effect on

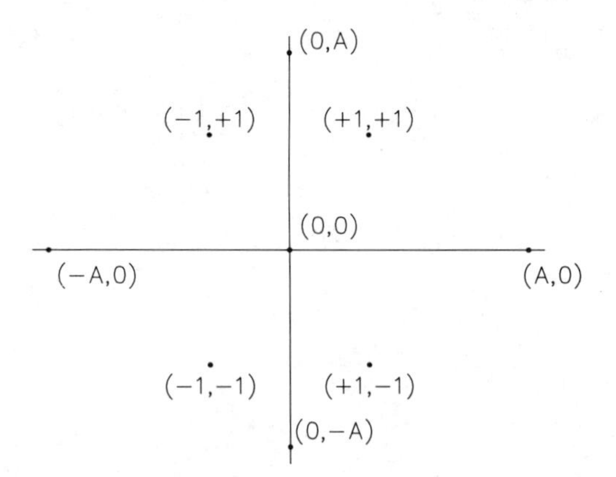

FIGURE 5.56. Central composite design.

cure speed and that there was a strong linear interaction between temperature and humidity. The next step in the experimental design sequence is to determine the optimum combination of humidity and temperature that will minimize the cure time. Step 1 has already been completed: In this example, it is assumed the researchers did a thorough job of narrowing the list of significant factors down to two. Part of step 2 has also been completed: They ran an experiment to investigate a region and have analyzed the data. The analysis indicates that a first-order model may be adequate. Now, the data are used to develop the first-order model and to determine which direction we want to test next.

The equation for the first-order model is

$$Y = \beta_0 + \beta_1 X_1 + \beta_2 X_2 + \epsilon$$

for two independent variables. ϵ is the sum of squares of the error and the β terms are estimated using least squares. Only a portion of the 2^3 design will be used. This will be relabeled in "response surface" notation. The data from Section 5.6.3 indicate that high temperature and low humidity are beneficial for quick cure so the lower right four cells are appropriate for the initial model development. The data for these cells are shown in Table 5.95

The response surface analysis uses coded notation. The coded notation is easier to use in the calculations for estimating the coefficients β_0, β_1, and β_2. Factor A is designated as X_1 and factor B as X_2. The levels of the factors are also coded. The high levels are coded as $+1$ and the low levels as -1. This comes from the transference equations for X_1 and X_2. In our example,

$$X_1 = \frac{T - \overline{T}}{T_{\text{high}} - T_{\text{low}}} \qquad X_2 = \frac{H - \overline{H}}{H_{\text{high}} - H_{\text{low}}}$$

Table 5.95 Data Table for Initial Response Surface Design on Adhesive Cure Speed[a]

Humidity (B)	Temperature (A) 75		Temperature (A) 85	
70	7 6	(6.5)	7 8	(7.5)
50	4 4	(4.0)	2 3	(2.5)

[a]The mean values of the two responses in each cell are shown circled.

Using these equations, the factors and their levels become:
Temperature:

$$X_1 = -1 \quad \text{when } T = 75°\text{F}$$

$$X_1 = +1 \quad \text{when } T = 85°\text{F}$$

therefore

$$X_1 = 0.2T - 16$$

Humidity [relative (RH)]:

$$X_2 = -1 \quad \text{when } H = 50\% \text{ RH}$$

$$X_2 = +1 \quad \text{when } H = 70\% \text{ RH}$$

therefore

$$X_2 = 0.1H - 6$$

The responses of the experimentation with the coded factor levels are shown in Table 5.96. Using these data, the estimated values of the β coefficients can be found and the equation for the response in the experimental region can be identified. The estimated values for the β coefficients are calculated from the following equations:

$$\beta_0 = \frac{\Sigma \bar{Y}}{n}$$

$$\beta_1 = \frac{\Sigma X_1 \bar{Y}}{\Sigma X_1^2}$$

$$\beta_2 = \frac{\Sigma X_2 \bar{Y}}{\Sigma X_2^2}$$

where n is the total number of cells (treatment combinations).

Table 5.96 Response Surface Notation Labeling of Combinations

$\bar{Y}$	X_1	X_2
4.0	−1	−1
6.5	−1	1
2.5	1	−1
7.5	1	1

For this example,

$$\beta_0 = \frac{(4.0 + 6.5 + 2.5 + 7.5)}{4} = 5.13$$

$$\beta_1 = \frac{(-4.0 - 6.5 + 2.5 + 7.5)}{4} = -0.13$$

$$\beta_2 = \frac{(-4.0 + 6.5 - 2.5 + 7.5)}{4} = 1.88$$

These values can be plugged into the equation for the response in the experimental region:

$$Y_{est} = 5.13 - 0.13X_1 + 1.88X_2 + \epsilon$$

In the initial experimentation, it is often assumed that ϵ is negligible compared to the other values in the equation to simplify the calculations. In subsequent experimentation, the error term can be estimated by the sum of squares method using replicates of the response at the center point. Assuming that ϵ is negligible, simplifies the equation to

$$Y_{est} = 5.13 - 0.13X_1 + 1.88X_2$$

Once the equation is determined for estimating Y, estimated contour lines of the response should be drawn within the experimental region. This is accomplished by choosing a value for Y and then rearranging the resulting equation to get X_2 as a function of X_1. The equation will take the form

$$X_2 = mX_1 + b$$

where m = slope of the line defined by the equation
$b = y$ intercept

The results of the experimentation ranged from 2 to 8 s. Therefore, it makes sense to draw lines for $Y = 2$, 4, 6, and 8 s.

For $Y = 8$, the contour line is calculated from

$$8 = 5.13 - 0.13X_1 + 1.88X_2$$

or

$$X_2 = 0.067X_1 + 1.53$$

Similarly, contour line equations can be found for $Y = 2$, 4, and 6. Once

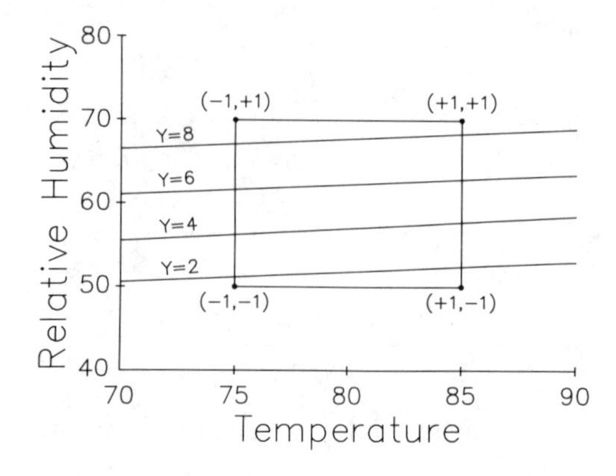

FIGURE 5.57. Response plot for temperature/relative humidity experiment.

the equations are known, the contour lines can be plotted resulting in the graph shown in Figure 5.57.

The direction of the greatest improvement, is called the path of steepest ascent. Although the word "ascent" is used, the path can actually be in any direction. Ascent refers to improvement in the response. The path of steepest ascent is a line perpendicular to the contour lines in the direction of improvement. This is shown in Figure 5.58.

Using simple algebra, the slope of the path of steepest ascent can be determined. We know the equations of the contour lines. As stated earlier, these equations are in the form

$$X_2 = mX_1 + b$$

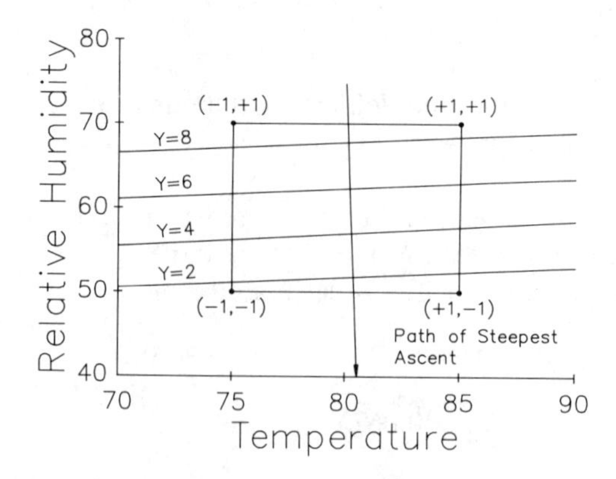

FIGURE 5.58. Path of steepest ascent for temperature/relative humidity experiment.

The slopes of all four contour lines are equal. The slope of a line perpendicular to the contour lines is by mathematical definition equal to $(-1/m)$. In this example, the slope of the perpendicular is

$$(-1/0.067) = -15$$

The path of steepest ascent is a line drawn through the center of the experimental region with a slope of -15. A region for further experimentation can be determined by moving along this path in the direction of improvement. Knowing the results from the initial experimentation and the path of steepest ascent, the center point of the second experimental region can be determined. In this case, a good selection might be

$$X_1 = 0.1 \qquad X_2 = -15X_1 = -1.5$$

The surrounding points for the second experiment might be selected as shown in Figure 5.59. Using the equations for X_1 and X_2, the coded values can be transformed back into the temperature and humidity values at which the experiment will be run. For example, the center point would be

$$X_1 = 0.2T - 16 \qquad X_2 = 0.1H - 6$$

$$T = \frac{X_1 + 16}{0.2} \qquad H = \frac{X_2 + 6}{0.1}$$

$$T = 80.5°\text{F} \qquad H = 45\%\ \text{RH}$$

In this example, the optimal point at which to run the process is very close to the initial experimental region. This is known because the response is cure time, which can never get below a value of zero. Being this close to the optimal point, the second designed experiment here should be a central composite design to determine if the contour lines in this region fit

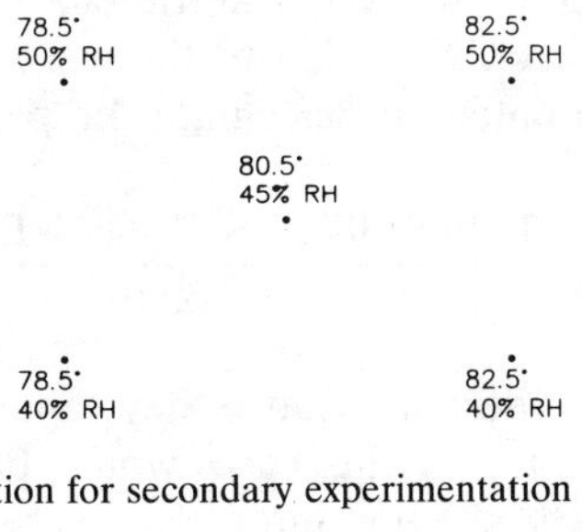

78.5°
50% RH

82.5°
50% RH

80.5°
45% RH

78.5°
40% RH

82.5°
40% RH

FIGURE 5.59. Level selection for secondary experimentation on temperature and relative humidity.

a higher-order (nonlinear) equation. In other cases, you may have to run more than just two first-order experiments before finding a region to optimize with a second-order experiment.

5.11 EVOP

EVOP means evolutionary operation. It is an experimental method, designed by Dr. Box (1957), that uses an experimental region within the current range of operating conditions of a process. Using this method, improvements in output and quality can be achieved without interrupting the process and, usually, without producing bad parts. A 2^2 factorial is most often used in this method with a center point added. The center point, or reference point, usually corresponds to the known or typical settings of the operating parameters. EVOP experiments usually only look at one or two factors. These experiments are used after the critical control factors have been determined in preliminary experimentation.

The terminology used with EVOP designs follows.

Reference Point: The center point of the factorial. The known or typical operating parameters of the process.
Cycle: One complete set of runs with one run at each set of conditions.
Phase: All the cycles that use the same settings of the same factors that led to a significant improvement in the process output.

When running the experiment, the cycles need to be repeated at least once (to get an estimate of experimental error). However, in most cases, the cycles have to be repeated a number of times to detect an effect.

EVOP Example

Suppose that a manufacturer of a lawn fertilizer wants to find a way to increase the yield of their reaction kettles. Their two critical reaction parameters are the temperature in the kettle and the time the raw materials are left in the kettle. Their kettle charge weight (raw materials) is held constant at 5000 lb. They record the weight of usable product that is produced from each batch and calculate the percent yield as

$$\text{Yield} = \frac{\text{pounds of good product produced}}{5000 \text{ lb}}$$

They selected an EVOP design because they cannot afford to manufacture bad product (bad product, in this case, would be hazardous waste) and they cannot afford any production interruptions because they are currently

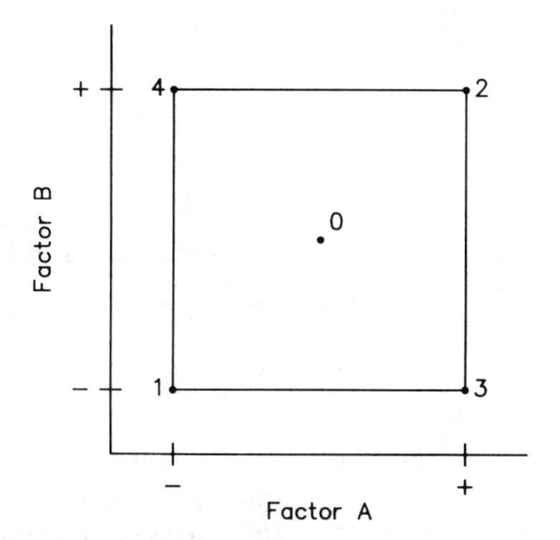

FIGURE 5.60. EVOP design with time and temperature factors.

running close to capacity. They used their current conditions of temperature and time, as center point of their design. The design is shown in Figure 5.60. Their cycle consisted of five runs, labeled zero through four, in the form of a 2^2 factorial with a center point added.

Before running any experiment, it is useful to have a data collection and analysis worksheet. Hicks (1982) proposes a fairly simple format as shown in Table 5.97.

The data collected for the yield experiments are shown in the next three worksheets, Tables 5.98, 5.99, and 5.100. Note that the $f_{K,n}$ values for a five point design are

	Cycle					
	2	3	4	5	6	7
$f_{5,n}$	0.30	0.35	0.37	0.38	0.39	0.40

The $f_{5,n}$ values are derived from the estimate of standard deviation:

$$s = \frac{\overline{R}}{d_2}$$

where $\overline{R}$ = average range
d_2 = constant that depends on the sample size
K = 5 (in our example, we have five data points)
d_2 = 2.326 (see appendix A-5)

Table 5.97 EVOP Worksheet

Cycle name: n = Date:
Response description:

Calculation of Averages

Operating Conditions	(0)	(1)	(2)	(3)	(4)
a. Previous cycle sum	(from previous cycle worksheet row e)				
b. Previous cycle average	(from previous cycle worksheet row f)				
c. New observations	(response data from experimentation)				
d. Differences	(b − c)				
e. New sums	(a + c)				
f. New averages	(e/n)				

Calculation of Standard Deviation

g. Previous sums s =	(from previous cycle worksheet row k)
h. Previous average s =	(from previous cycle worksheet row l)
i. Range =	(range of the values in row d)
j. New s = range $\times f_{K,n}$	$(i \times f_{K,n})$
k. New sum s =	(g + j)
l. New average = $\dfrac{\text{new sum } s}{(n-1)}$	$\dfrac{k}{(n-1)}$

Calculation of Effect	Calculation of Error Limit
$\text{Effect}_A = \frac{1}{2}(Y_2 + Y_3 - Y_4 - Y_1)$	For new average = $\dfrac{2}{\sqrt{n}}s$
$\text{Effect}_B = \frac{1}{2}(Y_4 + Y_2 - Y_1 - Y_3)$	For new effects = $\dfrac{2}{\sqrt{n}}s$
$\text{Effect}_{A\times B} = \frac{1}{2}(Y_1 + Y_2 - Y_3 - Y_4)$	For change in means = $\dfrac{1.78}{\sqrt{n}}s$

Change in mean effect = $\frac{1}{5}(Y_1 + Y_2 + Y_3 + Y_4 - 4Y_0)$

The population mean σ_p is then calculated using the equation

$$\sigma_p = \sqrt{\frac{n-1}{n}} \times s$$

where n is the cycle number. So, for this example,

$$\sigma_p = \sqrt{\frac{n-1}{n}} \times \frac{\bar{R}}{2.326}$$

Table 5.98 EVOP Work Sheet for Cycle 1

Cycle name: $n = 1$ Date: 6-9-91

Response description: Reactor Yield (%)

Calculation of Averages

Operating Conditions	(0)	(1)	(2)	(3)	(4)
Previous cycle sum	—	—	—	—	—
Previous cycle average	—	—	—	—	—
New Observations	95.0	94.5	96.0	95.0	95.5
Differences	—	—	—	—	—
New sums	95.0	94.5	96.0	95.0	95.5
New averages	95.0	94.5	96.0	95.0	95.5

Calculation of Standard Deviation

Note: Cannot do with only one cycle

Previous sums s =
Previous average s =
Range =
New s = range $\times f_{K,n}$ =
New sum s =
New average =

Calculation of Effect

Effect of temperature = 0.5
Effect of time = 1.0
$T \times t$ interaction = 0.0
Change in mean effect = 0.2

Calculation of Error Limit

Note: Cannot do with only one cycle

For new average
For new effects
For change in means

For $n = 2$ cycles and $K = 5$ data points,

$$\sigma_p = \sqrt{\frac{2-1}{2}} \times \frac{\overline{R}}{2.326} = 0.30\overline{R}$$

Similarly, for three cycles,

$$\sigma_p = \sqrt{\frac{3-1}{3}} \times \frac{\overline{R}}{2.326} = 0.35\overline{R}$$

Table 5.99 EVOP Work Sheet for Cycle 2

Cycle name: $n = 2$ Date: 6-16-91
Response description: Reactor yield (%)

Calculation of Averages

Operating Conditions	(0)	(1)	(2)	(3)	(4)
Previous cycle sum	95.0	94.5	96.0	95.0	95.5
Previous cycle average	95.0	94.5	96.0	95.0	95.5
New observations	93.0	92.0	94.0	92.5	94.5
Differences	2.0	2.5	2.0	2.5	1.0
New sums	188.0	186.5	190.0	187.5	190.0
New averages	94.0	93.25	95.0	93.75	95.0

Calculation of Standard Deviation

Previous sums s = __
Previous average s = __
Range = 1.5
New s = 0.45
New sum s = 0.45
New Average = 0.45

Calculation of Effect

Effect of temperature = 0.25
Effect of time = 1.5
$T \times t$ interaction = 0.25
Change in mean effect = 0.2

Calculation of Error Limit

For new average = 0.64
for new effects = 0.64
For change in means = 0.57

With the worksheet for the second cycle (Table 5.99) complete, the effects must be compared to the error limits. Because none of the effects is greater than its error limits, no conclusions can be drawn and another cycle must be run.

After this third cycle (Table 5.100), the effect of time (2.25) is greater than the error limits for effects (1.36), so the effect is considered to be significant. Because the sign is positive, increasing the time will increase the yield. As a result of the experiment, the time should be changed in the process to the high level, and if further improvement is desired, another EVOP design can be used with the existing temperature and the new time as the center point. If the effect were strongly significant, the next design can be performed at levels that aren't equal or adjacent to those in the first design as long as the design is moved in the direction indicated by the first set of results.

The "change in the mean" effect is a measure of the curvature of the plane formed with the five points. In some cases, a significant amount of

Table 5.100 EVOP Work Sheet for Cycle 3

Cycle name: $n = 3$ Date: 6-23-91
Response description: Reactor Yield (%)

Calculation of Averages

Operating Conditions	(0)	(1)	(2)	(3)	(4)
Previous cycle sum	188.0	186.5	190.0	187.5	190.0
Previous cycle average	94.0	93.23	95.0	93.75	95.0
New observations	94.0	92.0	96.0	93.0	96.5
Differences	0	1.25	−1.0	0.75	−1.5
New sums	282.0	278.5	286.0	280.5	286.5
New Averages	94.0	92.8	95.3	93.5	95.5

Calculation of Standard Deviation

Previous sums $s = 0.45$
Previous average $s = 0.45$
Range = 2.75
New sum $s = 0.96$
New sum $s = 1.41$
New average = 0.71

Calculation of Effect

Effect of temperature = 0.25
Effect of time = 2.25
$T \times t$ interaction = 0.225
Change in mean effect = 0.22

Calculation of Error Limit

For new average = 1.36
For new effects = 1.36
For change in means = 1.21

curvature may be found, which may mean that the center point is actually either a maximum or minimum.

5.12 HANDLING MISSING DATA

Although most of the examples of experimental designs discussed in this work seem fairly straightforward, they are not always realistic. They do not address a very important and unfortunately, frequently occurring problem —missing data. Many times samples are lost or destroyed or simply cannot be produced under the prescribed conditions, which leaves the experimenter with unequal sample sizes from cell to cell or, in an extreme case, missing cells. This section is included to give the experimenter some general rules of thumb so that the experiment will still yield meaningful results.

Randomized Designs / Randomized Block Designs / Factorial Designs

Missing Data Points:

If, after completion of the experiment, a data point is missing (causing unequal sample sizes) there are a number of different techniques that can be used. If the planned sample size for each cell is greater than or equal to 3, the easiest approach is to take the average of the two or more values available and use the average as the missing data point.

An example is shown in Table 5.101. The cell A_2B_3 has only two data points, so the average of the two will be used for the third value:

$$\text{Average} = \frac{4 + 5}{2} = 4.5$$

The values used for cell A_2B_3 would be 4, 5, and 4.5.

Another approach would be to abandon the traditional ANOVA method and revert back to strictly a test of means. A standard t test will indicate whether there is an effect of using one factor versus another and will accommodate unequal sample sizes. This approach is particularly useful when sample sizes are small ($n = 2$) and it is not possible to use an average value.

For an example, use the data in Table 5.101 again and find the totals for all the rows and columns as shown in Table 5.102.

Table 5.101 3^3 Experiment with Missing Data Point

Factor B	Factor A		
Level	1	2	3
3	5, 4, 4	4, 5	4, 3, 4
2	6, 6, 6	5, 6, 5	5, 5, 6
1	7, 5, 5	7, 6, 6	7, 8, 8

Table 5.102 Totals of Rows and Columns for Running t Tests to Evaluate Main Effects

Factor B	Factor A			
Level	1	2	3	Row
3	5, 4, 4	4, 5	4, 3, 4	33, $n = 8$
2	6, 6, 6	5, 6, 5	5, 5, 6	50, $n = 9$
1	7, 5, 5	7, 6, 6	7, 8, 8	59, $n = 9$
Column	48	44	50	
	$n = 9$	$n = 8$	$n = 9$	

Table 5.103 3^3 Experiment with Cell with Lost Data

	Factor A		
Factor B	1	2	3
3	$\bar{x} = 5.9$ $s = 0.4$	$\bar{x} = 4.3$ $s = 0.5$	$\bar{x} = 3.9$ $s = 0.2$
2	$\bar{x} = 6.2$ $s = 0.5$	$\bar{x} = 5.9$ $s = 0.7$	$\bar{x} = 6.0$ $s = 0.4$
1	$\bar{x} = 7.0$ $s = 0.7$	$\bar{x} = 7.5$ $s = 0.7$	Lost Data

By running t tests on pairs from the rows or columns, it is still possible to evaluate the main effects. Interactions are not as easily calculated, but can be viewed graphically.

Missing Cells

In cases where $n = 1$ and/or the entire cell was either destroyed or not produced, the best alternative may be to rerun that cell along with one or two others on a separate day. Then compare the results of the replicate runs to the original data. If the results of the replicate runs are approximately equivalent for both the mean and the standard deviation of the data for their corresponding cells in the initial runs, then plug in the data collected during the replicate runs for the empty cell into the original table of results. If a t test and an F test show the mean and the variation of the replicated runs to be significantly different from the original set, then normalize the new data as accurately as possible before plugging it into the experiment.

Table 5.103 shows the data from a 3^3 full factorial experiment. Due to an equipment malfunction during the experiment, the results from cell A_3B_1 were invalid.

The steps to recover as much information as possible from these results are as follows:

1. Rerun the missing cell with one or two others (for this example, cells A_1B_1 and A_3B_3 were selected):

Data	Cell A_1B_1	$\bar{x} = 6.8$	$s = 0.6$
	Cell A_3B_3	$\bar{x} = 3.6$	$s = 0.3$
	Cell A_3B_1	$\bar{x} = 8.1$	$s = 0.8$

2. Run F test and t tests on cells A_1B_1 and A_3B_3 to determine if they are the same or different than the original data for these two cells.
3. If they are the same, plug in the new cell A_3B_1 data and do the analysis as always.
4. If they are not the same, normalize the new cell A_3B_1 data before entering it into the table. In this example, they are not the same so the multiplier to normalize the data must be calculated:

$$\text{Cell } A_1B_1 \quad \frac{7.0}{6.8} = 1.029$$

$$\text{Cell } A_3B_3 \quad \frac{3.9}{3.6} = 1.083$$

$$\text{Average} = \frac{1.029 + 1.083}{2} = 1.056$$

To normalize, multiply all values in new cell A_3B_1 by 1.056 before entering them into the original grid for analysis.

Fractional Factorials and Screening Designs

Unequal sample sizes are not a problem with these types of experiments because average values are used for much of the analysis and conventional ANOVA isn't needed. However, missing treatment combinations are a problem that will require additional experimentation. The best approach, as in the case of missing cells in full factorial designs, is to rerun the missing combination along with two others. Once the combinations are run, the same methods are used, as with missing cells in the last section. Compare the new data with the original data and either plug it in as is or attempt to "normalize" it if a blocking effect appears to exist.

Mixture Designs

Again, unequal samples sizes are not typically a problem because traditional ANOVA is not used in the analysis. Use the guidelines (averaging existing data or rerunning and normalizing) outlined for missing cells in the last two sections.

General Comments

Although there are some basic rules of thumb outlined here, the experimenter must keep the following in mind:

Any digression from the recommended experimental method could lead to erroneous results and incorrect conclusions.

Any digression from the recommended experimental method should be described in detail along with the review of the results.

It is always *safer* to rerun the experiment than to use the methods described here. Response surface, EVOP, and 2^2 factorials should always be rerun completely if data are missing.

Use these rules of thumb only to salvage some of the results from an expensive experiment, NOT as a way to cut down on the number of data points needed.

Always perform a follow-up experiment to confirm your results and conclusions before putting them into practice.

5.13 PREVENTIVE MEASURES

An old adage states "an ounce of prevention is worth a pound of cure." Companies generally follow this advice in several areas:

- Total productive maintenance (also known as total preventive maintenance or TPM) techniques reduce or eliminate the potential causes of equipment downtime.
- There are tremendous benefits to regularly maintaining equipment compared to the costs and lost production time caused by breakdowns.
- Some companies offer employee wellness programs that aim to prevent health problems.
- Safety programs are strictly prevention-minded. Even after an accident, the goal is to prevent a recurrence.

Just like these examples, a TQM effort should use prevention techniques; techniques to prevent quality problems up front, to prevent quality failure relapses once a process is in control, and to prevent backsliding once improvements have been made. In summary, they assure that gains have permanence by changing the process to incorporate the improvements. This section covers three measures for the prevention of quality problems and for preventing quality relapses. These measures are process audits, risk analysis, and concurrent engineering.

Process audits inspect the process to ensure that it has remained in control or that proper reactions occurred when the process went out of control. Audits are useful to check that the improvements made with SPI techniques remain in place. Without audits, it is too easy to let the process slide back into old habits. Audits help institutionalize the gains so the improved process becomes the new status quo.

Risk analysis employs a technique known as failure mode and effect analysis (FMEA). In FMEA, numerical values are assigned to the severity, probability of occurrence, and the probability of detection for all potential quality problems within a product or process. These three values are used to calculate a risk factor and prioritize the risk potentials. Preventive measures can then be put in place for those potential quality problems with high risk factors.

Concurrent engineering works to prevent quality problems with a new product or new process starting in the initial design stages. Concurrent engineering uses a combination of FMEA and DOE techniques. Employing these improvement tools in the design, construction, and start-up phases for a new product or process can identify potential quality problems. This allows problems to be corrected earlier and, usually, more easily, so that start-up costs and time are minimized.

5.13.1 Process Audits

Once SPC has been put in place and process improvements have been made through SPI techniques, companies face perhaps the most difficult tasks of all: the tasks of ensuring that the changes (process improvements) are made with permanence and the improved level is set as the new baseline for continuous improvement. All too often, as companies move to improve another area or process, the focus on the previous area or process is abruptly dropped. If this occurs before the improvements are ingrained in the culture of the operation and a new, "better" status quo is created, the improvements will not be permanent. When the focus shifts, the natural tendency of all employees from operators to managers is to assume (or hope) that the changes made will stay in place at the improved level. However, they won't remain in place without some type of periodic review process to help measure the improvement process. Without this review, or self-audit, the operation may gradually drift back toward old methods and habits. The process audit is an improvement tool to help ensure that planned changes are really made, that gains *are held*, and that we have checkpoints in the continuous improvement journey.

The dictionary defines an audit as an examination. An examination in school measures where the student is and can identify areas where a student and or a group of students need improvement. An audit is just like an examination in school. It describes where the process is and identifies area that need to be reinforced.

Process Audit

A process audit uses product information to verify how the process is operating. The audit checks more than just the product; it checks procedural and equipment improvements as well.

Planning the process audit involves four steps:

1. *Develop the audit criteria.* Criteria should include:
   Ensuring that the operating procedures are up to date and are being followed. This includes an audit of both methods and operating parameters.
   Verifying that the control (or precontrol) charts are filled out correctly, that any out-of-control conditions are reacted to with the corrective action noted on the chart, and that control limits have been recalculated when appropriate.
   Calculating the process capability to determine if it has improved (or gotten worse). The process capability can be calculated from control chart data for audit purposes.
   Verifying that measurement equipment has been calibrated as scheduled.
   Checking to ensure measurement system error has been recognized and quantified.
   Verifying that preventative maintenance has been done on the equipment as scheduled.
   In addition to these, the audit criteria should include a check on specific improvements made.
   The process audit criteria may also include management measurements on the overall process such as yields, cycle times, productivity measures, and percentage uptime of equipment. These are not the normal measures associated with SPC or SPI, but they will indicate if the process has changed. It is important to know why changes have occurred, whether good or bad, so corrections to problems can be initiated or positive changes can be incorporated into the process.
2. *Develop a data collection format.* As with all data collection tasks, a process audit data collection form should be organized so that it is easy to use. Wherever possible, the data collection sheet should contain questions that can be checked "yes" or "no" and should lead the auditor through the audit. Figure 5.61 shows a simple data collection sheet for a process audit.
3. *Decide who will conduct the audit.* The audit should be independent and impartial, conducted by individuals who do not have a vested interest in either success or failure. The process audit may be done by an individual, but is best done by a team. Many companies use their engineering staffs to conduct process audits. If engineers are given processes to audit that they work with every day, the benefits of an independent audit may be lost, particularly when auditing the operating procedures. If an engineer is "too" familiar with the

TQ SELF AUDIT

A. MANUFACTURING AUDIT CHECKLIST

PART NUMBER ______ DATE: ______
ATTRIBUTE ______ AUDITED: ______
OPERATION ______ REVIEWED BY: ______
MACHINE NO.(S) ______
INSPECTION GAUGE ______

SAMPLE-SIZE ______ -FREQUENCY ______

B/P SPEC-NOMINAL ______ -TOLERANCE ______

1. CONTROL CHART MODE OR RUN MODE (CIRCLE ONE)
CONTROL CHART INFORMATION

$\bar{X}$	R	P
$\bar{\bar{X}}$ ______	$\bar{R}$ ______	$\bar{P}$ ______
$UCL_{\bar{X}}$ ______	UCL_R ______	UCL_P ______
$LCL_{\bar{X}}$ ______	LCL_R ______	LCL_P ______

DATA CONTROL LIMITS CALCULATED ______ DATE TO REVIEW ______
WERE ALL OUT OF CONTROL CONDITIONS REACTED TO AND DOCUMENTED?
CPK ______ YES__NO__

RUN CHART INFORMATION

ESTIMATED DATE TO CONVERT TO A CONTROL CHART ______

INDIVIDUAL	RANGE
UPPER ACTION LIMIT ______	UPPER ACTION LIMIT ______
LOWER ACTION LIMIT ______	LOWER ACTION LIMIT ______

DATE ACTION LIMITS ISSUED ______

2. SETUP INSTRUCTIONS AVAILABLE? YES__NO__N/A__
SETUP INSTRUCTIONS COMPLETE? YES__NO__N/A__
DATE LAST REVISED ______

3. OPERATION SOP AVAILABLE? YES__NO__
DATE LAST REVISED ______

4. GAUGES, MSP AVAILABLE? YES__NO__
DATE GAUGE LAST CALIBRATED ______ DATE TO BE RECALIBRATED ______
GAUGE(S) IN GOOD WORKING ORDER? YES__NO__

CORRECTIVE ACTION

ACTION REQUIRED	RI	COMPLETE DATE

FIGURE 5.61. Process audit data collection form.

process, he or she may "know" why the actual operating procedures are different from the written procedures, or the engineer may assume something is still being done because it was done in the past. Neither may be the case. To ensure the audit is independent, it is best to assign audit responsibilities to those that do not work in the assigned areas daily. If the teams don't know the area thoroughly, the members will be more apt to rely on the operating procedures and other documentation to get a clear picture of the process. This makes for a more accurate audit. (This also has other advantages in that it exposes people to other processes. Engineers or others may learn things that can improve their own processes.)

In forming the audit team in plant operations, first line manufacturing supervisors and operators should always be included in the process audit team. Again, the team make-up should be as independent as possible (removed from the process) to ensure a thorough audit. For example, a manufacturing supervisor and an engineer may be teamed up with one or two operators to audit a process outside of their normal responsibilities. Having operators involved adds another dimension to the audit. The operator can determine whether or not the operating procedures are written in language and terminology that is understandable to the employees that have to use them. This approach also has additional benefits. It promotes team-building between the supervisor, the engineer, and the operators. The team will tend to band together, especially if they are auditing an area that none of the members is especially familiar with.

4. *Select the frequency of the audits.* The frequency of the process auditing depends on the process and the product or products being manufactured. There is no need for weekly audits if the process is producing high volumes of the same product continuously. Quarterly audits should be adequate to ensure that appropriate SPC techniques are in place and that improvements developed with the SPI tools are still in place. If the audits start showing many discrepancies, then their frequency should be increased.

   Processes that manufacture a variety of products usually need to be audited more frequently. These processes should also be audited periodically for each of the products or product families that goes through the operation. With this approach, however, those parts of the process audit common to all products may be conducted more frequently than necessary. An approach to help make audits with common paths more effective is to separate the process audit into product specific topics (e.g., operating procedure for each product or product family) and into a second set of common audit items.

The recommended minimum audit frequency for each product or product family is one year. Set up an audit schedule that lists which products will be audited each month for the year. Some flexibility is needed because some products may not run in the month specified on the schedule. The audits for these products would, of course, have to be postponed until the next time they were manufactured.

Once the preliminary work is completed, the actual audit will be easy to conduct. The audit is not the end of the process, however. Unless action is taken using the audit results, the audit will only be a paperwork exercise. The audit team should review audit results with the manager responsible for the area. This enables the manager to understand the audit results so that action items, based on the audit results, can be established and supported. The data collection form shown in Figure 5.61 includes a section at the bottom for identification of action items. Always clearly indicate individuals responsible for completing the action items so accountability is understood by everyone.

Management must commit to the use of audits and to the priority with which identified action items are completed. The facility TQM Steering Committee or Management Council should review process audits in their monthly meetings. The Steering Committeee need not get into specific details of each audit. Instead, an audit summary form should be submitted monthly to the Steering Committee for review. The audit summary form, as shown in Figure 5.62, should include what audits were conducted, why any were missed, and what actions were taken or remain to be taken based on problem areas found in the audit.

Management must understand that the process audit is not a replacement for SPC or SPI techniques. An audit will not make improvements to the process and will not identify why improvements were not maintained. The process audit is a tool in the SPI toolbox and is not intended to be used as a hammer. If improvements are not made or not maintained, management must look at the culture and systems they created: It may be that they (management) do not facilitate change despite a great deal of effort, on the shop floor, to identify areas for improvement.

5.13.2 Risk Analysis

Risk analysis is another technique that can be used to prevent quality problems, even though it is more commonly thought of in a safety or environmental assurance context. Risk analysis is also known as failure

MANUFACTURING QUARTERLY TQM AUDIT SUMMARY

QUARTER ______ YEAR ______

	NO. OF PARTS AUDITED	PERCENT COMPLETION
P/L 24		
P/L 27		
P/L 28		
TOTAL		

____ ALL KANBAN CARDS CURRENT

____ SOPS UP–TO–DATE

____ ALL CHARTS IN PLACE – PRECONTROL, CONTROL, OR RUN CHARTS

____ ALL SET–UP FORMS CURRENT

____ VISUAL STANDARDS IN PLACE AND CURRENT

____ GAUGES HAVE UP–TO–DATE CALIBRATION STICKERS

PARTS REQUIRING ACTION

PART NUMBER	ACTION REQUIRED	R. I.	COMPLETION DATE

FIGURE 5.62. Audit summary form

mode and effect analysis (FMEA). It is a subjective technique that calculates the risk of an event detrimental to quality by its severity, potential frequency of occurrence, and ease of detection. This allows potential problem areas to be highlighted and corrections made to high risk areas (the vital few) to prevent problems from occurring.

All engineers review processes, at least intuitively or subjectively, for potential failure modes prior to starting up a new process or introducing a new product into the process. FMEA brings discipline and some degree of objectivity into this review process. It methodically covers the process step by step and component by component. An FMEA has the following sequential steps:

- Identify component or method failures that could cause a quality or safety incident.
- Develop a ranking of the potential for these failures based on a frequency of occurrence, severity, and the probability of detection.
- Evaluate the preventative measures in place and make recommendations to correct inadequacies.
- Document the results for use in future improvement activities.

The general procedure for conducting an FMEA follows:

1. *Select a review team and team leader.* An FMEA should include a team of four to six participants. Representatives from cross-functional areas such as manufacturing, engineering, and maintenance should participate. If a manufacturing process is involved, include at least one of the manufacturing operators, if possible. The team leader does not have to be the person most familiar with the process. In fact, it may be a better team effort if the process "expert" only participates in the team and does not act as leader.
2. *Define the process boundaries.* An FMEA on an entire process would be extremely complex, and could possibly cause some potential failure modes to be overlooked. The process should be broken down into a series of subprocesses and an FMEA conducted on each for ease of analysis. Some companies use separate review teams for each subprocess. Use process flow diagrams to help identify the boundaries of each subprocess to be studied.
3. *Brainstorm potential failure modes.* The brainstorming should focus on the process under study and on the potential process failure modes that will affect the product. Remember, the product from this process will usually be the incoming materials for another process.
   The brainstorming could follow either of two approaches. The approach could be brainstorming on all potential failure modes or it could be a series of directed brainstormings on each specific area: equipment and components, methods, materials, people, and the measurement system. The directed brainstormings will limit the team's scope allowing greater concentration on the specific area. It may uncover more potential failure modes than brainstorming on all areas at once. In either case, the goal of the brainstorming is to

uncover all failure modes that could even remotely occur without addressing (at this time) whether they will actually occur.
The brainstorming results should be organized and transferred onto the FMEA data collection form (Figure 5.63). Use a cause-and-effect diagram to aid in grouping related failure modes.

4. *List all potential effects of each failure.* Next to each of the potential failure modes on the FMEA form, list the potential effects for each failure, describing what will result if this failure occurs. The failure could impact other components in the system, leading to a domino effect. It could impact the whole process. It could obviously affect the customer, whether it be an internal customer (the next system or manufacturing process) or an external (paying) customer. The description of the potential effects should be as specific as possible.
5. *List the potential causes of each failure.* The potential causes will also be listed next to the potential failure modes on the FMEA form. These are the possible reasons why the failure could occur. This information is important later in the FMEA process to help direct the improvement efforts.
6. *List the current control.* For each of the potential causes of failure, list the controls that are in place to prevent each cause from occurring, to detect the cause of failure, or to detect the failure mode.
7. *Estimate the frequency or probability of occurrence.* The frequency or probability of occurrence for each cause of failure is rated from 1 to 10. Table 5.104 shows an example ranking scale for probability and frequency. This table is only an example of rankings. Each company or team should develop their own rankings. The ranking system used must remain constant throughout the FMEA.

   In estimating the occurrence probability, consideration must be given to those controls designed to *prevent* the cause of failure from occurring. (Neither the controls intended to detect failure after occurrence nor the severity of a failure occurring, should enter into this estimate of occurrence probability.)
8. *Estimate the severity.* For each of the effects of failure, rank the seriousness of the failure, if it had occurred, from 1 to 10. Table 5.105 shows a possible ranking scheme for severity. Again, each company should establish their own standardized ranking scale and criteria, especially for quality problems that effect their final customers. Teams working on internal processes could establish their own rankings of the severity of quality problems on their internal customers.
9. *Estimate the detection ranking.* The detection ranking is the probability of detecting a defect or quality problem before it is sent to the

POTENTIAL
FAILURE MODE AND EFFECTS ANALYSIS
(PROCESS FMEA)

PAGE 3 OF 5
FMEA TEAM MRB, MBM, KBM
SUPERVISOR EMH
DATE: ______

PROCESS MIXING
OUTSIDE SUPPLIERS AFFECTED ______

PART NAME/ PART NUMBER	PROCESS FUNCTION	POTENTIAL FAILURE MODE	POTENTIAL EFFECT(S) OF FAILURE	POTENTIAL CAUSE(S) OF FAILURE	EXISTING CONDITIONS: CURRENT CONTROLS	OCCURRENCE	SEVERITY	DETECTION	(A.P.M.) FMEA PRIORITY NUMBER	RECOMMENDED ACTION(S) AND STATUS	RESULTING: ACTION(S) TAKEN	OCCURRENCE	SEVERITY	DETECTION	(A.P.M.) FMEA PRIORITY NUMBER	RESPONSIBLE ACTIVITY
101		MATERIAL CONTAMINATION	UNUSEABLE MATERIAL	• IMPROPER CLEANING IN PRODUCT CHANGE OVER	• OPERATOR	6	4	2	48							
				• ROOF LEAKING AROUND HOOD	• VISUAL	3	1	1	3							
				• RUSTY ROLLS	• VISUAL	2	3	1	6							
				• CONTAMINATED MASTER BATCH	• BURTO SPC VISUAL	3	10	10	(300)	DISCUSS WITH BURTO						
				• PLASTIC BAG FELL INTO MIX	• VISUAL	2	10	1	20							
				• BROWN RELEASE PAPER (FROM MAT'L TURNMILL) FALLS INTO MIX	• VISUAL	2	10	1	20							
				• CONTAMINATED RE-MILL/TAILINGS	• VISUAL	3	10	3	90							
				• USE WRONG RE-MILL/TAILINGS	• ID TICKET	1	10	10	100							
				• FULL GREASE TROUGHS	• VISUAL	10	5	2	100							
				• MATERIAL TRAPPED UNDER SIDE GUARDS	• VISUAL	3	3	2	18							

FIGURE 5.63. FMEA data collection form.

Table 5.104 FMEA Ranking Scale for Probability and Frequency

Ranking	Criteria
1	Essentially no chance of occurrence (but still "possible")—once every 10 or more years or much greater than $\pm 6s$ probability
2	Remote chance of occurrence—once every 3–5 years or less than 2 occurrences in 10^9 events ($\pm 6s$) probability
3	Very low probability—once every 1–3 years or 6 occurrences in 10^8 events ($\pm 5s$)
4	Low probability—once every 6 months to 1 year or 6 in 10^6 ($\pm 4s$)
5	Moderately low probability—once every 1–3 months or 3 in 1000 ($\pm 3s$)
6	Moderate probability—once every 2 weeks to 1 month or 1 in 100 ($\pm 2.5s$)
7	Moderately high probability—once every 1–2 weeks or 5 in 100 ($\pm 2s$)
8	High probability—once every week or 13 in 100 ($\pm 1.5s$)
9	Very high probability—once every 2–3 days or 3 in 10 ($\pm 1s$)
10	Extremely high probability—once per day or more than 3 in 10 ($< \pm 1s$)

customer. Table 5.106 shows one ranking scheme. This, again, should be customized for each company.

When considering the detection ranking, assume that the failure mode has occurred. Then the capabilities of the controls designed to detect the failure (and to prevent shipment of defectives) are assessed.

Table 5.105 FMEA Ranking Scale for Severity

Ranking	Criteria
1	If failure occurred, cannot be detected by the customer and would not impact customer's process or product
2	Very low severity—cannot be detected by the customer but may cause a small change in their process or product
3	Low severity—enough for the customer to notice and may cause a small change in their process or product
4	Moderately low severity—minor nuisance to the customer but can be corrected for in their process
5	Moderate severity—minor nuisance to the customer with some performance degradation to the customer's process
6	Moderately high severity—subsystem degradation enough to initiate customer complaint
7	High severity—high degree of customer dissatisfaction due to impact on their process or product
8	High severity—unit not fit for use or inoperable
9	Very high severity—failure causes noncompliance with federal regulations
10	Extremely high severity—failure would create safety hazard to customer

Table 5.106 Typical FMEA Ranking Scheme

Ranking	Criteria
1	Remote chance that the defect will not be detected—the defect is obvious
2	Very low chance—100% automated inspection in place for a simple property or SPC used with $C_{pk} \gg 2.0$
3	Low chance—automated inspection in place
4	Moderately low chance—SPC is used to predict (detect) defects, $C_{pk} > 1.5$
5	Moderate chance—SPC used to "detect" defects, $C_{pk} < 1.5$
6	Moderately high chance—detection done through AQL sampling plan
7	High chance—100% manual inspection
8	Very high chance—100% manual inspection of property that is not easy to detect or 200% inspection is used
9	Extremely high chance—visual inspection
10	No method developed to detect defects or product is not inspected

10. *Calculate the "risk".* This is not a statistical risk calculation. It is a relative ranking method used to prioritize the items with the greatest risk to focus improvement efforts. The calculation for the "risk" is

$$\text{risk} = \text{occurrence} \times \text{severity} \times \text{detection}$$

where the highest possible risk is 1000 and the lowest is 1. Do this for all of the causes of failure.

11. *Determine recommended actions.* The FMEA team should use the Pareto principle to identify those causes of failure with the highest risk. These will be the first items targeted for corrective actions, although the team should also consider improving those causes of failure with very high occurrence rankings. A cutoff point may be set where all items with a risk greater than a preset "danger" level (such as 150 or 200) must be corrected before the process is put into operation.

To reduce the risk, the improvement effort can focus on:

- Reducing the probability or frequency of occurrence.
- Reducing the severity of failure occurring.
- Improving the detection methods.

The improvement efforts may focus on only one of these areas or the efforts may strive for some improvement in all three to reduce the overall risk. The team should establish responsible individuals and set a due data for each of the items slated for corrective measures or improvement.

12. *Follow-up on actions.* The team should review actions taken and then revise the occurrence, severity, and detection rankings. The new risk number can be calculated from the new rankings to determine if the actions were effective in reducing the risk to an acceptable level. When all the ratings are below the danger level, the team may elect to disband. Of course, they may also elect to continue the improvement process by working down their Pareto of risks that are unsatisfactory. It is recommended that each FMEA team review their progress with management before they disband.

5.13.3 Process Start-Up

Figure 5.64 shows the typical learning curve for companies starting up a new process or product in the process industries. Unfortunately, this learning curve is generally accepted as a necessary part of creating new business. Management tolerates low yields, poor quality, late shipments, and wasted manpower during the time the new process is on the learning curve. Few people have recognized that "learning curve" is synonymous with "waste" (Figure 5.65); even fewer people understand that the learning curve is an evil that is not necessary. Some of the same tools described earlier for process improvement can be applied to new processes to minimize waste, which results in improved profits and improved customer satisfaction. The use of these tools may delay the initial runs of materials for sales purposes due to the heavy investment in planning, but will result

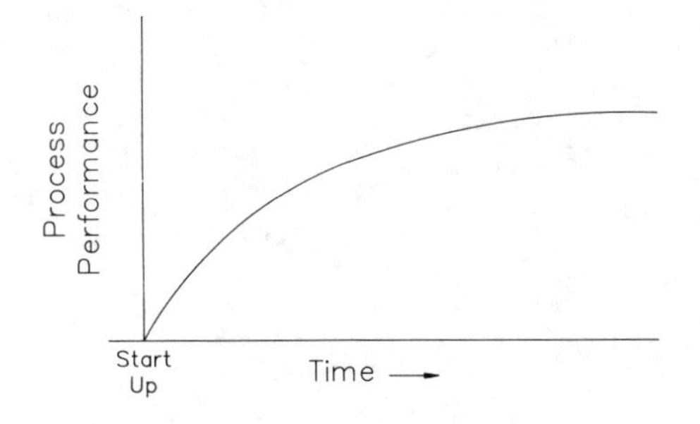

FIGURE 5.64. Typical start-up curve for the process industries.

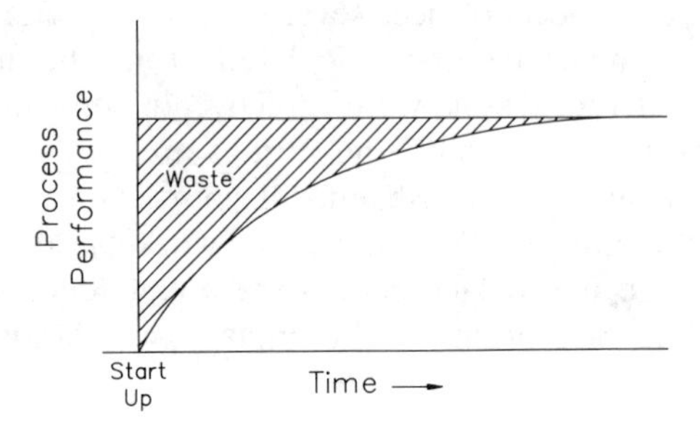

FIGURE 5.65. Waste in the typical start-up.

in an acceptable process more quickly and with less waste than the conventional approach. This can be seen in Figure 5.66.

The major process improvement techniques used for process start-up are brainstorming, failure mode and effect analysis (FMEA), and design of experiments (DOE). These improvement tools are initiated in the design phase. It is important to start using these tools as soon as possible because changes and improvements to the process are less costly the earlier they are uncovered. These techniques, also known as concurrent engineering, can convert many of the design engineers' "gut feelings" or opinions on product or process design parameters to facts. In some cases, it may not be possible to gather all of the facts, but the use of these techniques taps the

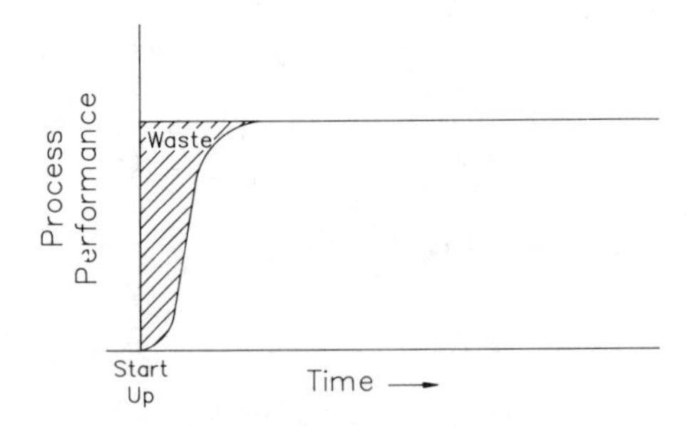

FIGURE 5.66. Waste reduction in concurrent engineering.

knowledge of an entire team and result in a consensus of opinions. The team approach yields a more powerful and comprehensive plan than individuals can possibly develop working alone.

Failure Mode and Effect Analysis (FMEA)

Any new product should undergo a design (FMEA) once the preliminary product design is available. The product design FMEA looks at potential failure modes for the product itself. The FMEA brainstorming session explores what can go wrong with the product that can effect the function of the product in its application. The failure modes that are identified are examined for severity, occurrence, and detection. Design changes can then be made to correct those items that have a high associated risk.

Another brainstorming session should be held on ideas to improve the manufacture of the product. This could also be examined in an FMEA format by focusing the brainstorming on those aspects of the product design that would make the product difficult to manufacture. These "failure modes" could be addressed by improvements to the product design. If they cannot be easily acted on, at least, the process design engineer has valuable input on potential problems that the process must be designed to overcome. For example, if a product is heat sensitive, the process can be modified in the design stage to take the heat sensitivity into account.

Mixture Design of Experiments

For new products, such as chemicals or materials, it is desirable to run a series of mixture design DOEs on the product prior to the detailed process design. The goal of the DOEs may be to identify the maximum (or minimum) variation that can be tolerated in each raw material without affecting the critical properties of the final product. The factors for the DOEs could include raw material weight, volume, or variations in the key properties of the raw material. (*Caution*: Some of these experiments may have been done in developing the product. It is not necessary to duplicate work done earlier in the design process.) The results from the DOEs may define the design requirement for the raw material feed equipment and may also provide valuable feedback to the material supplier by helping to determine the variation tolerable in the key properties and by laying the groundwork for material specifications.

Additional DOEs, in the pilot plant or laboratory, exploring the operating parameters for the process would also be valuable input to the process design engineer. Care must be taken here to consider scale-up factors when translating DOE results to full size equipment specifications.

Mistake-Proofing the Construction Phase

With the completion of the design, conduct another FMEA on the design itself. The focus of this brainstorming session would be to generate ideas on what could go wrong that would affect the construction or the operation of the process. This FMEA could be augmented with additional brainstorming on those aspects of the construction that might fall prey to subtle changes and therefore affect the process start-up or operation. By working up front to reduce the occurrence, severity, and detection of design errors, the construction and start-up will be smoother and quicker.

As-Built FMEA

Once the installation is complete or nearly complete, the FMEA team should be reassembled with the operators who will run the process. (As always, the core team should include some of the operators.) This enlarged team could conduct an FMEA on the process as installed. (The process as installed or constructed is often different than the original process design.) Use the results from this FMEA to correct errors and improve the process immediately while the construction crew is still available. If there is a delay in completing the FMEA and the construction crew is disbanded, corrections and improvements will take longer because the infamous final project "punchlist" often drags out for months before it is completed.

Safety Review

As soon as power is available, each piece of equipment should be run through its operating parameters independently first, and then in a system mode. Always take safety into account first when starting up a new piece of equipment. It is recommended that safety analysis, such as a safety FMEA or fault tree analysis, be done on each subprocess prior to full start-up. Some companies start safety analysis on both the product and process during the design phase.

Test for Stability and Capability

The initial start-up of the process should be carefully planned beforehand. First, allow the process to reach steady state and then perform an overall process capability study to determine if the process design criteria were met. If there are discrete operations within the process, capability studies on those individual operations or subprocesses should be conducted to understand the inherent variability in the process components. The results of the capability studies determine the next improvement steps. If the results indiciate the process is not stable, use the problem solving techniques described in Section 5.2 to identify and correct the special causes of variation. If the results show a stable process, conduct a series of designed experiments aimed at understanding the variation in the process and the direction to take to reduce the variation and improve the process.

In experimental strategy, the first DOE to be performed is usually a screening experiment. If the capability study indicated the process was stable but not capable, the objectives of the screening experiments are to minimize variation, optimize key product characteristics, and maximize yield. Once the process is stable and capable, additional DOEs should be performed to gain an understanding of the interactions of the key process variables and the effects of adjusting those variables on the process output. Full factorial, response surface, and EVOP designs are helpful at this stage.

An important resource of concurrent engineering often overlooked is the supplier to the process. Supplier parameters may be key to the development of a robust process. In the DOE strategy, consider the entire business continuum from supplier to customer. The DOEs could include important parameters from the supplier's process as well as from the customer's process.

The Improvement Payoff

The DOE portion of the startup process is not the phase of the project to take shortcuts. Often, there is pressure from sales and marketing to get the process running so that product can be shipped. The engineering/manufacturing start-up team should resist this pressure and explain the payoff the entire organization will receive from the completion of the DOEs. Use Figures 5.65 and 5.66 to put the explanation into language of management—money. Money, actually long term profits, is what continuous improvement is about. With a long term focus on continuously improving all processes, by everyone, all of the time, the bottom line can't help but grow.

6

Applications

6.1 OVERVIEW

Many managers and engineers narrowly view SPC and SPI as applicable strictly to manufacturing processes or those processes producing tangible products. But the "administration processes" that support the manufacturing organization can also benefit from the same SPC and SPI tools. The same techniques that are applied to control and improve manufacturing processes can be used to control and improve service processes and their products. This is true regardless of whether the services directly support manufacturing (e.g., maintenance, personnel, and materials control) or if the services support the product in the customers' uses or applications (e.g., customer service and technical support). All of these processes can use SPC and SPI tools, if applied correctly.

As discussed in Chapter 1, continuous improvement is the responsibility of everyone in the organization. If we view each function as a process, we can then see that function as adding value to an input (data, product, or services supplied) to create an output (the process "product") for the customers. Unfortunately, as we strive to add value to the input, we often add costs without value. The cost adders may be:

- Features the customers (our internal customers, such as the department receiving our work, as well as external customers) do not need or even want.
- The cost of reworking (reworking a product, a report, a test) before delivering a product.
- The cost of delivering a flawed or incorrect product—an output that multiplies the cost adders by driving them into the next process.

At the same time that the support service costs are being reduced through improvement efforts, the quality of the support services will itself be improved. This provides another marketplace advantage. A company can provide an excellent quality product at competitive prices, yet fare poorly in the marketplace if they cannot provide excellent service to their customers as well. It's no different than a restaurant example most of us can relate to: Even though the chef may create an excellent meal, if the service is poor, we do not return to that restaurant. We'll try other restaurants until we find one that provides both an excellent meal and excellent service. That will be the restaurant we repeatedly take our customers and our families to. We're looking for high quality in the total transaction, not just part of it.

The combination of high quality products and excellent support services may provide the company with enough of a competitive advantage to allow them to charge a premium price, just like an excellent restaurant can. That company can base their unique sales proposition on the fact that their product provides the lowest overall cost of ownership to the customer, even at an apparent premium price.

However, we must never underestimate our competition. They will also be working to improve their products and services. In order to stay ahead of the pack, a company needs to involve all of their employees, both in manufacturing and in the support services, in their SPC and SPI efforts using a team process. That's commitment to TQM.

The remainder of this chapter will provide some suggestions for practical applications of the basic tools plus SPC and SPI in the support services for those companies committed to improvement in all of their activities. These lists of applications for the various support functions are not intended to be complete. They are provided to help stimulate thinking about where basic tools, SPC, and SPI techniques may be applied to improve their service (administration) organizations.

6.2 CUSTOMER SERVICE

Statistical Tool	Application	Ways to Use
Pareto diagrams	Complaints analysis	Use the Pareto diagram as a tool to help reduce complaints. Pareto diagrams can be made by: type of complaint (e.g., cosmetic, dimensions,

Statistical Tool	Application	Ways to Use
		physical properties) cause (e.g., poor workmanship, employee error, equipment malfunction) formulation customer age of the part in the product life cycle (i.e., is it a prototype or a newly introduced part or an existing part?)
	Returned products analysis	Use the same types of Pareto diagrams as for complaint analysis.
	Order errors analysis	Pareto diagrams can be made by: type of error cause by employee
	Invoice errors analysis	Use the same types of diagrams as for order errors analysis.
	Customer survey response analysis	Plot the responses from the highest rated question to the lowest. In this case, the areas to work on are the lowest rated questions, not the highest, as in standard Pareto diagrams.
Concentration diagrams	Order errors	Use an actual order entry form or a printout of the computer screen for order entry. Place an × on the form at the location of each error. This will identify data entry areas that are unclear or difficult to do so that they can be improved.
	Invoice errors	Use the same approach as for order errors.

Statistical Tool	Application	Ways to Use
Flowcharts	Order entry process	Flowchart the entire order entry process from receipt of order up to the order being put into the manufacturing schedule. This should include any feedback steps to the customer. Areas of duplicate effort or other areas that can be improved can often be identified just by making a flowchart of the process.
	Invoicing process	Flowchart the invoicing process from the shipment of the goods up to the mailing of the invoice.
	Process for collecting information to respond to customer inquiries	A flowchart can help the customer service representative to quickly identify the department or the individual with the needed information.
Run charts and trend charts	On-time shipments	Plot (on a weekly or monthly basis) the percentage of orders that are shipped on the original date confirmed to the customer. Partial orders should not be counted as on time.
	On-time delivery	Plot (on a weekly or monthly basis) the percentage of orders, that are delivered on the agreed date. If a company does not have control over the transportation method, an on-time delivery may be defined as delivery on the agreed date ± 1 day.
	On-time service	On-time service is the ability to meet the customer's

Statistical Tool	Application	Ways to Use
		initially requested delivery date no matter how unrealistic this may be. Improving this measure is an indication that cycle times are being reduced or that customer's expectations and supplier's capabilities are matched more closely. Plot this on a weekly or monthly basis.
	Complaints	Plot (monthly or quarterly) the number of complaints received per time period.
	Response time to customer information inquiries	Plot (on a weekly basis) the average response time or the worst response time for that week. The customer service representative needs to log in customer inquiries and log out responses to track this metric.
	Response time for requests for quotations (RFQs)	Plot (on a weekly or monthly basis) the average response time or the worst response time for the time period chosen.
	Time of order placement to receipt of payment	This measure looks at the overall business cycle to see if it is improving. The average time per order could be plotted on a quarterly basis.
	Customer survey responses	Plot the trends in the data collected in customer satisfaction surveys.
Control charts	Same applications as trend charts	Some companies may wish to use attribute control charts to identify the occurrence of a special cause of

Statistical Tool	Application	Ways to Use
		variation. However, companies must learn not to just accept a stable or controlled process. They must strive to improve their processes. Improvement may cause out-of-control conditions on the "good" side. If the improvement rate is slow, simply recalculate the control limits periodically. If the improvement is rapid, use run charts instead of control charts until the improvement rate flattens and the process stability is again achieved.
FMEA	Order entry process	Analyze the risks for potential errors in the order entry process. Change the process to reduce those areas with high risks for errors.
	Invoicing process	Analyze the invoicing process for areas with high risks for potential errors.
Problem solving process	For addressing complaints or any other area identified to be a potential problem.	

6.3 MATERIALS CONTROL

Statistical Tool	Application	Ways to Use
Pareto diagram	Raw material outages	Make diagrams by product and by cause for those raw materials outages that occur most frequently.

Statistical Tool	Application	Ways to Use
	Trucker performance	Pareto of on-time performance of different trucking firms (either for delivering or shipping goods).
Trend charts	Raw material inventory levels	Plot inventory levels (on a weekly or monthly basis) to look for patterns or to track improvement.
	Work in-process (WIP) inventory levels	Same as for raw materials.
	Finished goods inventory levels	Same as for raw materials.
	Inventory accuracy	Measure accuracy of physical counts of inventory versus quantity shown to be in inventory. Calculate the percentage of the actual difference from the inventory control amount and plot on a monthly basis.
	Daily shipments—dollar values or quantities	Look for trends in the shipments made during the month. Are most shipments made on Friday or on the last week of the fiscal month? Knowing these trends should help to plan staffing levels.
	Number of moves made	In a warehouse, plot the number of moves made each day. Again, this may help plan staffing or indicate a need to level the warehousing process if there are large peaks or valleys.
	Response time to material requests from the shop floor	Plot the average time from request for material to delivery from the warehouse

Statistical Tool	Application	Ways to Use
		on a daily or weekly basis. Work to improve the response time.
Control charts	The same applications as trend charts; use trend charts if you're looking to improve the process and control charts if you want to maintain it (control it) as is.	

6.4 MAINTENANCE

Statistical Tool	Application	Ways to Use
Pareto diagrams	Equipment downtime	Make diagrams of hours of downtime by piece of equipment to indicate which pieces of equipment require the most effort.
	Process downtime	Pareto the hours of downtime for the processes in an operation, or make a diagram for the pieces of equipment that cause the most downtime in a process.
	Maintenance costs	Make diagrams of maintenance costs by piece of equipment, process, or department. Look at this on a monthly or quarterly basis.
	Maintenance effort	Pareto the hours spent on maintenance by process line or by department.

Statistical Tool	Application	Ways to Use
	Parts cost	Make diagrams of those pieces of equipment or processes that cost the most for replacement. Construct diagrams of all maintenance replacement parts, supplies, and services costs to see where the money is being spent and to ensure that adequate controls are in place.
	Parts or supply outages	Pareto parts and/or maintenance supplies by those that run out most frequently.
Flowcharts	Work order process	Flowchart the process for the work order process from inception to the communication that it is complete.
	Maintenance item purchasing process	Charting the process should start with recognition of the need and end with the part or supply in-house.
Concentration diagrams	Work orders	Make concentration diagrams of where work orders are received from in the plant to see if there is any apparent concentration of effort; if so, work to find the root cause.
Trend charts	Time to complete work orders	Plot weekly or monthly trends of the average time or maximum time it takes to complete a work order. This should be a work order for a routine equipment repair and not a work order for a major capital project or a safety item.

Statistical Tool	Application	Ways to Use
	Number of work orders	Graphing this on a weekly or monthly basis may show trends in overall equipment performance. A downward trend may indicate that equipment is running more reliably but it also may be an indication that little attention is being paid to the condition of the equipment.
	Time to failure	Plot the time to failure for key pieces of equipment. The *x* axis would be the failure number and the *y* axis would be the time since the last start-up from a repair. This information can be used to establish preventive maintenance timing.
	Ratio of hours spent on preventive maintenance to hours spent on emergency repairs	Plot this on a monthly or quarterly basis. The higher the ratio, the better. A downward trend is dangerous, indicating emergency repairs are overtaking preventive maintenance efforts.
Control charts	Use for the same applications as the trend charts. Use control chart where the goal is to maintain the performance level.	
Normalized control chart	Time to complete repairs	Plot the time to complete a repair, normalized to the standard time for that repair. Data bases of stan-

Statistical Tool	Application	Ways to Use
		dard times for common repairs are publically available.

6.5 MARKETING AND SALES

Statistical Tool	Application	Ways to Use
Pareto diagrams	Sales	There are many ways to use Pareto diagrams to analyze sales. Among them are by: plant or store area customer customer sex product line population density customer age customer earnings
	Inquiries from ads	Make diagrams by the number of inquiries from ads in different magazines or from different mailings to determine which is the most effective.
Trend charts	Sales parameters	Use these charts to analyze trends in: claims mandatory recalls returns amount of field service amount of repeat service required

Statistical Tool	Application	Ways to Use
		customer retention ad responses sales forecast accuracy bingo card responses
DOE	Responses to ads	Conduct a designed experiment on those factors that get the most responses, the highest number of quality responses, or the highest amount of sales. Among the factors studied could be magazine, ad positioning, size of ad, and colors used.
	Responses to direct mailings	Again, look for the factors that give the most responses, highest number of quality responses, or the largest amount of sales. Some of these factors might be type of envelope, type of enclosures, who in the organization signs the letter, number of pages enclosed, and the effect of a gift (e.g., a quarter) enclosed.
Correlation analysis (scatter diagram)	Number of quotes versus sales	Determine if there are relationships between the number of quotes from a given period and the sales volumes from those quotes after six months and/or one year. If there is a relationship, this would help the company plan early for future sales requirements.
	Visitors to display booths at trade shows versus sales	Look at this measure to help determine the worth of participating at a particular trade show.

6.6 HUMAN RESOURCES (PERSONNEL)

Statistical Tool	Application	Ways to Use
Check sheets	New employee training	Establish a check sheet on all of the topics a new employee is to be trained in. Once the training is complete, have the employee sign the check sheet for record keeping.
	Interviewing	Use the check sheet to aid the interviewer in what topics to cover.
Pareto diagrams	Absenteeism	Look at absenteeism by: plant area or department day of the week month individual job title shift
	Grievance activity	Analyze grievances by: plant area or department shift
Trend charts	Absenteeism	Look for trends in absenteeism over time.
	Recruitment time	Keep trend charts for the length of time it takes to recruit candidates for a position. In large firms, have separate charts for different jobs. The x axis on the trend chart would be $1, 2, 3, \ldots$ for the first individual hired, the second, and so forth. The y axis would be the time.
Control charts	Absenteeism	Use an attribute control chart to show that absenteeism has (or hasn't) stayed in control.

Statistical Tool	Application	Ways to Use
Correlation analysis (scatter diagram)	Test scores versus performance	Have existing employees take an employment test and correlate their scores to their job performance. Use this correlation to help predict how job applicants will perform based on their test results.
DOE	Recruiting process	Conduct a DOE on those factors that affect the recruiting process. One response to study would be the recruitment time. Factors may include advertising source, type of ad, size of ad, personnel agency.

6.7 PURCHASING

Statistical Tool	Application	Ways to Use
Flowcharts	All purchasing processes	Construct separate flowcharts of the processes for buying raw materials, capital goods, services, maintenance/repair organization (MRO) items, and departmental operating supplies. Look for differences between the processes and make them similar where possible. Try to streamline the processes.
Brainstorming	Negotiations	Use the brainstorming approach in preparation for purchasing negotiations. The brainstorming goal could be simply how to get the best price or it may be

Statistical Tool	Application	Ways to Use
		more complex, such as what does the seller want from the deal.
Trend charts	Supplier on-time delivery performance	Plot the supplier performance to a delivery window of ± 1–2 days or $+0$, -3 days. These data should be fed back to the supplier's management. It could be part of a monthly or quarterly "report card" sent to each key supplier.
	Material quality	Plot the percentage of parts or material that meets the quality standards upon receipt. This should also be part of any report card sent back to key suppliers.
	Supplier response to inquiries	This chart would look for trends in the length of time it takes for a supplier to respond to an inquiry. The inquiry could be a request for information or a request for a quote.
Control charts	The same applications where trend charts are used could also be monitored with control charts.	
Normalized control charts	Purchased prices versus standard	Normalized purchased material or part price to the standard costs. Look for abnormal trends in pricing.
Correlation analyses	Purchased price versus price indices	Analyze the purchased price versus national price indices for that industry over a period of time. Ask the supplier what they're

Statistical Tool	Application	Ways to Use
		doing right if their prices are running below the indices or wrong if their prices are rising at rates above the indices.
Problem solving process	Supplier problems	Train suppliers in your problem-solving approach so that they have an organized process and so that they will be speaking the same language your people do.

6.8 QUALITY DEPARTMENT

Statistical Tool	Application	Ways to Use
Pareto diagrams	Output by inspection	Plot the average number of inspections made by each inspector per day or in a week. Learn from the inspectors with high outputs what they are doing differently from the others. If it is something positive, train the others. If they are cutting corners, retrain them.
	Test equipment downtime	Make diagrams of test equipment downtime to allow effort to be focused on the equipment with the most downtime.
	Test equipment recalibration	Plot which test equipment requires recalibration most frequently.
Trend charts	Completion time for services	Graph the average time on a weekly or monthly basis for QA to complete a request for services. Look for

Statistical Tool	Application	Ways to Use
		a positive trend that shows continual improvement in the QA process.
	Completion time for inspections	Look for trends in the average time it takes for the inspectors to complete an audit or inspection. Plot this on a weekly or monthly basis.
	First pass lot acceptance	Plot on a weekly or monthly basis the percentage of lots that pass QA inspection the first time through. Do not count lots that were previously rejected and then reworked.
	Quality department costs	Monthly, graph QA costs. As product quality improves, there should be a downward trend in QA costs. Some companies break down their quality costs into three areas —prevention costs, detection costs, and rework costs. These would include all costs in these areas, not just the costs for the quality department. Most companies just implementing a TQM approach should look for detection and rework costs to trend downward with prevention costs trending upward at first, and then levelling off.
	Time to recalibration	To understand the amount of time to expect between recalibrations for a piece of equipment, graph the

Statistical Tool	Application	Ways to Use
		length of time since the last calibration on the y axis and the occurrence number on the x axis. Two responses could be studied with these data. The first is to use the data to set a calibration schedule to prevent the instrument from going out of calibration. The second is to improve the calibration process or the equipment itself and use the graph to monitor the improvement.
Problem solving techniques	Measurement system	Use the problem-solving techniques to attack the causes of errors in the measurement system.
	Determining the best measurement method	Use brainstorming and cause-and-effect diagrams to identify the best method to measure a product or a part.

6.9 SAFETY AND ENVIRONMENTAL

Improvement Tool	Application	Way to Use
Concentration diagram	Accident location	Use a plot of the facility and mark on it where each accident occurs. You may want to use a code such as an × for each OSHA recordable, ○ for each lost-time accident, △ for each accident requiring first aid, and a □ for each inci-

Improvement Tool	Application	Way to Use
		dent that occurred but didn't require first aid.
	Environmental incident	Keep a separate concentration diagram of the facility showing where environmental incidents (e.g., spills) occur.
	Injury location	Draw a front and back view of the human body. Mark on it where injuries occur. The improvement activities can then concentrate on eliminating the types of accidents that cause the most frequent injuries.
Histograms	Accident analysis	Accidents can be put in a number of categories for analysis: by cause (such as unsafe equipment, unsafe act, unsafe procedure, unsafe conditions, or a combination of these) by department by type of injury (such as strains, cuts, bruises, or broken bones) by shift by time of the day by time of the week by time of the month by the length of time the employee has been on the job by years of service by age by employee
	Environmental incident	These can be categorized similarly to accidents: by cause (such as equipment malfunction,

Improvement Tool	Application	Way to Use
		incorrect procedure, or operator error) by type (such as chemical spills, emissions to atmosphere, or releases to the river) by department by shift by time of the day
Pareto diagrams	Use for the same type of analysis as histograms	
Trend charts	OSHA incident rate (OSHA IR)	Plot the OSHA IR on a monthly basis for the facility.
	Lost workday case incident rate	Plot these more severe accidents on a monthly basis.
	Number of accidents	Plot on a weekly or monthly basis.
	Number of near misses (incidents)	Plot on a weekly or monthly basis.
Control charts	Use attribute charts for the same applications as shown for trend charts	
Brainstorming	Ideas on improving safety	
	On unsafe conditions or potential unsafe conditions	
FMEA	New processes	Conduct FMEAs for both safety and environmental concerns on new processes. They should be conducted on the preliminary design of the process and on the process as installed (before start-up).
	Existing processes	Here also, FMEAs can be conducted for both safety

Improvement Tool	Application	Way to Use
		and environmental concerns to prioritize improvement areas.
Problem solving techniques	Accident investigation	Use problem-solving techniques to uncover the underlying cause of the accident and to identify preventive measures.

6.10 MANAGEMENT

Improvement Tool	Application	Ways to Use
Pareto diagram	People time by project	Pareto the department work by project.
	Your time	Pareto the time you spend by activity.
Trend chart	Employee suggestion participation rate	Chart percent of employees submitting improvement suggestions per month.
	Operating profit	Chart operating profits by month for the operating unit.
Control chart	$\overline{X}$ and R chart for standard cost of sales	Track percent COS per week. Out-of-control points will indicate product mix changes from an historical basis.
	Forecast accuracy with a CUSUM signal chart	Track difference in actual to forecast.
	P-chart for product returns	Track product returns as a percent of product sales dollars or units.
	Internal damage of rework rates	Track as a percent of manufacturing cost per unit of time. If the control chart always shows this measure-

Improvement Tool	Application	Ways to Use
		ment to be out of control, use a trend chart instead.
Flowchart	The planning process	Flowchart the planning process from start to finish.
PERT chart	The planning cycle	Use the planning process flowchart (above) to develop a time-based PERT chart and determine the planning cycle critical path and shortest possible time.
Problem-solving tools	The planning process	Use the problem-solving cycle to identify areas to improve the process. Establish project teams to work through those improvement potentials.

A

SPC

A.1 STANDARD DEVIATIONS ON THE NORMAL CURVE

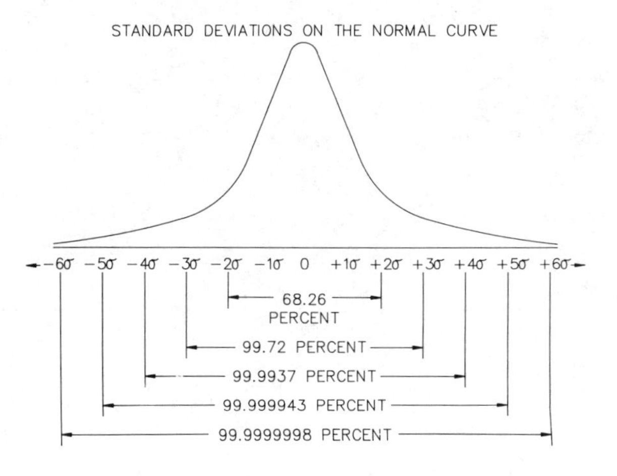

FIGURE A.1. Area under the normal curve for up to ± 6s.

A.2 PROCESS CAPABILITY MEASURES

The process capability ratio C_r and the process capability index C_{pk} are two indicators of process capability. They indicate how the inherent, common cause variation in the process output compares to the specification. They are calculated as shown.

C_r

$$C_r = \frac{\text{variation in process output distribution}}{\text{specification total tolerance range}}$$

Variation in process output distribution = $6 \times s$

Specification total tolerance range = upper specification limit (USL) minus lower specification limit (LSL) = USL − LSL

$$C_r = \frac{6s}{\text{USL} - \text{LSL}}$$

C_{pk} *(Two-Sided Specification)*

$$C_{pk} = \text{minimum of } \{C_{pu}, C_{pl}\}$$

where

C_{pu} = compares the upper half of the variation in the process output distribution versus the upper specification limit (USL)

C_{pl} = compares the upper half of the variation in the process output distribution versus the lower specification limit (LSL)

$$C_{pu} = \frac{\text{upper specification limit} - \text{process output mean}}{\text{One-half of the variation in the process output}}$$

$$= \frac{\text{USL} - \bar{x}}{1/2(6s)}$$

$$= \frac{\text{USL} - \bar{x}}{3s}$$

$$C_{pl} = \frac{\text{process output mean} - \text{lower specification limit}}{\text{One-half of the variation in the process output distribution}}$$

$$= \frac{\bar{x} - \text{LSL}}{1/2(6s)}$$

$$= \frac{\bar{x} - \text{LSL}}{3s}$$

C_{pk} (One-Sided Specification)

If the specification is for a minimum value, use:

$$C_{pk} = \frac{\text{process output mean} - \text{specification limit (SL)}}{\text{One-half of the variation in the process output distribution}}$$

$$C_{pk} = \frac{\bar{x} - \text{LS}}{1/2(6s)}$$

$$C_{pk} = \frac{\bar{x} - \text{LS}}{(3s)}$$

If the specification is for a maximum value, use:

$$C_{pk} = \frac{\text{specification limit} - \text{process output mean}}{\text{One-half of the variation in the process output distribution}}$$

$$C_{pk} = \frac{\text{SL} - \bar{x}}{1/2(6s)}$$

$$C_{pk} = \frac{\text{SL} - \bar{x}}{(3s)}$$

A.3 C_{pk} CONVERSION TABLE

C_{pk}	ppm Out of Specification
2.0	0.002
1.67	0.57
1.33	67
1.2	320
1.1	960
1.0	2,700
0.9	7,000
0.75	24,400
0.5	133,600

Note: These values assume the distribution is centered within the specification. If the distribution is NOT centered within the specification, LESS product will be out of specification.

A.4 CONTROL CHART FORMULAS

	CL	UCL	LCL
$\bar{x}$ & R			
$\bar{x}$	$\bar{\bar{x}}$	$\bar{\bar{x}} + A_2\bar{R}$	$\bar{\bar{x}} - A_2\bar{R}$
R	$\bar{R}$	$D_4\bar{R}$	$D_3\bar{R}$
p	$\bar{p}$	$\bar{p} + 3\sqrt{\bar{p}(1-\bar{p})/n}$	$\bar{p} - 3\sqrt{\bar{p}(1-\bar{p})/n}$
np	$\overline{np}$	$\overline{np} + 3\sqrt{\overline{np}\left(1 - \left(\overline{np}/n\right)\right)}$	$\overline{np} - 3\sqrt{\overline{np}\left(1 - \left(\overline{np}/n\right)\right)}$
c	$\bar{c}$	$\bar{c} + 3\sqrt{\bar{c}}$	$\bar{c} - 3\sqrt{\bar{c}}$
u	$\bar{u}$	$\bar{u} + 3\sqrt{\bar{u}/n}$	$\bar{u} - 3\sqrt{\bar{u}/n}$
$\bar{x}$ & s			
$\bar{x}$	$\bar{\bar{x}}$	$\bar{\bar{x}} + A_3\bar{s}$	$\bar{\bar{x}} - A_3\bar{s}$
s	$\bar{s}$	$B_4\bar{s}$	$B_3\bar{s}$
x & R_m			
x	$\bar{x}$	$\bar{x} + E_2\bar{R}_m$	$\bar{x} - E_2\bar{R}_m$
R_m	$\bar{R}_m$	$D_4\bar{R}_m$	$D_3\bar{R}_m$
$\bar{x}$, R_m, &R_w			
x	$\bar{\bar{x}}$	$\bar{\bar{x}} + 3\bar{R}_m/d_2$	$\bar{\bar{x}} - 3\bar{R}_m/d_2$
R_m	$\bar{R}_m$	$D_4\bar{R}_m$	$D_3\bar{R}_m$
R_w	$\bar{R}_w$	$D_4\bar{R}_w$	$D_3\bar{R}_w$
	Modified Control Charts		
	Upper center line = upper spec limit $- 3\bar{R}/d_2$		
	Lower center line = lower spec limit $+ 3\bar{R}/d_2$		
	UCL = upper center line $+ A_2\bar{R}$		
	LCL = lower center line $- A_2\bar{R}$		
	Normalized $\bar{x}$ & R		
$\bar{x}$	0	$+A_2$	$-A_2$
R	1	D_4	D_3
	Nominal $\bar{x}_V$ & R_V		
$\bar{x}_V$	$\bar{\bar{x}}_V$	$\bar{\bar{x}}_V + A_2\bar{R}_V$	$\bar{\bar{x}}_V - A_2\bar{R}_V$
R_V	$\bar{R}_V$	$D_4\bar{R}_V$	$D_3\bar{R}_V$
	Group Control Chart $\bar{x}_p$ & R_p		
$\bar{x}_p$	$\bar{\bar{x}}_p$	$\bar{\bar{x}} + A_2\bar{R}_p$	$\bar{\bar{x}} - A_2\bar{R}_p$
R_p	$\bar{R}_p$	$D_4\bar{R}_p$	$D_3\bar{R}_p$

A.5 CONSTANTS FOR CONTROL CHART FORMULAS

		$\bar{x}$ & R Charts		
		Factors for $\bar{x}$-Chart Control Limits,	Factors for R-Chart Control Limits	
Subgroup Size	Factor for Estimate Standard Deviation, d_2	A_2	D_3	D_4
2	1.128	1.880	—	3.267
3	1.693	1.023	—	2.574
4	2.059	0.729	—	2.282
5	2.326	0.577	—	2.114
6	2.534	0.483	—	2.004
7	2.704	0.419	0.076	1.924
8	2.847	0.373	0.136	1.864
9	2.970	0.337	0.184	1.816
10	3.078	0.308	0.223	1.777
11	3.173	0.285	0.256	1.744
12	3.258	0.266	0.283	1.717
13	3.336	0.249	0.307	1.693
14	3.407	0.235	0.328	1.672
15	3.472	0.223	0.347	1.653
16	3.532	0.212	0.363	1.637
17	3.588	0.203	0.378	1.622
18	3.640	0.194	0.391	1.608
19	3.689	0.187	0.403	1.597
20	3.735	0.180	0.415	1.585
21	3.778	0.173	0.425	1.575
22	3.819	0.167	0.434	1.566
23	3.858	0.162	0.443	1.557
24	3.895	0.157	0.451	1.548
25	3.931	0.153	0.459	1.541

Copyright ASTM. Reprinted with permission.

$\bar{x}$ & s Charts			x & R_m Charts			
Factors for $\bar{x}$-Chart Control Limit,	Factors for s-Chart Control Limits		Factors for x-Chart Control Limits,	Factors for R_m-Chart Control Limits		Subgroup Size
A_3	B_3	B_4	E_2	D_3	D_4	n
2.659	—	3.267	2.660	—	3.267	2
1.954	—	2.568	1.772	—	2.574	3
1.628	—	2.266	1.457	—	2.282	4
1.427	—	2.089	1.290	—	2.114	5
1.287	0.030	1.970	1.184	—	2.004	6
1.182	0.118	1.882	1.109	0.076	1.924	7
1.099	0.185	1.815	1.054	0.136	1.864	8
1.032	0.239	1.761	1.010	0.184	1.816	9
0.975	0.284	1.716	0.975	0.223	1.777	10
0.927	0.321	1.679	0.946	0.256	1.744	11
0.886	0.354	1.646	0.921	0.283	1.717	12
0.850	0.382	1.618	0.899	0.307	1.693	13
0.817	0.406	1.594	0.881	0.328	1.672	14
0.789	0.428	1.572	0.864	0.347	1.653	15
0.763	0.448	1.552	0.849	0.363	1.637	16
0.739	0.466	1.534	0.836	0.378	1.622	17
0.718	0.482	1.518	0.824	0.391	1.608	18
0.698	0.497	1.503	0.813	0.403	1.597	19
0.680	0.510	1.490	0.803	0.415	1.585	20
0.663	0.523	1.477	0.794	0.425	1.575	21
0.647	0.534	1.466	0.785	0.434	1.566	22
0.633	0.545	1.455	0.778	0.443	1.557	23
0.619	0.555	1.445	0.770	0.451	1.548	24
0.606	0.565	1.435	0.763	0.459	1.541	25

A.6 VALUES OF d_E, d_O, AND d_M FOR MEASUREMENT SYSTEM ANALYSIS

In measurement systems studies, the value of d_2 varies depending on which variation (repeatability, reproducibility, and material) is being calculated. The table that follows shows the matrix of d_2 values. To use this table, use the rules outlined here:

Repeatability d_E:

m = number of trials

g = number of samples times the number of operators

Reproducibility d_o:

m = number of operators
g = 1 because there is only one range calculation associated with the reproducibility calculation

Material d_M:

m = number of samples
g = number of operators

	m													
g	2	3	4	5	6	7	8	9	10	11	12	13	14	15
1	1.41	1.91	2.24	2.48	2.67	2.83	2.96	3.08	3.18	3.27	3.35	3.42	3.49	3.55
2	1.28	1.81	2.15	2.40	2.60	2.77	2.91	3.02	3.13	3.22	3.30	3.38	3.45	3.51
3	1.23	1.77	2.12	2.38	2.58	2.75	2.89	3.01	3.11	3.21	3.29	3.37	3.43	3.50
4	1.21	1.75	2.11	2.37	2.57	2.74	2.88	3.00	3.10	3.20	3.28	3.36	3.43	3.49
5	1.19	1.74	2.10	2.36	2.56	2.73	2.87	2.99	3.10	3.19	3.28	3.35	3.42	3.49
6	1.18	1.73	2.09	2.35	2.56	2.73	2.87	2.99	3.10	3.19	3.27	3.35	3.42	3.49
7	1.17	1.73	2.09	2.35	2.55	2.72	2.87	2.99	3.10	3.19	3.27	3.35	3.42	3.48
8	1.17	1.72	2.08	2.35	2.55	2.72	2.87	2.98	3.09	3.19	3.27	3.35	3.42	3.48
9	1.16	1.72	2.08	2.34	2.55	5.72	2.86	2.98	3.09	3.18	3.27	3.35	3.42	3.48
10	1.16	1.72	2.08	2.34	2.55	2.72	2.86	2.98	3.09	3.18	3.27	3.34	3.42	3.48
11	1.16	1.71	2.08	2.34	2.55	2.72	2.86	2.98	3.09	3.18	3.27	3.34	3.41	3.48
12	1.15	1.71	2.07	2.34	2.55	2.72	2.85	2.98	3.09	3.18	3.27	3.34	3.41	3.48
13	1.15	1.71	2.07	2.34	2.55	2.71	2.85	2.98	3.09	3.18	3.27	3.34	3.41	3.48
14	1.15	1.71	2.07	2.34	2.54	2.71	2.85	2.98	3.08	3.18	3.27	3.34	3.41	3.48
15	1.15	1.71	2.07	2.34	2.54	2.71	2.85	2.98	3.08	3.18	3.26	3.34	3.41	3.48
∞	1.128	1.693	2.059	2.326	2.534	2.704	2.847	2.970	3.078	3.173	3.258	3.336	3.407	3.472

Table reprinted with permission of: Automotive Industry Action Group, *Measurement Systems Analysis Reference Manual*, Troy, MI 1990.

B

Hypothesis Testing and Test Statistics

B.1 TYPES OF HYPOTHESIS TESTING

There are four types of hypothesis tests commonly used:

1. Test on a single mean (sample vs. population)

$$H_0: \mu = 12{,}000 \text{ lb/in.}^2$$

2. Test on two means (population vs. population)

$$H_0: \mu_1 = \mu_2$$

3. Test on a single variance (sample vs. population)

$$H_0: \sigma^2 = 1200 \text{ lb/in.}^2$$

4. Test on two variances (population vs. population)

$$H_0: \sigma_1^2 = \sigma_2^2$$

B.2 TYPES OF TESTS OF SIGNIFICANCE

Tests of significance use the characteristics of the normal curve to compare groups of data. There are four tests of significance:

Z test t test	Used for comparing either a sample mean to a population mean OR a population mean to a population mean.
F test	Used for comparing the variances of two populations.
χ^2 test	Used for comparing the variance of a sample to a population variance.

For each hypothesis test statistic (t, Z, F, and χ^2) we choose a value for α that is our confidence level for the test or the level of type I error that we will accept.

After completing the test statistic calculation, the calculated value is compared to the table value.

If test statistic (calculated) is greater than table value

Then we *reject* the null hypothesis

If test statistic (calculated) is less than or equal to table value

Then we do not reject the null hypothesis. We *theoretically* cannot accept the null hypothesis, but *practically* we do accept.

Test on a Single Mean

Use the t test if $n < 30$ and σ^2 is *un*known; use the Z test if $n \geq 30$ and σ^2 is known:

$$H_0\colon \mu = 5.0$$

Test on Two Means

Use the t test if σ_1^2 and σ_2^2 are *un*known; use the Z test if σ_1^2 and σ_2^2 are known:

$$H_0\colon \mu_1 = \mu_2$$

Test on a Single Variance

Use the χ^2-test statistic:

$$H_0\colon \sigma^2 = 1.89$$

Test on Two Variances

Use the F-test statistic:

$$H_0\colon \sigma_1^2 = \sigma_2^2$$

B.3 *t* TABLE

To use the *t* table:

1. Find the degrees of freedom (DF) in the left column:

$$DF = n - 1 \quad \text{or} \quad DF = n_1 + n_2 - 2$$

2. Choose the α level.
3. Calculate the probability based on α or $\alpha/2$ depending on the test:

$$\text{Probability} = (1 - \alpha) \times 100 \quad \text{or} \quad (1 - \alpha/2) \times 100$$

(See chart following this list.)

4. Match the row and column to get the *t* (table) value.

If $\alpha, \alpha/2 =$	Then use column labeled
0.005	99.5
0.01	99
0.025	97.5
0.05	95
0.10	90
0.20	80
0.30	70

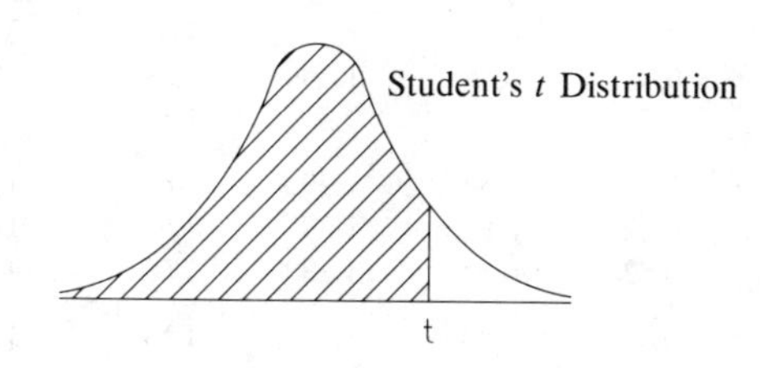

t Table

DF	Percentile Point						
	70	80	90	95	97.5	99	99.5
1	0.73	1.38	3.08	6.31	12.71	31.82	63.66
2	0.62	1.06	1.89	2.92	4.30	6.96	9.92
3	0.58	0.98	1.64	2.35	3.18	4.54	5.84
4	0.57	0.94	1.53	2.13	2.78	3.75	4.60
5	0.56	0.92	1.48	2.01	2.57	3.36	4.03
6	0.55	0.91	1.44	1.94	2.45	3.14	3.71
7	0.55	0.90	1.42	1.90	2.36	3.00	3.50
8	0.55	0.89	1.40	1.86	2.31	2.90	3.36

t Table *(Continued)*

DF	Percentile Point						
	70	80	90	95	97.5	99	99.5
9	0.54	0.88	1.38	1.83	2.26	2.82	3.25
10	0.54	0.88	1.37	1.81	2.23	2.76	3.17
11	0.54	0.88	1.36	1.80	2.20	2.72	3.11
12	0.54	0.87	1.36	1.78	2.18	2.68	3.06
13	0.54	0.87	1.35	1.77	2.16	2.65	3.01
14	0.54	0.87	1.34	1.76	2.14	2.62	2.98
15	0.54	0.87	1.34	1.75	2.13	2.60	2.95
16	0.54	0.86	1.34	1.75	2.12	2.58	2.92
17	0.53	0.86	1.33	1.74	2.11	2.57	2.90
18	0.53	0.86	1.33	1.73	2.10	2.55	2.88
19	0.53	0.86	1.33	1.73	2.09	2.54	2.86
20	0.53	0.86	1.32	1.72	2.09	2.53	2.84
21	0.53	0.86	1.32	1.72	2.08	2.52	2.83
22	0.53	0.86	1.32	1.72	2.07	2.51	2.82
23	0.53	0.86	1.32	1.71	2.07	2.50	2.81
24	0.53	0.86	1.32	1.71	2.06	2.49	2.80
25	0.53	0.86	1.32	1.71	2.06	2.48	2.79
26	0.53	0.86	1.32	1.71	2.06	2.48	2.78
27	0.53	0.86	1.31	1.70	2.05	2.47	2.77
28	0.53	0.86	1.31	1.70	2.05	2.47	2.76
29	0.53	0.85	1.31	1.70	2.04	2.46	2.76
30	0.53	0.85	1.31	1.70	2.04	2.46	2.75
40	0.53	0.85	1.30	1.68	2.02	2.42	2.70
50	0.53	0.85	1.30	1.67	2.01	2.40	2.68
60	0.53	0.85	1.30	1.67	2.00	2.39	2.66
80	0.53	0.85	1.29	1.66	1.99	2.37	2.64
100	0.53	0.84	1.29	1.66	1.98	2.36	2.63
200	0.52	0.84	1.29	1.65	1.97	2.34	2.60
500	0.52	0.84	1.28	1.65	1.96	2.33	2.59
∞	0.52	0.84	1.28	1.64	1.96	2.33	2.58

B.4 *Z* TABLE

To use the *Z* table:

1. Choose an α value.
2. Calculate $(1 - \alpha)$ or $(1 - \alpha/2)$ depending on the type of test used.
3. Look in the body of the table for the value corresponding to the $(1 - \alpha)$ or $(1 - \alpha/2)$ value.
4. Move to the top row and the left column corresponding to that point.
5. The *Z* value = (left column) $\times$ (top row) intersection.

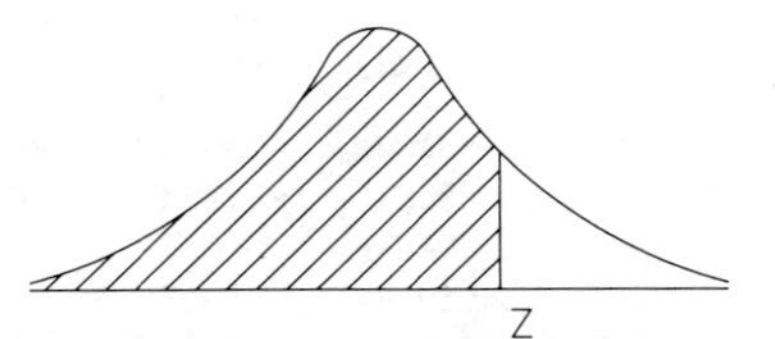

Z Table

Areas under the Normal Curve[a] (Proportion of Total Area under the Curve from $-\infty$ to Designated *Z* Value)

Z	0.09	0.08	0.07	0.06	0.05	0.04	0.03	0.02	0.01	0.00
− 3.5	0.00017	0.00017	0.00018	0.00019	0.00019	0.00020	0.00021	0.00022	0.00022	0.00023
− 3.4	0.00024	0.00025	0.00026	0.00027	0.00028	0.00029	0.00030	0.00031	0.00033	0.00034
− 3.3	0.00035	0.00036	0.00038	0.00039	0.00040	0.00042	0.00043	0.00045	0.00047	0.00048
− 3.2	0.00050	0.00052	0.00054	0.00056	0.00058	0.00060	0.00062	0.00064	0.00066	0.00069
− 3.1	0.00071	0.00074	0.00076	0.00079	0.00082	0.00085	0.00087	0.00090	0.00094	0.00097
− 3.0	0.00100	0.00104	0.00107	0.00111	0.00114	0.00118	0.00122	0.00126	0.00131	0.00135
− 2.9	0.0014	0.0014	0.0015	0.0015	0.0016	0.0016	0.0017	0.0017	0.0018	0.0019
− 2.8	0.0019	0.0020	0.0021	0.0021	0.0022	0.0023	0.0023	0.0024	0.0025	0.0026
− 2.7	0.0026	0.0027	0.0028	0.0029	0.0030	0.0031	0.0032	0.0033	0.0034	0.0035
− 2.6	0.0036	0.0037	0.0038	0.0039	0.0040	0.0041	0.0043	0.0044	0.0045	0.0047
− 2.5	0.0048	0.0049	0.0051	0.0052	0.0054	0.0055	0.0057	0.0059	0.0060	0.0062
− 2.4	0.0064	0.0066	0.0068	0.0069	0.0071	0.0073	0.0075	0.0078	0.0080	0.0082
− 2.3	0.0084	0.0087	0.0089	0.0091	0.0094	0.0096	0.0099	0.0102	0.0104	0.0107
− 2.2	0.0110	0.0113	0.0116	0.0119	0.0122	0.0125	0.0129	0.0132	0.0136	0.0139
− 2.1	0.0143	0.0146	0.0150	0.0154	0.0158	0.0162	0.0166	0.0170	0.0174	0.0179
− 2.0	0.0183	0.0188	0.0192	0.0197	0.0202	0.0207	0.0212	0.0217	0.0222	0.0228
− 1.9	0.0233	0.0239	0.0244	0.0250	0.0256	0.0262	0.0268	0.0274	0.0281	0.0287
− 1.8	0.0294	0.0301	0.0307	0.0314	0.0322	0.0329	0.0336	0.0344	0.0351	0.0359
− 1.7	0.0367	0.0375	0.0384	0.0392	0.0401	0.0409	0.0418	0.0427	0.0436	0.0446
− 1.6	0.0455	0.0465	0.0475	0.0485	0.0495	0.0505	0.0516	0.0526	0.0537	0.0548
− 1.5	0.0559	0.0571	0.0582	0.0594	0.0606	0.0618	0.0630	0.0643	0.0655	0.0668
− 1.4	0.0681	0.0694	0.0708	0.0721	0.0735	0.0749	0.0764	0.0778	0.0793	0.0808

Z Table *(Continued)*

Z	0.09	0.08	0.07	0.06	0.05	0.04	0.03	0.02	0.01	0.00
−1.3	0.0823	0.0838	0.0853	0.0869	0.0885	0.0901	0.0918	0.0934	0.0951	0.0968
−1.2	0.0985	0.1003	0.1020	0.1038	0.1057	0.1075	0.1093	0.1112	0.1131	0.1151
−1.1	0.1170	0.1190	0.1210	0.1230	0.1251	0.1271	0.1292	0.1314	0.1335	0.1357
−1.0	0.1379	0.1401	0.1423	0.1446	0.1469	0.1492	0.1515	0.1539	0.1562	0.1587
−0.9	0.1611	0.1635	0.1660	0.1685	0.1711	0.1736	0.1762	0.1788	0.1814	0.1841
−0.8	0.1867	0.1894	0.1922	0.1949	0.1977	0.2005	0.2033	0.2061	0.2090	0.2119
−0.7	0.2148	0.2177	0.2207	0.2236	0.2266	0.2297	0.2327	0.2358	0.2389	0.2420
−0.6	0.2451	0.2483	0.2514	0.2546	0.2578	0.2611	0.2643	0.2676	0.2709	0.2743
−0.5	0.2776	0.2810	0.2843	0.2877	0.2912	0.2946	0.2981	0.3015	0.3050	0.3085
−0.4	0.3121	0.3156	0.3192	0.3228	0.3264	0.3300	0.3336	0.3372	0.3409	0.3446
−0.3	0.3483	0.3520	0.3557	0.3594	0.3632	0.3669	0.3707	0.3745	0.3783	0.3821
−0.2	0.3859	0.3897	0.3936	0.3974	0.4013	0.4052	0.4090	0.4129	0.4168	0.4207
−0.1	0.4247	0.4286	0.4325	0.4364	0.4404	0.4443	0.4483	0.4522	0.4562	0.4602
−0.0	0.4641	0.4681	0.4721	0.4761	0.4801	0.4840	0.4880	0.4920	0.4960	0.5000

Z	0.00	0.01	0.02	0.03	0.04	0.05	0.06	0.07	0.08	0.09
+0.0	0.5000	0.5040	0.5080	0.5120	0.5160	0.5199	0.5239	0.5279	0.5319	0.5359
+0.1	0.5398	0.5438	0.5478	0.5517	0.5557	0.5596	0.5636	0.5675	0.5714	0.5753
+0.2	0.5793	0.5832	0.5871	0.5910	0.5948	0.5987	0.6026	0.6064	0.6103	0.6141
+0.3	0.6179	0.6217	0.6255	0.6293	0.6331	0.6368	0.6406	0.6443	0.6480	0.6517
+0.4	0.6554	0.6591	0.6628	0.6664	0.6700	0.6736	0.6772	0.6808	0.6844	0.6879
+0.5	0.6915	0.6950	0.6985	0.7019	0.7054	0.7088	0.7123	0.7157	0.7190	0.7224
+0.6	0.7257	0.7291	0.7324	0.7357	0.7389	0.7422	0.7454	0.7486	0.7517	0.7549
+0.7	0.7580	0.7611	0.7642	0.7673	0.7704	0.7734	0.7764	0.7794	0.7823	0.7852
+0.8	0.7881	0.7910	0.7939	0.7967	0.7995	0.8023	0.8051	0.8079	0.8106	0.8133
+0.9	0.8159	0.8186	0.8212	0.8238	0.8264	0.8289	0.8315	0.8340	0.8365	0.8389
+1.0	0.8413	0.8438	0.8461	0.8485	0.8508	0.8531	0.8554	0.8577	0.8599	0.8621
+1.1	0.8643	0.8665	0.8686	0.8708	0.8729	0.8749	0.8770	0.8790	0.8810	0.8830
+1.2	0.8849	0.8869	0.8888	0.8907	0.8925	0.8944	0.8962	0.8980	0.8997	0.9015

+1.3	0.9032	0.9049	0.9066	0.9082	0.9099	0.9115	0.9131	0.9147	0.9162	0.9177
+1.4	0.9192	0.9207	0.9222	0.9236	0.9251	0.9265	0.9279	0.9292	0.9306	0.9319
+1.5	0.9332	0.9345	0.9357	0.9370	0.9382	0.9394	0.9406	0.9418	0.9429	0.9441
+1.6	0.9452	0.9463	0.9474	0.9484	0.9495	0.9505	0.9515	0.9525	0.9535	0.9545
+1.7	0.9554	0.9564	0.9573	0.9582	0.9591	0.9599	0.9608	0.9616	0.9625	0.9633
+1.8	0.9641	0.9649	0.9656	0.9664	0.9671	0.9678	0.9686	0.9693	0.9699	0.9706
+1.9	0.9713	0.9719	0.9726	0.9732	0.9738	0.9744	0.9750	0.9756	0.9761	0.9767
+2.0	0.9773	0.9778	0.9783	0.9788	0.9793	0.9798	0.9803	0.9808	0.9812	0.9817
+2.1	0.9821	0.9826	0.9830	0.9834	0.9838	0.9842	0.9846	0.9850	0.9854	0.9857
+2.2	0.9861	0.9864	0.9868	0.9871	0.9875	0.9878	0.9881	0.9884	0.9887	0.9890
+2.3	0.9893	0.9896	0.9898	0.9901	0.9904	0.9906	0.9909	0.9911	0.9913	0.9916
+2.4	0.9918	0.9920	0.9922	0.9925	0.9927	0.9929	0.9931	0.9932	0.9934	0.9936
+2.5	0.9938	0.9940	0.9941	0.9943	0.9945	0.9946	0.9948	0.9949	0.9951	0.9952
+2.6	0.9953	0.9955	0.9956	0.9957	0.9959	0.9960	0.9961	0.9962	0.9963	0.9964
+2.7	0.9965	0.9966	0.9967	0.9968	0.9969	0.9970	0.9971	0.9972	0.9973	0.9974
+2.8	0.9974	0.9975	0.9976	0.9977	0.9977	0.9978	0.9979	0.9979	0.9980	0.9981
+2.9	0.9981	0.9982	0.9983	0.9983	0.9984	0.9984	0.9985	0.9985	0.9986	0.9986
+3.0	0.99865	0.99869	0.99874	0.99878	0.99882	0.99886	0.99889	0.99893	0.99896	0.99900
+3.1	0.99903	0.99906	0.99910	0.99913	0.99915	0.99918	0.99921	0.99924	0.99926	0.99929
+3.2	0.99931	0.99934	0.99936	0.99938	0.99940	0.99942	0.99944	0.99946	0.99948	0.99950
+3.3	0.99952	0.99953	0.99955	0.99957	0.99958	0.99960	0.99961	0.99962	0.99964	0.99965
+3.4	0.99966	0.99967	0.99969	0.99970	0.99971	0.99972	0.99973	0.99974	0.99975	0.99976
+3.5	0.99977	0.99978	0.99978	0.99979	0.99980	0.99981	0.99981	0.99982	0.99983	0.99983

[a]Adapted from E. L. Grant, *Statistical Quality Control* (2nd ed.), McGraw-Hill, New York, 1952, Table A, pp. 510–511. Reproduced by permission of the publisher.

B.5 χ^2 TABLE

To use the χ^2 table:

1. Choose an α value.
2. Determine the degrees of freedom.

$$DF = n - 1$$

3. Find the row corresponding to the DF and the column corresponding to the α or $\alpha/2$ value. The intersection of those is the χ^2 (table) value.

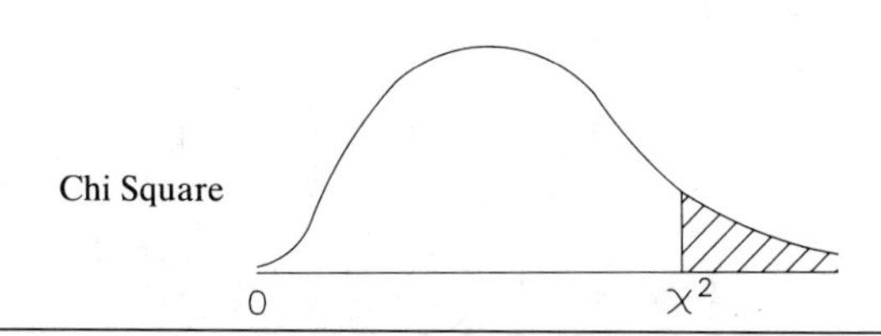

χ^2 Table[a]

ν	0.99	0.98	0.95	0.90	0.80	0.20	0.10	0.05	0.02	0.01	0.001
	Probability										
1	0.0^3157	0.0^3628	0.00393	0.0158	0.0642	1.642	2.706	3.841	5.412	6.635	10.827
2	0.0201	0.0404	0.103	0.211	0.446	3.219	4.605	5.991	7.824	9.210	13.815
3	0.115	0.185	0.352	0.584	1.005	4.642	6.251	7.815	9.837	11.341	16.268
4	0.297	0.429	0.711	1.064	1.649	5.989	7.779	9.488	11.668	13.277	18.465
5	0.554	0.752	1.145	1.610	2.343	7.289	9.236	11.070	13.388	15.086	20.517
6	0.872	1.134	1.635	2.204	3.070	8.558	10.645	12.592	15.033	16.812	22.457
7	1.239	1.564	2.167	2.833	3.822	9.803	12.017	14.067	16.622	18.475	24.322
8	1.646	2.032	2.733	3.490	4.594	11.030	13.362	15.507	18.168	20.090	26.125
9	2.088	2.532	3.325	4.168	5.380	12.242	14.684	16.919	19.679	21.666	27.877
10	2.558	3.059	3.940	4.865	6.179	13.442	15.987	18.307	21.161	23.209	29.588
11	3.053	3.609	4.575	5.578	6.989	14.631	17.275	19.675	22.618	24.725	31.264
12	3.571	4.178	5.226	6.304	7.807	15.812	18.549	21.026	24.054	26.217	32.909
13	4.107	4.765	5.892	7.042	8.634	16.985	19.812	22.362	25.472	27.688	34.528
14	4.660	5.368	6.571	7.790	9.467	18.151	21.064	23.685	26.873	29.141	36.123
15	5.229	5.985	7.261	8.547	10.307	19.311	22.307	24.996	28.259	30.578	37.697
16	5.812	6.614	7.962	9.312	11.152	20.465	23.542	26.296	29.633	32.000	39.252
17	6.408	7.255	8.672	10.085	12.002	21.615	24.769	27.587	30.995	33.409	40.790
18	7.015	7.906	9.390	10.865	12.857	22.760	25.989	28.869	32.346	34.805	42.312
19	7.633	8.567	10.117	11.651	13.716	23.900	27.204	30.144	33.687	36.191	43.820
20	8.260	9.237	10.851	12.443	14.578	25.038	28.412	31.410	35.020	37.566	45.315
21	8.897	9.915	11.591	13.240	15.445	26.171	29.615	32.671	36.343	38.932	46.797
22	9.542	10.600	12.338	14.041	16.314	27.301	30.813	33.924	37.659	40.289	48.268

χ^2 Table[a] *(Continued)*

ν	Probability										
	0.99	0.98	0.95	0.90	0.80	0.20	0.10	0.05	0.02	0.01	0.001
23	10.196	11.293	13.091	14.848	17.187	28.429	32.007	35.172	38.968	41.638	49.728
24	10.856	11.992	13.848	15.659	18.062	29.553	33.196	36.415	40.270	42.980	51.179
25	11.524	12.697	14.611	16.473	18.940	30.675	34.382	37.652	41.566	44.314	52.620
26	12.198	13.409	15.379	17.292	19.820	31.795	35.563	38.885	42.856	45.642	54.052
27	12.879	14.125	16.151	18.114	20.703	32.912	36.741	40.113	44.140	46.963	55.476
28	13.565	14.847	16.928	18.939	21.588	34.027	37.916	41.337	45.419	48.278	56.893
29	14.256	15.574	17.708	19.768	22.475	35.139	39.087	42.557	46.693	49.588	58.302
30	14.953	16.306	18.493	20.599	23.364	36.250	40.256	43.773	47.962	50.892	59.703

[a]Abridged from R. A. Fisher and F. Yates, *Statistical Tables for Biological, Agricultural and Medical Research*, Oliver & Boyd, London, 1953, Table IV. Reproduced by permission of the authors and publishers. For larger values of ν, the expression $\sqrt{2\chi^-} - \sqrt{2\nu - 1}$ may be used as a normal deviate with unit variance, remembering that the probability for χ^2 corresponds with that of a single tail of a normal curve.

B.6 *F* TABLE

To use the *F* table:

1. Choose an α value.
2. Calculate $(1 - \alpha)$ or $(1 - \alpha/2)$ depending on the test.
3. Calculate the degrees of freedom for both samples:

$$DF_1 = n_1 - 1$$
$$DF_2 = n_2 - 1$$

4. Look at the *F*-statistic calculation to determine which sample is in the numerator (top) and which is in the denominator (bottom).
5. Find DF (numerator), DF (denominator), and $(1 - \alpha)$ values on the table.
6. The intersection of all of those values is the *F* (table) value

Note:

If $\alpha, \alpha/2$ =	Then use rows labeled
0.01	0.99
0.05	0.95
0.10	0.90
0.25	0.75

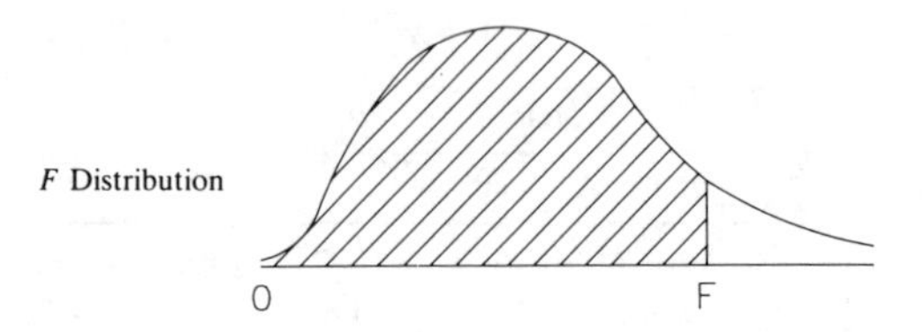

F Table[a]

DF for Denom.	$1-\alpha$	DF for Numerator 1	2	3	4	5	6	7	8	9	10	11	12
1	0.75	5.83	7.50	8.20	8.58	8.82	8.98	9.10	9.19	9.26	9.32	9.36	9.41
	0.90	39.9	49.5	53.6	55.8	57.2	58.2	58.9	59.4	59.9	60.2	60.5	60.7
	0.95	161	200	216	225	230	234	237	239	241	242	243	244
2	0.75	2.57	3.00	3.15	3.23	3.28	3.31	3.34	3.35	3.37	3.38	3.39	3.39
	0.90	8.53	9.00	9.16	9.24	9.29	9.33	9.35	9.37	9.38	9.39	9.40	9.41
	0.95	18.5	19.0	19.2	19.2	19.3	19.3	19.4	19.4	19.4	19.4	19.4	19.4
	0.99	98.5	99.0	99.2	99.2	99.3	99.3	99.4	99.4	99.4	99.4	99.4	99.4
3	0.75	2.02	2.28	2.36	2.39	2.41	2.42	2.43	2.44	2.44	2.44	2.45	2.45
	0.90	5.54	5.46	5.39	5.34	5.31	5.28	5.27	5.25	5.24	5.23	5.22	5.22
	0.95	10.1	9.55	9.28	9.12	9.10	8.94	8.89	8.85	8.81	8.79	8.76	8.74
	0.99	34.1	30.8	29.5	28.7	28.2	27.9	27.7	27.5	27.3	27.2	27.1	27.1
4	0.75	1.81	2.00	2.05	2.06	2.07	2.08	2.08	2.08	2.08	2.08	2.08	2.08
	0.90	4.54	4.32	4.19	4.11	4.05	4.01	3.98	3.95	3.94	3.92	3.91	3.90
	0.95	7.71	6.94	6.59	6.39	6.26	6.16	6.09	6.04	6.00	5.96	5.94	5.91
	0.99	21.2	18.0	16.7	16.0	15.5	15.2	15.0	14.8	14.7	14.5	14.4	14.4
5	0.75	1.69	1.85	1.88	1.89	1.89	1.89	1.89	1.89	1.89	1.89	1.89	1.89
	0.90	4.06	3.78	3.62	3.52	3.45	3.40	3.37	3.34	3.32	3.30	3.28	3.27
	0.95	6.61	5.79	5.41	5.19	5.05	4.95	4.88	4.82	4.77	4.74	4.71	4.68
	0.99	16.3	13.3	12.1	11.4	11.0	10.7	10.5	10.3	10.2	10.1	9.96	9.89
6	0.75	1.62	1.76	1.78	1.79	1.79	1.78	1.78	1.77	1.77	1.77	1.77	1.77
	0.90	3.78	3.46	3.29	3.18	3.11	3.05	3.01	2.98	2.96	2.94	2.92	2.90
	0.95	5.99	5.14	4.76	4.53	4.39	4.28	4.21	4.15	4.10	4.06	4.03	4.00
	0.99	13.7	10.9	9.78	9.15	8.75	8.47	8.26	8.10	7.98	7.87	7.79	7.72
7	0.75	1.57	1.70	1.72	1.72	1.71	1.71	1.70	1.70	1.69	1.69	1.69	1.68
	0.90	3.59	3.26	3.07	2.96	2.88	2.83	2.78	2.75	2.72	2.70	2.68	2.67
	0.95	5.59	4.74	4.35	4.12	3.97	3.87	3.79	3.73	3.68	3.64	3.60	3.57
	0.99	12.2	9.55	8.45	7.85	7.46	7.19	6.99	6.84	6.72	6.62	6.54	6.47
8	0.75	1.54	1.66	1.67	1.66	1.66	1.65	1.64	1.64	1.64	1.63	1.63	1.62
	0.90	3.46	3.11	2.92	2.81	2.73	2.67	2.62	2.59	2.56	2.54	2.52	2.50
	0.95	5.32	4.46	4.07	3.84	3.69	3.58	3.50	3.44	3.39	3.35	3.31	3.28
	0.99	11.3	8.65	7.59	8.01	6.63	6.37	6.18	6.03	5.91	5.81	5.73	5.67
9	0.75	1.51	1.62	1.63	1.63	1.62	1.61	1.60	1.60	1.59	1.59	1.58	1.58
	0.90	3.36	3.01	2.81	2.69	2.61	2.55	2.51	2.47	2.44	2.42	2.40	2.38
	0.95	5.12	4.26	3.86	3.63	3.48	3.37	3.29	3.23	3.18	3.14	3.10	3.07
	0.99	10.6	8.02	6.99	6.42	6.06	5.80	5.61	5.47	5.35	5.26	5.18	5.11
10	0.75	1.49	1.60	1.60	1.59	1.59	1.58	1.57	1.56	1.56	1.55	1.55	1.54
	0.90	3.28	2.92	2.73	2.61	2.52	2.46	2.41	2.38	2.35	2.32	2.30	2.28
	0.95	4.96	4.10	3.71	3.48	3.33	3.22	3.14	3.07	3.02	2.98	2.94	2.91
	0.99	10.0	7.56	6.55	5.99	5.64	5.39	5.20	5.06	4.94	4.85	4.77	4.71
11	0.75	1.47	1.58	1.58	1.57	1.56	1.55	1.54	1.53	1.53	1.52	1.52	1.51
	0.90	3.23	2.86	2.66	2.54	2.45	2.39	2.34	2.30	2.27	2.25	2.23	2.21
	0.95	4.84	3.98	3.59	3.36	3.20	3.09	3.01	2.95	2.90	2.85	2.82	2.79
	0.99	9.65	7.21	6.22	5.67	5.32	5.07	4.89	4.74	4.63	4.54	4.46	4.40
12	0.75	1.46	1.56	1.56	1.55	1.54	1.53	1.52	1.51	1.51	1.50	1.50	1.49
	0.90	3.18	2.81	2.61	2.48	2.39	2.33	2.28	2.24	2.21	2.19	2.17	2.15
	0.95	4.75	3.89	3.49	3.26	3.11	3.00	2.91	2.85	2.80	2.75	2.72	2.69
	0.99	9.33	6.93	5.95	5.41	5.06	4.82	4.64	4.50	4.39	4.30	4.22	4.16

F Table *(Continued)*

DF for Numerator											
15	20	24	30	40	50	60	100	120	200	500	∞
9.49	9.58	9.63	9.67	9.71	9.74	9.76	9.78	9.80	9.82	9.84	9.85
61.2	61.7	62.0	62.3	62.5	62.7	62.8	63.0	63.1	63.2	63.3	63.3
246	248	249	250	251	252	252	253	253	254	254	254
3.41	3.43	3.43	3.44	3.45	3.45	3.46	3.47	3.47	3.48	3.48	3.48
9.42	9.44	9.45	9.46	9.47	9.47	9.47	9.48	9.48	9.49	9.49	9.49
19.4	19.4	19.5	19.5	19.5	19.5	19.5	19.5	19.5	19.5	19.5	19.5
99.4	99.4	99.5	99.5	99.5	99.5	99.5	99.5	99.5	99.5	99.5	99.5
2.46	2.46	2.46	2.47	2.47	2.47	2.47	2.47	2.47	2.47	2.47	2.47
5.20	5.18	5.18	5.17	5.16	5.15	5.15	5.14	5.14	5.14	5.14	5.13
8.70	8.66	8.64	8.62	8.59	8.58	8.57	8.55	8.55	8.54	8.53	8.53
26.9	26.7	26.6	26.5	26.4	26.4	26.3	26.2	26.2	26.2	26.1	26.1
2.08	2.08	2.08	2.08	2.08	2.08	2.08	2.08	2.08	2.08	2.08	2.08
3.87	3.84	3.83	3.82	3.80	3.80	3.79	3.78	3.78	3.77	3.76	3.76
5.86	5.80	5.77	5.75	5.72	5.70	5.69	5.66	5.66	5.65	5.64	5.63
14.2	14.0	13.9	13.8	13.7	13.7	13.7	13.6	13.6	13.5	13.5	13.5
1.89	1.88	1.88	1.88	1.88	1.88	1.87	1.87	1.87	1.87	1.87	1.87
3.24	3.21	3.19	3.17	3.16	3.15	3.14	3.13	3.12	3.12	3.11	3.10
4.62	4.56	4.53	4.50	4.46	4.44	4.43	4.41	4.40	4.39	4.37	4.36
9.72	9.55	9.47	9.38	9.29	9.24	9.20	9.13	9.11	9.08	9.04	9.02
1.76	1.76	1.75	1.75	1.75	1.75	1.74	1.74	1.74	1.74	1.74	1.74
2.87	2.84	2.82	2.80	2.78	2.77	2.76	2.75	2.74	2.73	2.73	2.72
3.94	3.87	3.84	3.81	3.77	3.75	3.74	3.71	3.70	3.69	3.68	3.67
7.56	7.40	7.31	7.23	7.14	7.09	7.06	6.99	6.97	6.93	6.90	6.88
1.68	1.67	1.67	1.66	1.66	1.66	1.65	1.65	1.65	1.65	1.65	1.65
2.63	2.59	2.58	2.56	2.54	2.52	2.51	2.50	2.49	2.48	2.48	2.47
3.51	3.44	3.41	3.38	3.34	3.32	3.30	3.27	3.27	3.25	3.24	3.23
6.31	6.16	6.07	5.99	5.91	5.86	5.82	5.75	5.74	5.70	5.67	5.65
1.62	1.61	1.60	1.60	1.59	1.59	1.59	1.58	1.58	1.58	1.58	1.58
2.46	2.42	2.40	2.38	2.36	2.35	2.34	2.32	2.32	2.31	2.30	2.29
3.22	3.15	3.12	3.08	3.04	3.02	3.01	2.97	2.97	2.95	2.94	2.93
5.52	5.36	5.28	5.20	5.12	5.07	5.03	4.96	4.95	4.91	4.88	4.86
1.57	1.56	1.56	1.55	1.55	1.54	1.54	1.53	1.53	1.53	1.53	1.53
2.34	2.30	2.28	2.25	2.23	2.22	2.21	2.19	2.18	2.17	2.17	2.16
3.01	2.94	2.90	2.86	2.83	2.80	2.79	2.76	2.75	2.73	2.72	2.71
4.96	4.81	4.73	4.65	4.57	4.52	4.48	4.42	4.40	4.36	4.33	4.31
1.53	1.52	1.52	1.51	1.51	1.50	1.50	1.49	1.49	1.49	1.48	1.48
2.24	2.20	2.18	2.16	2.13	2.12	2.11	2.09	2.08	2.07	2.06	2.06
2.85	2.77	2.74	2.70	2.66	2.64	2.62	2.59	2.58	2.56	2.55	2.54
4.56	4.41	4.33	4.25	4.17	4.12	4.08	4.01	4.00	3.96	3.93	3.91
1.50	1.49	1.49	1.48	1.47	1.47	1.47	1.46	1.46	1.46	1.45	1.45
2.17	2.12	2.10	2.08	2.05	2.04	2.03	2.00	2.00	1.99	1.98	1.97
2.72	2.65	2.61	2.57	2.53	2.51	2.49	2.46	2.45	2.43	2.42	2.40
4.25	4.10	4.02	3.94	3.86	3.81	3.78	3.71	3.69	3.66	3.62	3.60
1.48	1.47	1.46	1.45	1.45	1.44	1.44	1.43	1.43	1.43	1.42	1.42
2.10	2.06	2.04	2.01	1.99	1.97	1.96	1.94	1.93	1.92	1.91	1.90
2.62	2.54	2.51	2.47	2.43	2.40	2.38	2.35	2.34	2.32	2.31	2.30
4.01	3.86	3.78	3.70	3.62	3.57	3.54	3.47	3.45	3.41	3.38	3.36

F Table *(Continued)*

DF for Denom.	$1-\alpha$	DF for Numerator 1	2	3	4	5	6	7	8	9	10	11	12
13	0.75	1.45	1.54	1.54	1.53	1.52	1.51	1.50	1.49	1.49	1.48	1.47	1.47
	0.90	3.14	2.76	2.56	2.43	2.35	2.28	2.23	2.20	2.16	2.14	2.12	2.10
	0.95	4.67	3.81	3.41	3.18	3.03	2.92	2.83	2.77	2.71	2.67	2.63	2.60
	0.99	9.07	6.70	5.74	5.21	4.86	4.62	4.44	4.30	4.19	4.10	4.02	3.96
14	0.75	1.44	1.53	1.53	1.52	1.51	1.50	1.48	1.48	1.47	1.46	1.46	1.45
	0.90	3.10	2.73	2.52	2.39	2.31	2.24	2.19	2.15	2.12	2.10	2.08	2.05
	0.95	4.60	3.74	3.34	3.11	2.96	2.85	2.76	2.70	2.65	2.60	2.57	2.53
	0.99	8.86	6.51	5.56	5.04	4.69	4.46	4.28	4.14	4.03	3.94	3.86	3.80
15	0.75	1.43	1.52	1.52	1.51	1.49	1.48	1.47	1.46	1.46	1.45	1.44	1.44
	0.90	3.07	2.70	2.49	2.36	2.27	2.21	2.16	2.12	2.09	2.06	2.04	2.02
	0.95	4.54	3.68	3.29	3.06	2.90	2.79	2.71	2.64	2.59	2.54	2.51	2.48
	0.99	8.68	6.36	5.42	4.89	4.56	4.32	4.14	4.00	3.89	3.80	3.73	3.67
16	0.75	1.42	1.51	1.51	1.50	1.48	1.48	1.47	1.46	1.45	1.45	1.44	1.44
	0.90	3.05	2.67	2.46	2.33	2.24	2.18	2.13	2.09	2.06	2.03	2.01	1.99
	0.95	4.49	3.63	3.24	3.01	2.85	2.74	2.66	2.59	2.54	2.49	2.46	2.42
	0.99	8.53	6.23	5.29	4.77	4.44	4.20	4.03	3.89	3.78	3.69	3.62	3.55
17	0.75	1.42	1.51	1.50	1.49	1.47	1.46	1.45	1.44	1.43	1.43	1.42	1.41
	0.90	3.03	2.64	2.44	2.31	2.22	2.15	2.10	2.06	2.03	2.00	1.98	1.96
	0.95	4.45	3.59	3.20	2.96	2.81	2.70	2.61	2.55	2.49	2.45	2.41	2.38
	0.99	8.40	6.11	5.18	4.67	4.34	4.10	3.93	3.79	3.68	3.59	3.52	3.46
18	0.75	1.41	1.50	1.49	1.48	1.46	1.45	1.44	1.43	1.42	1.42	1.41	1.40
	0.90	3.01	2.62	2.42	2.29	2.20	2.13	2.08	2.04	2.00	1.98	1.96	1.93
	0.95	4.41	3.55	3.16	2.93	2.77	2.66	2.58	2.51	2.46	2.41	2.37	2.34
	0.99	8.29	6.01	5.09	4.58	4.25	4.01	3.84	3.71	3.60	3.51	3.43	3.37
19	0.75	1.41	1.49	1.49	1.47	1.46	1.44	1.43	1.42	1.41	1.41	1.40	1.40
	0.90	2.99	2.61	2.40	2.27	2.18	2.11	2.06	2.02	1.98	1.96	1.94	1.91
	0.95	4.38	3.52	3.13	2.90	2.74	2.63	2.54	2.48	2.42	2.38	2.34	2.31
	0.99	8.18	5.93	5.01	4.50	4.17	3.94	3.77	3.63	3.52	3.43	3.36	3.30
20	0.75	1.40	1.49	1.48	1.46	1.45	1.44	1.42	1.42	1.41	1.40	1.39	1.39
	0.90	2.97	2.59	2.38	2.25	2.16	2.09	2.04	2.00	1.96	1.94	1.92	1.89
	0.95	4.35	3.49	3.10	2.87	2.71	2.60	2.51	2.45	2.39	2.35	2.31	2.28
	0.99	8.10	5.85	4.94	4.43	4.10	3.87	3.70	3.56	3.46	3.37	3.29	3.23
22	0.75	1.40	1.48	1.47	1.45	1.44	1.42	1.41	1.40	1.39	1.39	1.38	1.37
	0.90	2.95	2.56	2.35	2.22	2.13	2.06	2.01	1.97	1.93	1.90	1.88	1.86
	0.95	4.30	3.44	3.05	2.82	2.66	2.55	2.46	2.40	2.34	2.30	2.26	2.23
	0.99	7.95	5.72	4.82	4.31	3.99	3.76	3.59	3.45	3.35	3.26	3.18	3.12
24	0.75	1.39	1.47	1.46	1.44	1.43	1.41	1.40	1.39	1.38	1.38	1.37	1.36
	0.90	2.93	2.54	2.33	2.19	2.10	2.04	1.98	1.94	1.91	1.88	1.85	1.83
	0.95	4.26	3.40	3.01	2.78	2.62	2.51	2.42	2.36	2.30	2.25	2.21	2.18
	0.99	7.82	5.61	4.72	4.22	3.90	3.67	3.50	3.36	3.26	3.17	3.09	3.03
26	0.75	1.38	1.46	1.45	1.44	1.42	1.41	1.40	1.39	1.37	1.37	1.36	1.35
	0.90	2.91	2.52	2.31	2.17	2.08	2.01	1.96	1.92	1.88	1.86	1.84	1.81
	0.95	4.23	3.37	2.98	2.74	2.59	2.47	2.39	2.32	2.27	2.22	2.18	2.15
	0.99	7.72	5.53	4.64	4.14	3.82	3.59	3.42	3.29	3.18	3.09	3.02	2.96
28	0.75	1.38	1.46	1.45	1.43	1.41	1.40	1.39	1.38	1.37	1.36	1.35	1.34
	0.90	2.89	2.50	2.29	2.16	2.06	2.00	1.94	1.90	1.87	1.84	1.81	1.79
	0.95	4.20	3.34	2.95	2.71	2.56	2.45	2.36	2.29	2.24	2.19	2.15	2.12
	0.99	7.64	5.45	4.57	4.07	3.75	3.53	3.36	3.23	3.12	3.03	2.96	2.90

DF for Numerator											
15	20	24	30	40	50	60	100	120	200	500	∞
1.46	1.45	1.44	1.43	1.42	1.42	1.42	1.41	1.41	1.40	1.40	1.40
2.05	2.01	1.98	1.96	1.93	1.92	1.90	1.88	1.88	1.86	1.85	1.85
2.53	2.46	2.42	2.38	2.34	2.31	2.30	2.26	2.25	2.23	2.22	2.21
3.82	3.66	3.59	3.51	3.43	3.38	3.34	3.27	3.25	3.22	3.19	3.17
1.44	1.43	1.42	1.41	1.41	1.40	1.40	1.39	1.39	1.39	1.38	1.38
2.01	1.96	1.94	1.91	1.89	1.87	1.86	1.83	1.83	1.82	1.80	1.80
2.46	2.39	2.35	2.31	2.27	2.24	2.22	2.19	2.18	2.16	2.14	2.13
3.66	3.51	3.43	3.35	3.27	3.22	3.18	3.11	3.09	3.06	3.03	3.00
1.43	1.41	1.41	1.40	1.39	1.39	1.38	1.38	1.37	1.37	1.36	1.36
1.97	1.92	1.90	1.87	1.85	1.83	1.82	1.79	1.79	1.77	1.76	1.76
2.40	2.33	2.29	2.25	2.20	2.18	2.16	2.12	2.11	2.10	2.08	2.07
3.52	3.37	3.29	3.21	3.13	3.08	3.05	2.98	2.96	2.92	2.89	2.87
1.41	1.40	1.39	1.38	1.37	1.37	1.36	1.36	1.35	1.35	1.34	1.34
1.94	1.89	1.87	1.84	1.81	1.79	1.78	1.76	1.75	1.74	1.73	1.72
2.35	2.28	2.24	2.19	2.15	2.12	2.11	2.07	2.06	2.04	2.02	2.01
3.41	3.26	3.18	3.10	3.02	2.97	2.93	2.86	2.84	2.81	2.78	2.75
1.40	1.39	1.38	1.37	1.36	1.35	1.35	1.34	1.34	1.34	1.33	1.33
1.91	1.86	1.84	1.81	1.78	1.76	1.75	1.73	1.72	1.71	1.69	1.69
2.31	2.23	2.19	2.15	2.10	2.08	2.06	2.02	2.01	1.99	1.97	1.96
3.31	3.16	3.08	3.00	2.92	2.87	2.83	2.76	2.75	2.71	2.68	2.65
1.39	1.38	1.37	1.36	1.35	1.34	1.34	1.33	1.33	1.32	1.32	1.32
1.89	1.84	1.81	1.78	1.75	1.74	1.72	1.70	1.69	1.68	1.67	1.66
2.27	2.19	2.15	2.11	2.06	2.04	2.02	1.98	1.97	1.95	1.93	1.92
3.23	3.08	3.00	2.92	2.84	2.78	2.75	2.68	2.66	2.62	2.59	2.57
1.38	1.37	1.36	1.35	1.34	1.33	1.33	1.32	1.32	1.31	1.31	1.30
1.86	1.81	1.79	1.76	1.73	1.71	1.70	1.67	1.67	1.65	1.64	1.63
2.23	2.16	2.11	2.07	2.03	2.00	1.98	1.94	1.93	1.91	1.89	1.88
3.15	3.00	2.92	2.84	2.76	2.71	2.67	2.60	2.58	2.55	2.51	2.49
1.37	1.36	1.35	1.34	1.33	1.33	1.32	1.31	1.31	1.30	1.30	1.29
1.84	1.79	1.77	1.74	1.71	1.69	1.68	1.65	1.64	1.63	1.62	1.61
2.20	2.12	2.08	2.04	1.99	1.97	1.95	1.91	1.90	1.88	1.86	1.84
3.09	2.94	2.86	2.78	2.69	2.64	2.61	2.54	2.52	2.48	2.44	2.42
1.36	1.34	1.33	1.32	1.31	1.31	1.30	1.30	1.30	1.29	1.29	1.28
1.81	1.76	1.73	1.70	1.67	1.65	1.64	1.61	1.60	1.59	1.58	1.57
2.15	2.07	2.03	1.98	1.94	1.91	1.89	1.85	1.84	1.82	1.80	1.78
2.98	2.83	2.75	2.67	2.58	2.53	2.50	2.42	2.40	2.36	2.33	2.31
1.35	1.33	1.32	1.31	1.30	1.29	1.29	1.28	1.28	1.27	1.27	1.26
1.78	1.73	1.70	1.67	1.64	1.62	1.61	1.58	1.57	1.56	1.54	1.53
2.11	2.03	1.98	1.94	1.89	1.86	1.84	1.80	1.79	1.77	1.75	1.73
2.89	2.74	2.66	2.58	2.49	2.44	2.40	2.33	2.31	2.27	2.24	2.21
1.34	1.32	1.31	1.30	1.29	1.28	1.28	1.26	1.26	1.26	1.25	1.25
1.76	1.71	1.68	1.65	1.61	1.59	1.58	1.55	1.54	1.53	1.51	1.50
2.07	1.99	1.95	1.90	1.85	1.82	1.80	1.76	1.75	1.73	1.71	1.69
2.81	2.66	2.58	2.50	2.42	2.36	2.33	2.25	2.23	2.19	2.16	2.13
1.33	1.31	1.30	1.29	1.28	1.27	1.27	1.26	1.25	1.25	1.24	1.24
1.74	1.69	1.66	1.63	1.59	1.57	1.56	1.53	1.52	1.50	1.49	1.48
2.04	1.96	1.91	1.87	1.82	1.79	1.77	1.73	1.71	1.69	1.67	1.65
2.75	2.60	2.52	2.44	2.35	2.30	2.26	2.19	2.17	2.13	2.09	2.06

F Table *(Continued)*

DF for Denom.	$1 - \alpha$	DF for Numerator											
		1	2	3	4	5	6	7	8	9	10	11	12
30	0.75	1.38	1.45	1.44	1.42	1.41	1.39	1.38	1.37	1.36	1.35	1.35	1.34
	0.90	2.88	2.49	2.28	2.14	2.05	1.98	1.93	1.88	1.85	1.82	1.79	1.77
	0.95	4.17	3.32	2.92	2.69	2.53	2.42	2.33	2.27	2.21	2.16	2.13	2.09
	0.99	7.56	5.39	4.51	4.02	3.70	3.47	3.30	3.17	3.07	2.98	2.91	2.84
40	0.75	1.36	1.44	1.42	1.40	1.39	1.37	1.36	1.35	1.34	1.33	1.32	1.31
	0.90	2.84	2.44	2.23	2.09	2.00	1.93	1.87	1.83	1.79	1.76	1.73	1.71
	0.95	4.08	3.23	2.84	2.61	2.45	2.34	2.25	2.18	2.12	2.08	2.04	2.00
	0.99	7.31	5.18	4.31	3.83	3.51	3.29	3.12	2.99	2.89	2.80	2.73	2.66
60	0.75	1.35	1.42	1.41	1.38	1.37	1.35	1.33	1.32	1.31	1.30	1.29	1.29
	0.90	2.79	2.39	2.18	2.04	1.95	1.87	1.82	1.77	1.74	1.71	1.68	1.66
	0.95	4.00	3.15	2.76	2.53	2.37	2.25	2.17	2.10	2.04	1.99	1.95	1.92
	0.99	7.08	4.98	4.13	3.65	3.34	3.12	2.95	2.82	2.72	2.63	2.56	2.50
120	0.75	1.34	1.40	1.39	1.37	1.35	1.33	1.31	1.30	1.29	1.28	1.27	1.26
	0.90	2.75	2.35	2.13	1.99	1.90	1.82	1.77	1.72	1.68	1.65	1.62	1.60
	0.95	3.92	3.07	2.68	2.45	2.29	2.17	2.09	2.02	1.96	1.91	1.87	1.83
	0.99	6.85	4.79	3.95	3.48	3.17	2.96	2.79	2.66	2.56	2.47	2.40	2.34
200	0.75	1.33	1.39	1.38	1.36	1.34	1.32	1.31	1.29	1.28	1.27	1.26	1.25
	0.90	2.73	2.33	2.11	1.97	1.88	1.80	1.75	1.70	1.66	1.63	1.60	1.57
	0.95	3.89	3.04	2.65	2.42	2.26	2.14	2.06	1.98	1.93	1.88	1.84	1.80
	0.99	6.76	4.71	3.88	3.41	3.11	2.89	2.73	2.60	2.50	2.41	2.34	2.27
∞	0.75	1.32	1.39	1.37	1.35	1.33	1.31	1.29	1.28	1.27	1.25	1.24	1.24
	0.90	2.71	2.30	2.08	1.94	1.85	1.77	1.72	1.67	1.63	1.60	1.57	1.55
	0.95	3.84	3.00	2.60	2.37	2.21	2.10	2.01	1.94	1.88	1.83	1.79	1.75
	0.99	6.63	4.61	3.78	3.32	3.02	2.80	2.64	2.51	2.41	2.32	2.25	2.18

Reproduced by permission of the Biometrika Trustees.

B.7 RANDOM NUMBERS TABLE

To use the random numbers table:

1. Determine the number of random numbers needed and the range of numbers to select from.
2. Pick a point arbitrarily on the table to start (some people simply close their eyes and point with a pencil).
3. Move in any direction (up, down, left, right, or diagonally) the appropriate number of digits (e.g., two digits of the range was 1 to 100 or four digits if the range was 1 to 5,000).
4. Record the random number if it is within the appropriate range. Skip any number that is outside the range.
5. Continue moving in the same direction until you hit the edge of the table. At the edge, move one digit in any direction. This will be your next starting point.

DF for Numerator											
15	20	24	30	40	50	60	100	120	200	500	∞
1.32	1.30	1.29	1.28	1.27	1.26	1.26	1.25	1.24	1.24	1.23	1.23
1.72	1.67	1.64	1.61	1.57	1.55	1.54	1.51	1.50	1.48	1.47	1.46
2.01	1.93	1.89	1.84	1.79	1.76	1.74	1.70	1.68	1.66	1.64	1.62
2.70	2.55	2.47	2.39	2.30	2.25	2.21	2.13	2.11	2.07	2.03	2.01
1.30	1.28	1.26	1.25	1.24	1.23	1.22	1.21	1.21	1.20	1.19	1.19
1.66	1.61	1.57	1.54	1.51	1.48	1.47	1.43	1.42	1.41	1.39	1.38
1.92	1.84	1.79	1.74	1.69	1.66	1.64	1.59	1.58	1.55	1.53	1.51
2.52	2.37	2.29	2.20	2.11	2.06	2.02	1.94	1.92	1.87	1.83	1.80
1.27	1.25	1.24	1.22	1.21	1.20	1.19	1.17	1.17	1.16	1.15	1.15
1.60	1.54	1.51	1.48	1.44	1.41	1.40	1.36	1.35	1.33	1.31	1.29
1.84	1.75	1.70	1.65	1.59	1.56	1.53	1.48	1.47	1.44	1.41	1.39
2.35	2.20	2.12	2.03	1.94	1.88	1.84	1.75	1.73	1.68	1.63	1.60
1.24	1.22	1.21	1.19	1.18	1.17	1.16	1.14	1.13	1.12	1.11	1.10
1.55	1.48	1.45	1.41	1.37	1.34	1.32	1.27	1.26	1.24	1.21	1.19
1.75	1.66	1.61	1.55	1.50	1.46	1.43	1.37	1.35	1.32	1.28	1.25
2.19	2.03	1.95	1.86	1.76	1.70	1.66	1.56	1.53	1.48	1.42	1.38
1.23	1.21	1.20	1.18	1.16	1.14	1.12	1.11	1.10	1.09	1.08	1.06
1.52	1.46	1.42	1.38	1.34	1.31	1.28	1.24	1.22	1.20	1.17	1.14
1.72	1.62	1.57	1.52	1.46	1.41	1.39	1.32	1.29	1.26	1.22	1.19
2.13	1.97	1.89	1.79	1.69	1.63	1.58	1.48	1.44	1.39	1.33	1.28
1.22	1.19	1.18	1.16	1.14	1.13	1.12	1.09	1.08	1.07	1.04	1.00
1.49	1.42	1.38	1.34	1.30	1.26	1.24	1.18	1.17	1.13	1.08	1.00
1.67	1.57	1.52	1.46	1.39	1.35	1.32	1.24	1.22	1.17	1.11	1.00
2.04	1.88	1.79	1.70	1.59	1.52	1.47	1.36	1.32	1.25	1.15	1.00

[a]Abridged from E. S. Pearson and H. O. Hartley, eds., *Biometrika Tables for Statisticians*, Vol. 1, 2nd ed., Cambridge University Press, New York, 1958, Table 18. Reproduced by permission of E. S. Pearson and the trustees of *Biometrika*.

6. Repeat steps 3 through 5 until you have the number of random numbers needed.

Example

We want to take 32 samples from a 1200-piece lot. Our arbitrary point to start happened to be row 1, column 1, and we decided to move right. Our range is 1 to 1200, so we need four digit random numbers.

Moving across: 07 28 part 728 is our first sample
68 61 outside range—skip
81 38 outside range—skip
11 98 part 1198 is our second sample
34 74 outside range—skip

As we move across row 1, there are no other numbers that fall within the range so we might move one digit diagonally to 1. Then we would pick another direction to move. This would continue until 32 part numbers were chosen.

Random Numbers[a] (40 columns by 40 rows)

07	28	68	61	81	38	11	98	34	74	64	03	48	09	18	10	15	25	98	80
29	24	86	11	41	21	16	12	96	17	56	61	49	32	48	35	43	29	34	12
76	05	58	54	35	55	35	59	07	19	00	92	65	95	34	88	26	32	61	36
95	01	20	28	66	31	15	92	14	33	39	98	55	85	71	35	82	04	51	64
73	89	25	53	83	33	75	79	98	20	09	06	76	92	43	42	55	86	41	67
41	58	46	41	68	72	73	78	34	65	87	08	10	93	46	00	32	48	29	68
53	46	33	57	86	99	47	87	14	55	98	93	72	15	77	23	13	26	37	20
39	46	65	77	16	92	33	65	57	49	18	41	87	68	05	23	73	33	55	49
40	98	58	06	54	13	55	31	86	06	34	94	43	59	08	54	86	44	59	84
06	45	65	80	97	46	95	38	82	01	88	12	28	75	93	39	33	60	00	48
84	72	36	35	94	11	36	23	17	09	95	90	26	46	90	70	81	40	77	38
61	14	68	60	77	44	75	28	56	67	36	58	03	82	16	76	39	12	73	70
07	47	15	19	64	62	17	97	36	08	22	55	58	81	17	77	83	65	75	05
70	43	84	46	41	98	44	54	23	72	39	79	53	16	88	04	66	00	66	43
57	10	02	26	17	12	56	48	43	97	65	06	21	97	65	97	95	77	93	01
95	01	58	34	51	77	89	80	79	72	60	94	43	05	89	83	88	15	09	58
53	00	18	66	58	39	02	95	62	79	35	52	01	06	50	18	98	88	87	81
51	86	20	34	89	54	54	61	15	00	96	89	11	34	05	18	26	77	17	23
38	63	42	41	87	99	37	18	91	08	55	42	27	51	69	48	94	14	70	96
47	77	39	28	14	56	98	96	73	22	31	67	20	90	85	04	01	87	42	17
26	20	46	66	36	28	98	66	97	56	78	29	19	53	46	08	20	30	55	61
58	58	28	68	36	45	83	66	12	05	17	37	74	90	81	86	99	04	17	90
80	83	75	20	32	63	09	41	69	12	43	82	63	40	08	89	71	89	68	44
40	90	05	68	85	00	90	91	49	16	23	00	26	56	52	66	71	22	63	40
77	38	50	26	29	57	56	31	37	52	88	88	37	72	14	52	73	79	23	79
51	62	77	67	70	21	17	88	22	26	66	77	78	55	87	14	39	07	31	67
66	81	52	18	87	47	01	60	71	73	90	72	90	39	37	64	44	26	82	07
67	72	78	24	07	12	61	67	78	85	92	68	95	24	69	57	74	13	28	64
14	29	00	91	50	43	64	63	85	17	54	46	92	58	58	52	97	54	84	09
30	89	99	07	56	26	49	27	83	67	52	35	36	93	63	60	15	71	16	34
26	42	43	27	81	79	67	35	84	28	64	59	79	16	11	54	85	34	01	49
98	05	34	47	71	14	87	98	70	21	53	51	01	46	60	71	19	33	62	43
02	82	10	42	11	62	87	83	16	96	34	46	04	25	33	69	55	37	82	29
99	88	34	85	46	77	12	00	89	17	04	48	85	62	32	77	08	24	88	65
83	59	57	38	84	22	08	75	21	10	58	75	87	70	19	07	94	83	09	37
76	27	52	23	67	14	39	88	57	00	72	71	21	68	81	49	24	94	19	37
03	80	24	56	17	64	66	90	80	09	62	03	65	61	66	39	83	87	41	95
40	86	98	74	63	72	14	00	08	38	25	25	37	93	89	96	74	66	36	06
38	02	78	20	39	15	04	67	68	27	46	22	43	79	26	45	45	17	66	13
19	51	85	12	56	95	63	15	44	74	88	26	02	10	68	09	84	86	26	81

[a]Reprinted with permission from *DataMyte Handbook*, 4th ed., DataMyte Division of Allen-Bradley Inc. Copyright 1989, Allen-Bradley Inc., Milwaukee, WI.

C

ANOVA Formats

C.1 ANOVA GENERIC

Source of Variation	DF	Sum of Squares	Mean Square	F (calc)
Variation between (factors, blocks, interactions)	$n - 1$	$\sum \frac{Y_{i\cdot}^2}{n_i} - \frac{(Y_{\cdot\cdot})^2}{N}$ (= SST)	$\frac{\text{SST}}{\text{DF}} = \text{MST}$	$\frac{\text{MST}}{\text{MSE}}$
Variation within (error)	By subtraction	By subtraction (= SSE)	$\frac{\text{SSE}}{\text{DF}} = \text{MSE}$	
Total (sum of all variations)	$N - 1$	$\sum \sum Y_{ij}^2 - \frac{(Y_{\cdot\cdot})^2}{N}$		

N = total number of observations
n = number of factors, blocks, interactions
Y_{ij} = treatment i, jth observation
$Y_{i\cdot}$ = sum of observations for treatment i
$Y_{\cdot\cdot}$ = sum of all observations

C.2 CRD ANOVA

Source	DF	Sum of Squares	Mean Square	F
Treatments	$i-1$	$\sum \frac{Y_{i\cdot}^2}{n_i} - \frac{(Y_{\cdot\cdot})^2}{N}$	$\frac{SS_t}{i-1} = MS_t$	$\frac{MS_t}{MSE}$
Error	$N-i$	By subtraction	$\frac{SSE}{N-i} = MSE$	
Total	$N-1$	$\sum\sum Y_{ij}^2 - \frac{(Y_{\cdot\cdot})^2}{N}$		

i = number of treatments
N = total sample size (total number of data points)
$Y_{i\cdot}$ = sum of all observations for treatment i
$Y_{\cdot\cdot}$ = sum of all observations
Y_{ij} = treatment i, jth observation
n_i = sample size for treatment i

C.3 CRBD ANOVA

Source	DF	Sum of Squares	Mean Square	F
Treatments	$I-1$	$\frac{1}{J}\sum Y_{i\cdot}^2 - \frac{(Y_{\cdot\cdot})^2}{N}$	$\frac{SST}{DF_t}$	$\frac{MS_t}{MSE}$
Blocks	$J-1$	$\frac{1}{I}\sum Y_{\cdot j}^2 - \frac{(Y_{\cdot\cdot})^2}{N}$	$\frac{SSB}{DF_b}$	$\frac{MS_B}{MSE}$
Error	By subtraction	By subtraction SSE	$\frac{SSE}{DF_E} = MSE$	
Total	$N-1$	$\sum\sum Y_{ij}^2 - \frac{(Y_{\cdot\cdot})^2}{N}$		

I = number of treatments
J = number of blocks
N = total number of samples taken
$Y_{i\cdot}$ = sum of observations for treatment i
$Y_{\cdot j}$ = sum of observations for block j
$Y_{\cdot\cdot}$ = sum of all observations
Y_{ij} = treatment i, block j

C.4 LATIN SQUARE ANOVA

Source	DF	Sum of Squares	Mean Square	F
Treatment	$p - 1$	$\sum \frac{Y_{i..}^2}{p} - \frac{(Y...)^2}{p^2}$	$\frac{SS_T}{DF_T}$	$\frac{MS_T}{MSE}$
Block 1 (rows)	$p - 1$	$\sum \frac{Y_{.j.}^2}{p} - \frac{(Y...)^2}{p^2}$	$\frac{SS_{row}}{DF_r}$	$\frac{MS_r}{MSE}$
Block 2 (columns)	$p - 1$	$\sum \frac{Y_{..k}^2}{p} - \frac{(Y...)^2}{p^2}$	$\frac{SS_{column}}{DF_c}$	$\frac{MS_c}{MSE}$
Error	By subtraction	By subtraction SSE	$\frac{SSE}{DF_E} = MSE$	
Total	$p^2 - 1$	$\sum\sum\sum Y_{ijk}^2 - \frac{(Y...)^2}{p^2}$		

p = number of treatments and number of levels in each block
$Y_{i..}$ = sum of all observations for treatment i
$Y_{.j.}$ = sum of all observations for block j
$Y_{..k}$ = sum of all observations for block k
$Y...$ = sum of all observations
Y_{ijk} = treatment i at blocks j and k

C.5 TWO-FACTOR FACTORIAL ANOVA

Source	DF	Sum of Squares	Mean Square	F
Treatments	$ab - 1$	$\sum\sum \frac{Y_{ij.}^2}{n} - \frac{(Y...)^2}{N}$	$\frac{SS_T}{DF_T}$	$\frac{MS_T}{MSE}$
Factor A	$a - 1$	$\sum \frac{Y_{i..}^2}{nb} - \frac{(Y...)^2}{N}$	$\frac{SS_A}{DF_A}$	$\frac{MS_A}{MSE}$
Factor B	$b - 1$	$\sum \frac{Y_{.j.}^2}{na} - \frac{(Y...)^2}{N}$	$\frac{SS_B}{DF_B}$	$\frac{MS_B}{MSE}$
Interaction $A \times B$	$(a - 1) \times (b - 1)$	$SS_T - SS_A - SS_B$	$\frac{SS_{A \times B}}{DF_{A \times B}}$	$\frac{MS_{A \times B}}{MSE}$
Error	By subtraction	$SS_{tot} - SS_T$	$\frac{SS_E}{DF_E} = MSE$	
Total	$abn - 1$	$\sum\sum\sum Y_{ijk}^2 - \frac{(Y...)^2}{N}$		

a = levels of A
b = levels of B
n = sample size per cell
$SS_A + SS_B + SS_{A \times B} = SS_T$

C.6 2^3 FACTORIAL ANOVA

Source	DF	Sum of Squares	Mean Square	F
Factor A	1		$\frac{SS_A}{DF}$	$\frac{MS_A}{MSE}$
Factor B	1		$\frac{SS_B}{DF}$	$\frac{MS_B}{MSE}$
Factor C	1	Found	$\frac{SS_C}{DF}$	.
$A \times B$	1	using	$SS_{A \times B}$	.
$A \times C$	1	contrasts	$SS_{A \times C}$	.
$B \times C$	1		$SS_{B \times C}$	.
$A \times B \times C$	1		$SS_{A \times B \times C}$	$\frac{MS_{A \times B \times C}}{MSE}$
Error	$N - 8$	By subtraction	$\frac{SS_E}{N - 8} = MSE$	
Total	$N - 1$	$\sum\sum\sum Y_{ijk}^2 - \frac{(Y...)^2}{N}$		

To find contrasts, use the table of treatment combinations and calculate the average of each factor (or interaction) at its high level and at its low level. Take the difference of the two, square it, and divide by the number of combinations (8 in this case) times the number of observations of each:

$$SS_A = \frac{(\text{contrast})^2}{8n}$$

C.7 NEWMAN–KEULS METHOD: ANALYSIS OF MEANS

1. List the means from lowest to highest.
2. Take the error mean square value from the ANOVA table.
3. Obtain the standard error of the mean for each treatment using the equation

$$S_{x \cdot j} = \sqrt{\frac{\text{error mean square}}{\text{number of observations in } x_{.j}}}$$

4. After selecting a level for α (0.01 or 0.05), use the Tables in Appendix A.8 (studentized range tables) to find the values for the $k - 1$ ranges where DF_E is the degrees of freedom of the error mean square (from the ANOVA table) and p equals 2 through k where k equals the number of treatments.
5. Multiply the values for the ranges by the standard errors calculated in step 3 to find the least significant ranges (LSR).
6. Test the observed ranges between the means using the LSRs calculated in step 5. Choose the LSR corresponding to the number of treatments between means being tested.

   Note: Treatments that are found to exhibit significantly different effects on the response are designated by different capital letter:

Treatment 1	*A*		
Treatment 2		*B*	
Treatment 3		*B*	
Treatment 4			*C*

   1 is significantly different than 2–4; 2 and 3 are not significantly different from each other.

C.8 UPPER PERCENT POINTS OF STUDENTIZED RANGE q

Upper 5 Percent Points[a]

	p[b]																		
DF_E	2	3	4	5	6	7	8	9	10	11	12	13	14	15	16	17	18	19	20
1	18.0	26.7	32.8	37.2	40.5	43.1	45.4	47.3	49.1	50.6	51.9	53.2	54.3	55.4	56.3	57.2	58.0	58.8	59.6
2	6.09	8.28	9.80	10.89	11.73	12.43	13.03	13.54	13.99	14.39	14.75	15.08	15.38	15.65	15.91	16.14	16.36	16.57	16.77
3	4.50	5.88	6.83	7.51	8.04	8.47	8.85	9.18	9.46	9.72	9.95	10.16	10.35	10.52	10.69	10.84	10.98	11.12	11.24
4	3.93	5.00	5.76	6.31	6.73	7.06	7.35	7.60	7.83	8.03	8.21	8.37	8.52	8.67	8.80	8.92	9.03	9.14	9.24
5	3.61	4.54	5.18	5.64	5.99	6.28	6.52	6.74	6.93	7.10	7.25	7.39	7.52	7.64	7.75	7.86	7.95	8.04	8.13
6	3.46	4.34	4.90	5.31	5.63	5.89	6.12	6.32	6.49	6.65	6.79	6.92	7.04	7.14	7.24	7.34	7.43	7.51	7.59
7	3.34	4.16	4.68	5.06	5.35	5.59	5.80	5.99	6.15	6.29	6.42	6.54	6.65	6.75	6.84	6.93	7.01	7.08	7.16
8	3.26	4.04	4.53	4.89	5.17	5.40	5.60	5.77	5.92	6.05	6.18	6.29	6.39	6.48	6.57	6.65	6.73	6.80	6.87
9	3.20	3.95	4.42	4.76	5.02	5.24	5.43	5.60	5.74	5.87	5.98	6.09	6.19	6.28	6.36	6.44	6.51	6.58	6.65
10	3.15	3.88	4.33	4.66	4.91	5.12	5.30	5.46	5.60	5.72	5.83	5.93	6.03	6.12	6.20	6.27	6.34	6.41	6.47
11	3.11	3.82	4.26	4.58	4.82	5.03	5.20	5.35	5.49	5.61	5.71	5.81	5.90	5.98	6.06	6.14	6.20	6.27	6.33
12	3.08	3.77	4.20	4.51	4.75	4.95	5.12	5.27	5.40	5.51	5.61	5.71	5.80	5.88	5.95	6.02	6.09	6.15	6.21
13	3.06	3.73	4.15	4.46	4.69	4.88	5.05	5.19	5.32	5.43	5.53	5.63	5.71	5.79	5.86	5.93	6.00	6.06	6.11
14	3.03	3.70	4.11	4.41	4.64	4.83	4.99	5.13	5.25	5.36	5.46	5.56	5.64	5.72	5.79	5.86	5.92	5.98	6.03
15	3.01	3.67	4.08	4.37	4.59	4.78	4.94	5.08	5.20	5.31	5.40	5.49	5.57	5.65	5.72	5.79	5.85	5.91	5.96
16	3.00	3.65	4.05	4.34	4.56	4.74	4.90	5.03	5.15	5.26	5.35	5.44	5.52	5.59	5.66	5.73	5.79	5.84	5.90
17	2.98	3.62	4.02	4.31	4.52	4.70	4.86	4.99	5.11	5.21	5.31	5.39	5.47	5.55	5.61	5.68	5.74	5.79	5.84
18	2.97	3.61	4.00	4.28	4.49	4.67	4.83	4.96	5.07	5.17	5.27	5.35	5.43	5.50	5.57	5.63	5.69	5.74	5.79
19	2.96	3.59	3.98	4.26	4.47	4.64	4.79	4.92	5.04	5.14	5.23	5.32	5.39	5.46	5.53	5.59	5.65	5.70	5.75
20	2.95	3.58	3.96	4.24	4.45	4.62	4.77	4.90	5.01	5.11	5.20	5.28	5.36	5.43	5.50	5.56	5.61	5.66	5.71
24	2.92	3.53	3.90	4.17	4.37	4.54	4.68	4.81	4.92	5.01	5.10	5.18	5.25	5.32	5.38	5.44	5.50	5.55	5.59
30	2.89	3.48	3.84	4.11	4.30	4.46	4.60	4.72	4.83	4.92	5.00	5.08	5.15	5.21	5.27	5.33	5.38	5.43	5.48
40	2.86	3.44	3.79	4.04	4.23	4.39	4.52	4.63	4.74	4.82	4.90	4.98	5.05	5.11	5.17	5.22	5.27	5.32	5.36
60	2.83	3.40	3.74	3.98	4.16	4.31	4.44	4.55	4.65	4.73	4.81	4.88	4.94	5.00	5.06	5.11	5.15	5.20	5.24
120	2.80	3.36	3.69	3.92	4.10	4.24	4.36	4.47	4.56	4.64	4.71	4.78	4.84	4.90	4.95	5.00	5.04	5.09	5.13
∞	2.77	3.32	3.63	3.86	4.03	4.17	4.29	4.39	4.47	4.55	4.62	4.68	4.74	4.80	4.84	4.89	4.93	4.97	5.01

[a] From J. M. May, Extended and corrected tables of the upper percentage points of the studentized range, *Biometrika*, Vol. 39, 1952, pp. 192–193. Reproduced by permission of the trustees of *Biometrika*.

[b] p is the number of quantities (for example, means) whose range is involved. DF_E is the degrees of freedom in the error estimate.

Upper 1 Percent Points

n_2[a]	p[a] 2	3	4	5	6	7	8	9	10	11	12	13	14	15	16	17	18	19	20
1	90.0	135.0	164.0	186.0	202.0	216.0	227.0	237.0	246.0	253.0	260.0	266.0	272.0	227.0	282.0	286.0	290.0	294.0	298.0
2	14.0	19.0	22.3	24.7	26.6	28.2	29.5	30.7	31.7	32.6	33.4	34.1	34.8	35.4	36.0	36.5	37.0	37.5	37.9
3	8.26	10.6	12.2	13.3	14.2	15.0	15.6	16.2	16.7	17.1	17.5	17.9	18.2	18.5	18.8	19.1	19.3	19.5	19.8
4	6.51	8.12	9.17	9.96	10.6	11.1	11.5	11.9	12.3	12.6	12.8	13.1	13.3	13.5	13.7	13.9	14.1	14.2	14.4
5	5.70	6.97	7.80	8.42	8.91	9.32	9.67	9.97	10.24	10.48	10.70	10.89	11.08	11.24	11.40	11.55	11.68	11.81	11.93
6	5.24	6.33	7.03	7.56	7.97	8.32	8.61	8.87	9.10	9.30	9.49	9.65	9.81	9.95	10.08	10.21	10.32	10.43	10.54
7	4.95	5.92	6.54	7.01	7.37	7.68	7.94	8.17	8.37	8.55	8.71	8.86	9.00	9.12	9.24	9.35	9.46	9.55	9.65
8	4.74	5.63	6.20	6.63	6.96	7.24	7.47	7.68	7.87	8.03	8.18	8.31	8.44	8.55	8.66	8.76	8.85	8.94	9.03
9	4.60	5.43	5.96	6.35	6.66	6.91	7.13	7.32	7.49	7.65	7.78	7.91	8.03	8.13	8.23	9.32	8.41	8.49	8.57
10	4.48	5.27	5.77	6.14	6.43	6.67	6.87	7.05	7.21	7.36	7.48	7.60	7.71	7.81	7.91	7.99	8.07	8.15	8.22
11	4.39	5.14	5.62	5.97	6.25	6.48	6.67	6.84	6.99	7.13	7.25	7.36	7.46	7.56	7.65	7.73	7.81	7.88	7.95
12	4.32	5.04	5.50	5.84	6.10	6.32	6.51	6.67	6.81	6.94	7.06	7.17	7.26	7.36	7.44	7.52	7.59	7.66	7.73
13	4.26	4.96	5.40	5.73	5.98	6.19	6.37	6.53	6.67	6.79	6.90	7.01	7.10	7.19	7.27	7.34	7.42	7.48	7.55
14	4.21	4.89	5.32	5.63	5.88	6.08	6.26	6.41	6.54	6.66	6.77	6.87	6.96	7.05	7.12	7.20	7.27	7.33	7.39
15	4.17	4.83	5.25	5.56	5.80	5.99	6.16	6.31	6.44	6.55	6.66	6.76	6.84	6.93	7.00	7.07	7.14	7.20	7.26
16	4.13	4.78	5.19	5.49	5.72	5.92	6.08	6.22	6.35	6.46	6.56	6.66	6.74	6.82	6.90	6.97	7.03	7.09	7.15
17	4.10	4.74	5.14	5.43	5.66	5.85	6.01	6.15	6.27	6.38	6.48	6.57	6.66	6.73	6.80	6.87	6.94	7.00	7.05
18	4.07	4.70	5.09	5.38	5.60	5.79	5.94	6.08	6.20	6.31	6.41	6.50	6.58	6.65	6.72	6.79	6.85	6.91	6.96
19	4.05	4.67	5.05	5.33	5.55	5.73	5.89	6.02	6.14	6.25	6.34	6.43	6.51	6.58	6.65	6.72	6.78	6.84	6.89
20	4.02	4.64	5.02	5.29	5.51	5.69	5.84	5.97	6.09	6.19	6.29	6.37	6.45	6.52	6.59	6.65	6.71	6.76	6.82
24	3.96	4.54	4.91	5.17	5.37	5.54	5.69	5.81	5.92	6.02	6.11	6.19	6.26	6.33	6.39	6.45	6.51	6.56	6.61
30	3.89	4.45	4.80	5.05	5.24	5.40	5.54	5.65	5.76	5.85	5.93	6.01	6.08	6.14	6.20	6.26	6.31	6.36	6.41
40	3.82	4.37	4.70	4.93	5.11	5.27	5.39	5.50	5.60	5.69	5.77	5.84	5.90	5.96	6.02	6.07	6.12	6.17	6.21
60	3.76	4.28	4.60	4.82	4.99	5.13	5.25	5.36	5.45	5.53	5.60	5.67	5.73	5.79	5.84	5.89	5.93	5.98	6.02
120	3.70	4.20	4.50	4.71	4.87	5.01	5.12	5.21	5.30	5.38	5.44	5.51	5.56	5.61	5.66	5.71	5.75	5.79	5.83
∞	3.64	4.12	4.40	4.60	4.76	4.88	4.99	5.08	5.16	5.23	5.29	5.35	5.40	5.45	5.49	5.54	5.57	5.61	5.65

[a] p is the number of quantities (for example, means) whose range is involved. n_2 is the degrees of freedom in the error estimate.

D

Designed Experiments

D.1 LATIN SQUARE DESIGN MATRICES

3 × 3:

A	B	C
B	C	A
C	A	B

4 × 4:

A	B	C	D		A	B	C	D		A	B	C	D
B	A	D	C		B	A	D	C		B	D	A	C
C	D	B	A		C	D	A	B		C	A	D	B
D	C	A	B		D	C	B	A		D	C	B	A

5 × 5:

A	B	C	D	E
B	A	E	C	D
C	D	A	E	B
D	E	B	A	C
E	C	D	B	A

6 × 6:

A	B	C	D	E	F
B	F	D	C	A	E
C	D	E	F	B	A
E	C	A	B	F	D
F	E	B	A	D	C
D	A	F	E	C	B

D.2 TWO-LEVEL FACTORIAL DESIGNS

Treatment Combinations	A	B	C	D	E
1	−	−	−	−	−
2	+	−	−	−	−
3	−	+	−	−	−
$4 = 2^2$	+	+	−	−	−
5	−	−	+	−	−
6	+	−	+	−	−
7	−	+	+	−	−
$8 = 2^3$	+	+	+	−	−
9	−	−	−	+	−
10	+	−	−	+	−
11	−	+	−	+	−
12	+	+	−	+	−
13	−	−	+	+	−
14	+	−	+	+	−
15	−	+	+	+	−
$16 = 2^4$	+	+	+	+	−
17	−	−	−	−	+
18	+	−	−	−	+
19	−	+	−	−	+
20	+	+	−	−	+
21	−	−	+	−	+
22	+	−	+	−	+
23	−	+	+	−	+
24	+	+	+	−	+
25	−	−	−	+	+
26	+	−	−	+	+
27	−	+	−	+	+
28	+	+	−	+	+
29	−	−	+	+	+
30	+	−	+	+	+
31	−	+	+	+	+
$32 = 2^5$	+	+	+	+	+

D.3 RANDOM ORDERS FOR 2^f EXPERIMENTS

2^3					2^4					2^5				
3	6	1	6	2	3	9	11	10	3	21	12	22	12	28
5	1	7	1	3	1	15	1	7	5	18	20	5	24	10
1	4	5	5	1	11	16	12	11	10	3	2	29	13	31
8	2	4	3	5	5	12	4	5	1	11	16	18	4	1
6	3	6	8	4	8	10	14	8	14	26	7	11	15	3
7	5	8	4	6	4	5	3	4	16	17	32	17	25	5
4	8	2	2	7	12	13	15	15	8	31	3	8	14	8
2	7	3	7	8	15	1	10	14	13	1	5	23	28	15
					10	11	13	3	9	13	15	15	26	11
					2	4	6	2	11	23	22	14	22	26
					7	6	8	13	12	7	25	10	16	23
					9	8	7	1	6	20	9	16	27	19
					6	14	5	9	7	12	24	13	20	12
					14	3	2	12	4	16	31	30	5	27
					16	7	9	16	2	8	26	3	7	14
					13	2	16	6	15	25	30	32	6	17
										6	10	25	23	2
										14	17	7	21	7
										4	13	4	11	13
										19	18	27	17	29
										32	4	9	18	9
										5	8	12	10	32
										30	29	28	2	16
										15	19	19	31	18
										10	1	20	3	21
										24	6	31	29	24
										27	14	6	1	6
										2	11	1	19	22
										29	21	21	32	20
										9	28	26	8	4
										28	23	2	9	25
										22	27	24	30	30

D.4 TABLE FOR ANALYSIS OF 2^f FACTORIAL DESIGNS

Treatment	2^2				2^3				2^4							
Combinations	Mean	*A*	*B*	*AB*	*C*	*AC*	*BC*	*ABC*	*D*	*AD*	*BD*	*ABD*	*CD*	*ACD*	*BCD*	*ABCD*
1	+	−	−	+	−	+	+	−	−	+	+	−	+	−	−	+
2	+	+	−	−	−	−	+	+	−	−	+	+	+	+	−	−
3	+	−	+	−	−	+	−	+	−	+	−	+	+	−	+	−
4	+	+	+	+	−	−	−	−	−	−	−	−	+	+	+	+
5	+	−	−	+	+	−	−	+	−	+	+	−	−	+	+	−
6	+	+	−	−	+	+	−	−	−	−	+	+	−	−	+	+
7	+	−	+	−	+	−	+	−	−	+	−	+	−	+	−	+
8	+	+	+	+	+	+	+	+	−	−	−	−	−	−	−	−
9	+	−	−	+	−	+	+	−	+	−	−	+	−	+	+	−
10	+	+	−	−	−	−	+	+	+	+	−	−	−	−	+	+
11	+	−	+	−	−	+	−	+	+	−	+	−	−	+	−	+
12	+	+	+	+	−	−	−	−	+	+	+	+	−	−	−	−
13	+	−	−	+	+	−	−	+	+	−	−	+	+	−	−	+
14	+	+	−	−	+	+	−	−	+	+	−	−	+	+	−	−
15	+	−	+	−	+	−	+	−	+	−	+	−	+	−	+	−
16	+	+	+	+	+	+	+	+	+	+	+	+	+	+	+	+
17	+	−	−	+	−	+	+	−	−	+	+	−	+	−	−	+
18	+	+	−	−	−	−	+	+	−	−	+	+	+	+	−	−
19	+	−	+	−	−	+	−	+	−	+	−	+	+	−	+	−
20	+	+	+	+	−	−	−	−	−	−	−	−	+	+	+	+
21	+	−	−	+	+	−	−	+	−	+	+	−	−	+	+	−
22	+	+	−	−	+	+	−	−	−	−	+	+	−	−	+	+
23	+	−	+	−	+	−	+	−	−	+	−	+	−	+	−	+
24	+	+	+	+	+	+	+	+	−	−	−	−	−	−	−	−
25	+	−	−	+	−	+	+	−	+	−	−	+	−	+	+	−
26	+	+	−	−	−	−	+	+	+	+	−	−	−	−	+	+
27	+	−	+	−	−	+	−	+	+	−	+	−	−	+	−	+
28	+	+	+	+	−	−	−	−	+	+	+	+	−	−	−	−
29	+	−	−	+	+	−	−	+	+	−	−	+	+	−	−	+
30	+	+	−	−	+	+	−	−	+	+	−	−	+	+	−	−
31	+	−	+	−	+	−	+	−	+	−	+	−	+	−	+	−
32	+	+	+	+	+	+	+	+	+	+	+	+	+	+	+	+

2^5															
E	*AE*	*BE*	*ABE*	*CE*	*ACE*	*BCE*	*ABCE*	*DE*	*ADE*	*BDE*	*ABDE*	*CDE*	*ACDE*	*BCDE*	*ABCDE*
−	+	+	−	+	−	−	+	+	−	−	+	−	+	+	−
−	−	+	+	+	+	−	−	+	+	−	−	−	−	+	+
−	+	−	+	+	−	+	−	+	−	+	−	−	+	−	+
−	−	−	−	+	+	+	+	+	+	+	+	−	−	−	−
−	+	+	−	−	+	+	−	+	−	−	+	+	−	−	+
−	−	+	+	−	−	+	+	+	+	−	−	+	+	−	−
−	+	−	+	−	+	−	+	+	−	+	−	+	−	+	−
−	−	−	−	−	−	−	−	+	+	+	+	+	+	+	+
−	+	+	−	+	−	−	+	−	+	+	−	+	−	−	+
−	−	+	+	+	+	−	−	−	−	+	+	+	+	−	−
−	+	−	+	+	−	+	−	−	+	−	+	+	−	+	−
−	−	−	−	+	+	+	+	−	−	−	−	+	+	+	+
−	+	+	−	−	+	+	−	−	+	+	−	−	+	+	−
−	−	+	+	−	−	+	+	−	−	+	+	−	−	+	+
−	+	−	+	−	+	−	+	−	+	−	+	−	+	−	+
−	−	−	−	−	−	−	−	−	−	−	−	−	−	−	−
+	−	−	+	−	+	+	−	−	+	+	−	+	−	−	+
+	+	−	−	−	−	+	+	−	−	+	+	+	+	−	−
+	−	+	−	−	+	−	+	−	+	−	+	+	−	+	−
+	+	+	+	−	−	−	−	−	−	−	−	+	+	+	+
+	−	−	+	+	−	−	+	−	+	+	−	−	+	+	−
+	+	−	−	+	+	−	−	−	−	+	+	−	−	+	+
+	−	+	−	+	−	+	−	−	+	−	+	−	+	−	+
+	+	+	+	+	+	+	+	−	−	−	−	−	−	−	−
+	−	−	+	−	+	+	−	+	−	−	+	−	+	+	−
+	+	−	−	−	−	+	+	+	+	−	−	−	−	+	+
+	−	+	−	−	+	−	+	+	−	+	−	−	+	−	+
+	+	+	+	−	−	−	−	+	+	+	+	−	−	−	−
+	−	−	+	+	−	−	+	+	−	−	+	+	−	−	+
+	+	−	−	+	+	−	−	+	+	−	−	+	+	−	−
+	−	+	−	+	−	+	−	+	−	+	−	+	−	+	−
+	+	+	+	+	+	+	+	+	+	+	+	+	+	+	+

D.5 CONTRASTS FOR 2^f FACTORIAL DESIGNS

Coefficients for Effects in a 2^2 Factorial Experiment

Treatment Combination	Effect		
	A	*B*	*AB*
(1)	−	−	+
a	+	−	−
b	−	+	−
ab	+	+	+

Coefficients for Effects in a 2^3 Factorial Experiment

Treatment Combination	Effect							
	Total	*A*	*B*	*AB*	*C*	*AC*	*BC*	*ABC*
(1)	+	−	−	+	−	+	+	−
a	+	+	−	−	−	−	+	+
b	+	−	+	−	−	+	−	+
ab	+	+	+	+	−	−	−	−
c	+	−	−	+	+	−	−	+
ac	+	+	−	−	+	+	−	−
bc	+	−	+	−	+	−	+	−
abc	+	+	+	+	+	+	+	+

Table of Signs for Calculating Effects for a 2^4 Factorial

A	*B*	*C*	*D*	*AB*	*AC*	*AD*	*BC*	*BD*	*CD*	*ABC*	*ABD*	*ACD*	*BCD*	*ABCD*
−	−	−	−	+	+	+	+	+	+	−	−	−	−	+
+	−	−	−	−	−	−	+	+	+	+	+	+	−	−
−	+	−	−	−	+	+	−	−	+	+	−	−	+	−
+	+	−	−	+	−	−	−	−	+	−	+	+	+	+
−	−	+	−	+	−	+	−	+	−	+	−	+	+	−
+	−	+	−	−	+	−	−	+	−	−	+	−	+	+
−	+	+	−	−	−	+	+	−	−	−	−	+	−	+
+	+	+	−	+	+	−	+	−	−	+	+	−	−	−
−	−	−	+	+	+	−	+	−	−	−	+	+	+	−
+	−	−	+	−	−	+	+	−	−	+	−	−	+	+
−	+	−	+	−	+	−	−	+	−	+	+	+	−	+
+	+	−	+	+	−	+	−	+	−	−	−	−	−	−
−	−	+	+	+	−	−	−	−	+	+	+	−	−	+
+	−	+	+	−	+	+	−	−	+	−	−	+	−	−
−	+	+	+	−	−	−	+	+	+	−	+	−	+	−
+	+	+	+	+	+	+	+	+	+	+	−	+	+	+

D.6 COEFFICIENTS OF ORTHOGONAL POLYNOMIALS

		X											
k	Polynomial	1	2	3	4	5	6	7	8	9	10	$\sum \xi_j^2$	λ
3	Linear	−1	0	1								2	1
	Quadratic	1	−2	1								6	3
4	Linear	−3	−1	1	3							20	2
	Quadratic	1	−1	−1	1							4	1
	Cubic	−1	3	−3	1							20	$\frac{10}{3}$
5	Linear	−2	−1	0	1	2						10	1
	Quadratic	2	−1	−2	−1	2						14	1
	Cubic	−1	2	0	−2	1						10	$\frac{5}{6}$
	Quartic	1	−4	6	−4	1						70	$\frac{35}{12}$
6	Linear	−5	−3	−1	1	3	5					70	2
	Quadratic	5	−1	−4	−4	−1	5					84	$\frac{3}{2}$
	Cubic	−5	7	4	−4	−7	5					180	$\frac{5}{3}$
	Quartic	1	−3	2	2	−3	1					28	$\frac{7}{12}$
7	Linear	−3	−2	−1	0	1	2	3				28	1
	Quadratic	5	0	−3	−4	−3	0	5				84	1
	Cubic	−1	1	1	0	−1	−1	1				6	$\frac{1}{6}$
	Quartic	3	−7	1	6	1	−7	3				154	$\frac{7}{12}$
8	Linear	−7	−5	−3	−1	1	3	5	7			168	2
	Quadratic	7	1	−3	−5	−5	−3	1	7			168	1
	Cubic	−7	5	7	3	−3	−7	−5	7			264	$\frac{2}{3}$
	Quartic	7	−13	−3	9	9	−3	−13	7			616	$\frac{7}{12}$
	Quintic	−7	23	−17	−15	15	17	−23	7			2184	$\frac{7}{10}$
9	Linear	−4	−3	−2	−1	0	1	2	3	4		60	1
	Quadratic	28	7	−8	−17	−20	−17	−8	7	28		2772	3
	Cubic	−14	7	13	9	0	−9	−13	−7	14		990	$\frac{5}{6}$
	Quartic	14	−21	−11	9	18	9	−11	−21	14		2002	$\frac{7}{12}$
	Quintic	−4	11	−4	−9	0	9	4	−11	4		468	$\frac{3}{20}$
10	Linear	−9	−7	−5	−3	−1	1	3	5	7	9	330	2
	Quadratic	6	2	−1	−3	−4	−4	−3	−1	2	6	132	$\frac{1}{2}$
	Cubic	−42	14	35	31	12	−12	−31	−35	−14	42	8580	$\frac{5}{3}$
	Quartic	18	−22	−17	3	18	18	3	−17	−22	18	2860	$\frac{5}{12}$
	Quintic	−6	14	−1	−11	−6	6	11	1	−14	6	780	$\frac{1}{10}$

E

DOE: Screening Designs

E.1 EIGHT-RUN PLACKETT–BURMAN DESIGN

Run	Mean	*A*	*B*	*C*	*D*	*E*	*F*	*G*
1	+	+	+	+	−	+	−	−
2	+	−	+	+	+	−	+	−
3	+	−	−	+	+	+	−	+
4	+	+	−	−	+	+	+	−
5	+	−	+	−	−	+	+	+
6	+	+	−	+	−	−	+	+
7	+	+	+	−	+	−	−	+
8	+	−	−	−	−	−	−	−

E.2 EIGHT-RUN PLACKETT–BURMAN DESIGN REFLECTED AND REPLICATED

Run	Mean	*A*	*B*	*C*	*D*	*E*	*F*	*G*
1	+	+	+	+	−	+	−	−
2	+	−	+	+	+	−	+	−
3	+	−	−	+	+	+	−	+
4	+	+	−	−	+	+	+	−
5	+	−	+	−	−	+	+	+
6	+	+	−	+	−	−	+	+
7	+	+	+	−	+	−	−	+
8	+	−	−	−	−	−	−	−

(*continued*)

	Run	Mean	*A*	*B*	*C*	*D*	*E*	*F*	*G*
Reflected Original	9	+	−	−	−	+	−	+	+
	10	+	+	−	−	−	+	−	+
	11	+	+	+	−	−	−	+	−
	12	+	−	+	+	−	−	−	+
	13	+	+	−	+	+	−	−	−
	14	+	−	+	−	+	+	−	−
	15	+	−	−	+	−	+	+	−
	16	+	+	+	+	+	+	+	+
Replicated Original	R1	+	+	+	+	−	+	−	−
	R2	+	−	+	+	+	−	+	−
	R3	+	−	−	+	+	+	−	+
	R4	+	+	−	−	+	+	+	−
	R5	+	−	+	−	−	+	+	+
	R6	+	+	−	+	−	−	+	+
	R7	+	+	+	−	+	−	−	+
	R8	+	−	−	−	−	−	−	−
Replicated Reflected	R9	+	−	−	−	+	−	+	+
	R10	+	+	−	−	−	+	−	+
	R11	+	+	+	−	−	−	+	−
	R12	+	−	+	+	−	−	−	+
	R13	+	+	−	+	+	−	−	−
	R14	+	−	+	−	+	+	−	−
	R15	+	−	−	+	−	+	+	−
	R16	+	+	+	+	+	+	+	+

E.3 12-RUN PLACKETT–BURMAN DESIGN

Run	Mean	*A*	*B*	*C*	*D*	*E*	*F*	*G*	*H*	*I*	*J*	*K*
1	+	+	+	−	+	+	+	−	−	−	+	−
2	+	+	−	+	+	+	−	−	−	+	−	+
3	+	−	+	+	+	−	−	−	+	−	+	+
4	+	+	+	+	−	−	−	+	−	+	+	−
5	+	+	+	−	−	−	+	−	+	+	−	+
6	+	+	−	−	−	+	−	+	+	−	+	+
7	+	−	−	−	+	−	+	+	−	+	+	+
8	+	−	−	+	−	+	+	−	+	+	+	−
9	+	−	+	−	+	+	−	+	+	+	−	−
10	+	+	−	+	+	−	+	+	+	−	−	−
11	+	−	+	+	−	+	+	+	−	−	−	+
12	+	−	−	−	−	−	−	−	−	−	−	−

E.4 16-RUN PLACKETT–BURMAN DESIGN

Run	Mean	*A*	*B*	*C*	*D*	*E*	*F*	*G*	*H*	*I*	*J*	*K*	*L*	*M*	*N*	*O*
1	+	+	−	−	−	+	−	−	+	+	−	+	−	+	+	+
2	+	+	+	−	−	−	+	−	−	+	+	−	+	−	+	+
3	+	+	+	+	−	−	−	+	−	−	+	+	−	+	−	+
4	+	+	+	+	+	−	−	−	+	−	−	+	+	−	+	−
5	+	−	+	+	+	+	−	−	−	+	−	−	+	+	−	+
6	+	+	−	+	+	+	+	−	−	−	+	−	−	+	+	−
7	+	−	+	−	+	+	+	+	−	−	−	+	−	−	+	+
8	+	+	−	+	−	+	+	+	+	−	−	−	+	−	−	+
9	+	+	+	−	+	−	+	+	+	+	−	−	−	+	−	−
10	+	−	+	+	−	+	−	+	+	+	+	−	−	−	+	−
11	+	−	−	+	+	−	+	−	+	+	+	+	−	−	−	+
12	+	+	−	−	+	+	−	+	−	+	+	+	+	−	−	−
13	+	−	+	−	−	+	+	−	+	−	+	+	+	+	−	−
14	+	−	−	+	−	−	+	+	−	+	−	+	+	+	+	−
15	+	−	−	−	+	−	−	+	+	−	+	−	+	+	+	+
16	+	−	−	−	−	−	−	−	−	−	−	−	−	−	−	−

E.5 20-RUN PLACKETT–BURMAN DESIGN

Run	Mean	*A*	*B*	*C*	*D*	*E*	*F*	*G*	*H*	*I*	*J*	*K*	*L*	*M*	*N*	*O*	*P*	*Q*	*R*	*S*
1	+	+	+	−	−	+	+	+	+	−	+	−	+	−	−	−	−	+	+	−
2	+	+	−	−	+	+	+	+	−	+	−	+	−	−	−	−	+	+	−	+
3	+	−	−	+	+	+	+	−	+	−	+	−	−	−	−	+	+	−	+	+
4	+	−	+	+	+	+	−	+	−	+	−	−	−	−	+	+	−	+	+	−
5	+	+	+	+	+	−	+	−	+	−	−	−	−	+	+	−	+	+	−	−
6	+	+	+	+	−	+	−	+	−	−	−	−	+	+	−	+	+	−	−	+
7	+	+	+	−	+	−	+	−	−	−	−	+	+	−	+	+	−	−	+	+
8	+	+	−	+	−	+	−	−	−	−	+	+	−	+	+	−	−	+	+	+
9	+	−	+	−	+	−	−	−	−	+	+	−	+	+	−	−	+	+	+	+
10	+	+	−	+	−	−	−	−	+	+	−	+	+	−	−	+	+	+	+	−
11	+	−	+	−	−	−	−	+	+	−	+	+	−	−	+	+	+	+	−	+
12	+	+	−	−	−	−	+	+	−	+	+	−	−	+	+	+	+	−	+	−
13	+	−	−	−	−	+	+	−	+	+	−	−	+	+	+	+	−	+	−	+
14	+	−	−	−	+	+	−	+	+	−	−	+	+	+	+	−	+	−	+	−
15	+	−	−	+	+	−	+	+	−	−	+	+	+	+	−	+	−	+	−	−
16	+	−	+	+	−	+	+	−	−	+	+	+	+	−	+	−	+	−	−	−
17	+	+	+	−	+	+	−	−	+	+	+	+	−	+	−	+	−	−	−	−
18	+	+	−	+	+	−	−	+	+	+	+	−	+	−	+	−	−	−	−	+
19	+	−	+	+	−	−	+	+	+	+	−	+	−	+	−	−	−	−	+	+
20	+	−	−	−	−	−	−	−	−	−	−	−	−	−	−	−	−	−	−	−

E.6 28-RUN PLACKETT–BURMAN DESIGN

Run	Mean	*A*	*B*	*C*	*D*	*E*	*F*	*G*	*H*	*I*	*J*	*K*	*L*	*M*	*N*	*O*	*P*	*Q*	*R*	*S*	*T*	*U*	*V*	*W*	*X*	*Y*	*Z*	*a*
1	+	+	−	+	+	+	+	−	−	−	−	+	−	−	−	+	−	−	+	+	+	−	+	−	+	+	−	+
2	+	+	+	−	+	+	+	−	−	−	−	−	+	+	−	−	+	−	−	−	+	+	+	+	−	+	+	−
3	+	−	+	+	+	+	+	−	−	−	+	−	−	−	+	−	−	+	−	+	−	+	−	+	+	−	+	+
4	+	−	−	−	+	−	+	+	+	+	−	−	+	−	+	−	−	−	+	+	−	+	+	+	−	+	−	+
5	+	−	−	−	+	+	−	+	+	+	+	−	−	−	−	+	+	−	−	+	+	−	−	+	+	+	+	−
6	+	−	−	−	−	+	+	+	+	+	−	+	−	+	−	−	−	+	−	−	+	+	+	−	+	−	+	+
7	+	+	+	+	−	−	−	+	−	+	−	−	+	−	−	+	−	+	−	+	−	+	+	−	+	+	+	−
8	+	+	+	+	−	−	−	+	+	−	+	−	−	+	−	−	−	−	+	+	+	−	+	+	−	−	+	+
9	+	+	+	+	−	−	−	−	+	+	−	+	−	−	+	−	+	−	−	−	+	+	−	+	+	+	−	+
10	+	−	+	−	−	−	+	−	−	+	+	+	−	+	−	+	+	−	+	+	−	+	+	+	+	−	−	−
11	+	−	−	+	+	−	−	+	−	−	−	+	+	+	+	−	+	+	−	+	+	−	+	+	+	−	−	−
12	+	+	−	−	−	+	−	−	+	−	+	−	+	−	+	+	−	+	+	−	+	+	+	+	+	−	−	−
13	+	−	−	+	−	+	−	−	−	+	+	−	+	+	+	−	+	−	+	−	−	−	+	−	+	+	+	+
14	+	+	−	−	−	−	+	+	−	−	+	+	−	−	+	+	+	+	−	−	−	−	+	+	−	+	+	+
15	+	−	+	−	+	−	−	−	+	−	−	+	+	+	−	+	−	+	+	−	−	−	−	+	+	+	+	+
16	+	−	−	+	−	−	+	−	+	−	+	−	+	+	−	+	+	+	−	+	+	+	−	−	−	+	−	+
17	+	+	−	−	+	−	−	−	−	+	+	+	−	+	+	−	−	+	+	+	+	+	−	−	−	+	+	−
18	+	−	+	−	−	+	−	+	−	−	−	+	+	−	+	+	+	−	+	+	+	+	−	−	−	−	+	+
19	+	+	+	−	+	−	+	+	−	+	+	−	+	+	+	+	−	−	−	−	+	−	−	−	+	−	−	+
20	+	−	+	+	+	+	−	+	+	−	+	+	−	+	+	+	−	−	−	−	−	+	+	−	−	+	−	−
21	+	+	−	+	−	+	+	−	+	+	−	+	+	+	+	+	−	−	−	+	−	−	−	+	−	−	+	−
22	+	+	−	+	+	+	−	+	−	+	−	−	−	+	−	+	+	+	+	−	−	+	−	+	−	−	−	+
23	+	+	+	−	−	+	+	+	+	−	−	−	−	+	+	−	+	+	+	+	−	−	−	−	+	+	−	−
24	+	−	+	+	+	−	+	−	+	+	−	−	−	−	+	+	+	+	+	−	+	−	+	−	−	−	+	−
25	+	+	−	+	+	−	+	+	+	−	+	+	+	−	−	−	+	−	+	−	−	+	−	−	+	−	+	−
26	+	+	+	−	+	+	−	−	+	+	+	+	+	−	−	−	+	+	−	+	−	−	+	−	−	−	−	+
27	+	−	+	+	−	+	+	+	−	+	+	+	+	−	−	−	−	+	+	−	+	−	−	+	−	+	−	−
28	+	−	−	−	−	−	−	−	−	−	−	−	−	−	−	−	−	−	−	−	−	−	−	−	−	−	−	−

E.7 L4 TAGUCHI DESIGN

L4 Matrix

Treatment Combination	Factor 1	Factor 2	Factor 3
1	+1	+1	+1
2	+1	−1	−1
3	−1	+1	−1
4	−1	−1	+1

L4 Interactions Table

First Factor	Second Factor 1	2	3	Interaction Column
1	—	3	2	
2		—	1	
3			—	

E.8 L8 TAGUCHI DESIGN

L8 Matrix

Treatment Combination	Factor 1	2	3	4	5	6	7
1	+1	+1	+1	+1	+1	+1	+1
2	+1	+1	+1	−1	−1	−1	−1
3	+1	−1	−1	+1	+1	−1	−1
4	+1	−1	−1	−1	−1	+1	+1
5	−1	+1	−1	+1	−1	+1	−1
6	−1	+1	−1	−1	+1	−1	+1
7	−1	−1	+1	+1	−1	−1	+1
8	−1	−1	+1	−1	+1	+1	−1

L8 Interactions Table

First Factor	Second Factor 1	2	3	4	5	6	7	Interaction Column
1	—	3	2	5	4	7	6	
2		—	1	6	7	4	5	
3			—	7	6	5	4	
4				—	1	2	3	
5					—	3	2	
6						—	1	
7							—	

E.9 L12 TAGUCHI DESIGN

L12 DMatrix

Treatment Combination	Factor										
	1	2	3	4	5	6	7	8	9	10	11
1	+1	+1	+1	+1	+1	+1	+1	+1	+1	+1	+1
2	+1	+1	+1	+1	+1	−1	−1	−1	−1	−1	−1
3	+1	+1	−1	−1	−1	+1	+1	+1	−1	−1	−1
4	+1	−1	+1	−1	−1	+1	−1	−1	+1	+1	−1
5	+1	−1	−1	+1	−1	−1	+1	−1	+1	−1	+1
6	+1	−1	−1	−1	+1	−1	−1	+1	−1	+1	+1
7	−1	+1	−1	−1	+1	+1	−1	−1	+1	−1	+1
8	−1	+1	−1	+1	−1	−1	−1	+1	+1	+1	−1
9	−1	+1	+1	−1	−1	−1	+1	−1	−1	+1	+1
10	−1	−1	−1	+1	+1	+1	+1	−1	−1	+1	−1
11	−1	−1	+1	−1	+1	−1	+1	+1	+1	−1	−1
12	−1	−1	+1	+1	−1	+1	−1	+1	−1	−1	+1

Note: This design should not be used for experiments that involve looking for interactions.

E.10 L16 TAGUCHI DESIGN

L16 Matrix

Treatment Combination	Factor														
	1	2	3	4	5	6	7	8	9	10	11	12	13	14	15
1	+1	+1	+1	+1	+1	+1	+1	+1	+1	+1	+1	+1	+1	+1	+1
2	+1	+1	+1	+1	+1	+1	+1	−1	−1	−1	−1	−1	−1	−1	−1
3	+1	+1	+1	−1	−1	−1	−1	+1	+1	+1	+1	−1	−1	−1	−1
4	+1	+1	+1	−1	−1	−1	−1	−1	−1	−1	−1	+1	+1	+1	+1
5	+1	−1	−1	+1	+1	−1	−1	+1	+1	−1	−1	+1	+1	−1	−1
6	+1	−1	−1	+1	+1	−1	−1	−1	−1	+1	+1	−1	−1	+1	+1
7	+1	−1	−1	−1	−1	+1	+1	+1	+1	−1	−1	−1	−1	+1	+1
8	+1	−1	−1	−1	−1	+1	+1	−1	−1	+1	+1	+1	+1	−1	−1
9	−1	+1	−1	+1	−1	+1	−1	+1	−1	+1	−1	+1	−1	+1	−1
10	−1	+1	−1	+1	−1	+1	−1	−1	+1	−1	+1	−1	+1	−1	+1
11	−1	+1	−1	−1	+1	−1	+1	+1	−1	+1	−1	−1	+1	−1	+1
12	−1	+1	−1	−1	+1	−1	+1	−1	+1	−1	+1	+1	−1	+1	−1
13	−1	−1	+1	+1	−1	−1	+1	+1	−1	−1	+1	+1	−1	−1	+1
14	−1	−1	+1	+1	−1	−1	+1	−1	+1	+1	−1	−1	+1	+1	−1
15	−1	−1	+1	−1	+1	+1	−1	+1	−1	−1	+1	−1	+1	+1	−1
16	−1	−1	+1	−1	+1	+1	−1	−1	+1	+1	−1	+1	−1	−1	+1

16 Interaction Table

First Factor	Second Factor 1	2	3	4	5	6	7	8	9	10	11	12	13	14	15	Interaction Column
1	—	1	4	5	4	7	6	9	8	11	10	13	12	15	14	
2		—	1	6	7	4	5	10	11	8	9	14	15	12	13	
3			—	7	6	5	4	11	10	9	8	15	14	13	12	
4				—	1	2	3	12	13	14	15	8	9	10	11	
5					—	3	2	13	12	15	14	9	8	11	10	
6						—	1	14	15	12	13	10	11	8	9	
7							—	15	14	13	12	11	10	9	8	
8								—	1	2	3	4	5	6	7	
9									—	3	2	5	4	7	6	
10										—	1	6	7	4	5	
11											—	7	6	5	4	
12												—	1	2	3	
13													—	3	2	
14														—	1	
15															—	

F

Mixture Experiments

F.1 THREE-COMPONENT SIMPLEX DESIGN

Blend	Components		
	1	2	3
1	1	0	0
2	0	1	0
3	0	0	1
4	1/2	1/2	0
5	1/2	0	1/2
6	0	1/2	1/2
7	1/3	1/3	1/3
8	2/3	1/6	1/6
9	1/6	2/3	1/6
10	1/6	1/6	2/3

F.2 FOUR-COMPONENT SIMPLEX DESIGN

Blend	Components			
	1	2	3	4
1	1	0	0	0
2	0	1	0	0
3	0	0	1	0
4	0	0	0	1
5	1/2	1/2	0	0
6	1/2	0	1/2	0
7	1/2	0	0	1/2
8	0	1/2	1/2	0

(*continued*)

Blend	Components			
	1	2	3	4
9	0	1/2	0	1/2
10	0	0	1/2	1/2
11	1/4	1/4	1/4	1/4
12	5/8	1/8	1/8	1/8
13	1/8	5/8	1/8	1/8
14	1/8	1/8	5/8	1/8
15	1/8	1/8	1/8	5/8

F.3 FIVE-COMPONENT SIMPLEX DESIGN

Blend	Components				
	1	2	3	4	5
1	1	0	0	0	0
2	0	1	0	0	0
3	0	0	1	0	0
4	0	0	0	1	0
5	0	0	0	0	1
6	1/2	1/2	0	0	0
7	1/2	0	1/2	0	0
8	1/2	0	0	1/2	0
9	1/2	0	0	0	1/2
10	0	1/2	1/2	0	0
11	0	1/2	0	1/2	0
12	0	1/2	0	0	1/2
13	0	0	1/2	1/2	0
14	0	0	1/2	0	1/2
15	0	0	0	1/2	1/2
16	1/5	1/5	1/5	1/5	1/5
17	3/5	1/10	1/10	1/10	1/10
18	1/10	3/5	1/10	1/10	1/10
19	1/10	1/10	3/5	1/10	1/10
20	1/10	1/10	1/10	3/5	1/10
21	1/10	1/10	1/10	1/10	3/5

F.4 SIMPLEX SCREENING DESIGN POINTS: *n* COMPONENTS

Blend	1	2	3	·	·	·	x_n
1-1	1	0	0	·	·	·	0
1-2	0	1	0	·	·	·	0

(*continued*)

Blend	1	2	3	·	·	·	x_n
·	·	·	1	·			·
·	·	·	·		·		·
·	·	·	·	·	·	·	·
2-1	$(n+1)/2n$	$1/2n$	$1/2n$	·	·	·	$1/2n$
2-2	$1/2n$	$(n+1)/2n$	$1/2n$	·	·	·	$1/2n$
·	·	$1/2n$	$(n+1)/2n$	·			·
·	·	·	$1/2n$		·		·
·	·	·	·			·	·
2-n	$1/2n$	$1/2n$	$1/2n$	·	·	·	$(n+1)/2n$
3-1	n^{-1}	n^{-1}	n^{-1}	·	·	·	n^{-1}
4-1	0	$(n-1)^{-1}$	$(n-1)^{-1}$	·	·	·	$(n-1)^{-1}$
4-2	$(n-1)^{-1}$	0	$(n-1)^{-1}$	·	·	·	$(n-1)^{-1}$
·	·	$(n-1)^{-1}$	0	·			·
·	·	·	$(n-1)^{-1}$		·		·
·	·	·	·			·	·
4-n	$(n-1)^{-1}$	$(n-1)^{-1}$	$(n-1)^{-1}$	·	·	·	0

Note: Once the number of components has been selected, renumber the blends in sequential order.

F.5 SIX-COMPONENT SIMPLEX SCREENING DESIGN

	Components					
Blend	1	2	3	4	5	6
1	1	0	0	0	0	0
2	0	1	0	0	0	0
3	0	0	1	0	0	0
4	0	0	0	1	0	0
5	0	0	0	0	1	0
6	0	0	0	0	0	1
7	7/12	1/12	1/12	1/12	1/12	1/12
8	1/12	7/12	1/12	1/12	1/12	1/12
9	1/12	1/12	7/12	1/12	1/12	1/12
10	1/12	1/12	1/12	7/12	1/12	1/12
11	1/12	1/12	1/12	1/12	7/12	1/12
12	1/12	1/12	1/12	1/12	1/12	7/12
13	1/6	1/6	1/6	1/6	1/6	1/6
14	0	1/5	1/5	1/5	1/5	1/5
15	1/5	0	1/5	1/5	1/5	1/5
16	1/5	1/5	0	1/5	1/5	1/5
17	1/5	1/5	1/5	0	1/5	1/5
18	1/5	1/5	1/5	1/5	0	1/5
19	1/5	1/5	1/5	1/5	1/5	0

Glossary of Symbols

a_i	= minimum level of component i
a_j	= lower bound of component j
a_n	= lower bound of component n
A_2	= value dependent on the sample size that is multiplied by $\overline{R}$ to calculate control limits for an $\overline{X}$ chart
A_3	= value dependent on the sample size that is multiplied by $\bar{s}$ to calculate control limits for an $\overline{X}$ chart
AOQ	= average outgoing quality
AOQL	= average outgoing quality limit
AQL	= acceptable quality level
b	= y intercept value in a linear equation
b_i	= maximum level of component i
b_{ij}	= coefficients for mixture experiment equation
B_3	= value dependent on the sample size that is multiplied by $\bar{s}$ to calculate the LCL for an s chart
B_4	= value dependent on the sample size that is multiplied by $\bar{s}$ to calculate the UCL for an s chart
c	= number of defects or nonconformities
	= acceptable number of defects in a lot
c_k	= number of nonconformities in sample k
$\bar{c}$	= average number of defects or nonconformities
C	= the acceptable number of standard deviations that the process can differ from the target when it is being set up
	= when used as a subscript, indicates the cubic effect of a factor or an interaction
C_i	= component i in a mixture experiment

C_p = process capability index = (USL − LSL)/6s
C_{pk} = process capability index = minimum of $\{C_{pu}, C_{pl}\}$
C_{pl} = capability of the process against the lower specification limit (LSL) = $(\overline{X} - \text{LSL})/3s$
C_{pu} = capability of the process against the upper specification limit (USL) = $(\text{USL} - \overline{X})/3s$
C_r = process capability ratio = $6s/(\text{USL} - \text{LSL})$
CI = confidence interval
CL = centerline of a control chart

d_E = value of d_2 for calculating the repeatability error in a GR & R study
d_i = difference between data pair i
d_{ij} = edge length between vertices i and j in a mixture experiment
d_m = value of d_2 for calculating the materials variation in a GR & R study
d_0 = value of d_2 for calculating the reproducibility in a GR & R study
d_2 = factor based on the sample size used with $\overline{R}$ to estimate the sample standard deviation s
$\bar{d}$ = average difference between pairs of data
D_3 = value dependent on the sample size that is multiplied by $\overline{R}$ to calculate the LCL for an R chart
D_4 = value dependent on the sample size that is multiplied by $\overline{R}$ to calculate the UCL for an R chart
DF = degrees of freedom
DF_E = degrees of freedom of the error term

E_i = overall effect of factor i
E_{iH} = effect of factor i at its high level
E_{iL} = effect of factor i at its low level
E_2 = value dependent on the number of samples used to calculate the moving range R_m. This value is multiplied by $\overline{R}_m$ to calculate the control limits for an X chart

f = number of factors
$f(x)$ = equation for a curve based on x
F = test of significance for the variance
F_{calc} = calculated value of the F-test statistic = s_1^2/s_2^2
F_{table} = table value of the F-test statistic

H_a = alternate hypothesis to the null hypothesis H_0
H_0 = null hypothesis tested in a test of significance

i = number of treatments in a randomized block design
= factor, treatment, or component i when used as a subscript

j = number of observations within each treatment
= block or observation j when used as a subscript

k = number of samples taken for preparing the centerline and control limits of attribute charts
= number of check blends in a mixture experiment

L = when used as a subscript, indicates the linear effect of a factor or an interaction
L_s = lower alarm setpoint for process average setting
LCL = lower control limit
LSAL = lower signal alarm limit
LSL = lower specification limit
LSR = least significant range

m = slope of a line
MS = mean square value = SS/DF
MSE = mean square value of the error term = SSE/DF_E

n = sample or subgroup size
= EVOP cycle number
n_i = sample size for treatment i
n_k = total number of items or events in sample k
= number of samples in check blend k
np = number of defective items or events
$\overline{np}$ = average number of defective items or events
$(np)_k$ = number of nonconforming or defective items in sample k
N = size of a population
= total number of observations in a designed experiment
N_s = calculated sample size for use in process average setting

p = proportion or percentage of defective items or events
= upper 5% values of the studentized range
$\bar{p}$ = average proportion or percentage of defective items or events
P_a = probability of acceptance
PTV = percent of total variability taken up by the measurement

Q = when used as a subscript, indicates a quadratic effect of a factor or an interaction

r = number of trials
R = range = (highest value in a set of data) − (lowest value in that set)
$\bar{R}$ = average range
$\bar{\bar{R}}$ = overall or grand average range
$\bar{\bar{R}}_E$ = overall average range for calculating the repeatability in a GR & R study
$\bar{R}_m$ = average range for calculating the materials variation in a GR & R study
R_m = moving range calculated over two or more samples or subgroups

$\overline{R}_m$ = average value of R_m
R_0 = range for calculating the reproducibility in a GR & R study
$\overline{R}_p$ = process average range for a group control chart
R_w = within batch range
$\overline{R}_w$ = average value of R_w
R & R = repeatability and reproducibility

s = sample standard deviation
s^2 = sample variance
s_d = sample standard deviation of the differences between pairs of data
s_E = sample standard deviation for the repeatability calculated in a GR & R study
s_{eff} = sample standard deviation of an effect
s^2_{error} = variance due to error of predicted response to actual response for a mixture experiment
s_k = sample standard deviation of check blends
s^2_{LOF} = variance due to the lack of fit in a mixture experiment
s_m = sample standard deviation for the materials calculated in a GR & R study
s_0 = sample standard deviation for the reproducibility calculated in a GR & R study
s_p = pooled sample standard deviation
$S_{\bar{y}_{ij}}$ = standard error for each treatment
SF = signal factor for a CUSUM signal chart
S/N = signal-to-noise ratio in a measurement system study or in a designed experiment
SS = sum of squares
SSE = sum of squares of the error term

t = test of significance for sample means
t_{calc} = calculated value of the t-test statistic
t_{table} = table value of the t-test statistic
T = total number of runs in a designed experiment
TC = treatment combination
TSE = two sigma effect that an effect must be greater than to be significant at a 95% confidence level

u = proportion of defects or nonconformities per unit
$\bar{u}$ = average proportion of defects or nonconformities per unit

U_s = upper alarm setpoint for process average setting
UCL = upper control limit
USAL = upper signal alarm limit for a CUSUM chart
USL = upper specification limit

V = variation or, mathematically, variance

x = individual value
$\bar{x}$ = mean or arithmetic average
$\bar{\bar{x}}$ = overall or grand average
$\tilde{x}$ = median
x_i = proportion of component i in a mixture experiment
x_{ic} = centroid value for component i in a mixture experiment
x_j = proportion of pseudocomponent j in a mixture experiment
$\bar{\bar{x}}_p$ = overall process mean for a group control chart
x_v = nominalized value of $x = x -$ [nominal (target) value]
$\bar{x}_v$ = average value of x_v
X_i = transference value for factor i in a response surface analysis

y = dependent variable or response
Y = measured response
$\bar{Y}$ = average response
$\bar{\bar{Y}}$ = overall or grand average of the response
Y_i = response for treatment combination or factor i
Y_{ij} = response for treatment i, observation j
$Y_{i\cdot}$ = sum of responses for treatment i
$Y_{\cdot j}$ = sum of responses for block j
$Y_{\cdot\cdot}$ = sum of all responses for treatment i and block j
Y_{ijk} = response for treatment i, block j, block k
$Y_{i\cdot\cdot}$ = sum of responses for treatment i or factor i
$Y_{\cdot j\cdot}$ = sum of responses for block j or factor j
$Y_{\cdot\cdot k}$ = sum of responses for block k
$Y_{\cdot\cdot\cdot}$ = sum of all responses for i, j, k
$\bar{Y}_{iH}$ = average response at the high level of factor i
$\bar{Y}_{iL}$ = average response at the low level of factor i
$\bar{Y}_{mk}$ = measured average response for check blend k
Y_{pk} = predicted response for check blend k

z_j = pure component proportion of component j
Z = test of significance for population means
= the normal curve statistic
Z_{calc} = calculated value of the test statistic Z
Z_{table} = table value of the test statistic Z

α = the Type I risk of rejecting a hypothesis when it is true

β = the Type II risk of accepting a hypothesis when it is false
β_j = effect of block j
$\beta_0, \beta_1, \beta_2$ = coefficients for defining the response surface equation

ϵ = error term of response-surface-analysis curve fitting
ϵ_{ij} = random error term in equation for effects

ϵ_{ijk} = random error term in equation for effects

δ_k = effect of block k

μ = population mean
= common effect of all treatments in the effects equation
μ_d = difference between population mean values
μ_0 = population mean (used in t test)

σ = population standard deviation
σ^2 = population variance
σ_0^2 = population variance (used in t test)
σ_p = population standard deviation for EVOP
σ_x = population standard deviation for individual values
$\sigma_{\bar{x}}$ = population standard deviation for the mean values

τ_i = effect of treatment i

χ^2 = test of significance for two known population variances
χ^2_{calc} = calculated value of the χ^2-test statistic
χ^2_{table} = table value of the χ^2-test statistic

$f_{K,n}$ = factors for EVOP calculations for sample size K and cycle n; the factors are derived from d_2 values

Σ = sum of values following this symbol

References

A Seminar in Problem Solving and Decision Making, 1986. Madison, WI: Bud Erickson Associates.

Barrentine, L. B., 1989. Addressing sample variation. *Quality* June: 59.

Beauregard, M. R., R. J. Mikulak, R. McDermott, and M. J. DeLassus, 1990. *SPC in Action*. White Plains, NY: Quality Resources.

Box, G. E. P., 1957. Evolutionary operations: A method for increasing industrial productivity. *Applied Statistics* 6:81–101.

Box, G. E. P. and N. R. Draper, 1969. *Evolutionary Operations*. New York: John Wiley & Sons, Inc.

Box, G. E. P., W. G. Hunter, and J. S. Hunter, 1978. *Statistics for Experimenters*. New York: John Wiley & Sons, Inc.

Burr, I. W., 1949. A new method for approving a machine or process setting—Part I. *Industrial Quality Control* January: 12–18.

Burr, I. W., 1953. *Engineering Statistics and Quality Control*. New York: McGraw-Hill Book Company.

Charbonneau, H. C. and G. L. Webster, 1978. *Industrial Quality Control*. Englewood Cliffs, NJ: Prentice-Hall, Inc.

Chesmire, A. D., 1988. SPC charting for job shop and short-run applications. Internal Memo, Rogers Corporation, 12 February, Rogers, CT.

Cochran, W. G. and G. M. Cox, 1957. *Experimental Designs*. New York: John Wiley & Sons, Inc.

Cornell, J. A., 1973. Experiments with mixtures: A review. *Technometrics* 15(3):437–455.

Cornell, J. A., 1979. Experiments with mixtures: An update and bibliography. *Technometrics* 21(1):95–106.

Cornell, J. A., 1981. *Experiments with Mixtures: Designs, Models, and the Analysis of Mixture Data*. New York: John Wiley & Sons, Inc.

Cornell, J. A., 1990. *How to Apply Response Surface Methodology*, rev. ed. Milwaukee, WI: American Society for Quality Control.

Cox, D. R., 1966. *Planning of Experiments*. New York: John Wiley & Sons, Inc.

Crosby, P. B., 1979. *Quality is Free*. New York: McGraw-Hill Book Company.

Crosby, P. B., 1984. *Quality Without Tears*. New York: McGraw-Hill Book Company.

Daigle, M. J., 1990. Measurement error vs. blueprint specification. Internal Memo, Rogers Corporation, 15 November, South Windham, CT.

Daniel, C. and F. S. Wood, 1991. *Fitting Equations to Data*. New York: John Wiley & Sons, Inc.

Data Myte Handbook—A Practical Guide to Computerized Data Collection for Statistical Process Control, 4th ed. Minnetonka, WI: Data Myte Corporation.

Davies, O. L., ed., 1960. *The Design and Analysis of Industrial Experiments*. London: Oliver and Boyd.

DeLassus, M. J., 1990. Material presented at Rogers Corporation training session on Taguchi methods, Quinnebaug Valley Community College, Danielson, CT.

Deming, W. E., 1986. *Out of the Crisis*. Cambridge: Massachusetts Institute of Technology, Center for Advanced Engineering Study.

Beauregard, M. R., R. J. Mikulak, M. J. DeLassus, and B. A. Olson, *Designing for Experimental Success*, 1991. Vernon, CT: Resource Engineering, Inc.

Diamond, W. J., 1981. *Practical Experimental Design for Engineers and Scientists*. New York: Van Nostrand Reinhold, Inc.

Dixon, W. J. and F. J. Massey, Jr., 1969. *Introduction to Statistical Analysis*, 3rd ed. New York: McGraw-Hill Book Company.

Draper, N. P. and H. Smith, 1981. *Applied Regression Analysis*, 2nd ed. New York: John Wiley & Sons, Inc.

Duncan, A. J., 1986. *Quality Control and Industrial Statistics*, 5th ed. Homewood, IL: Richard D. Irwin, Inc.

Fenton, B. A. and D. M. Baars, 1990. The use of designed experimentation and supplier involvement to aid in process start-up. SAE Technical Paper Series, International Congress and Exposition in Detroit, MI, 26 February–2 March 1990.

Fischer, R. A., 1935. *The Design of Experiments*. Edinburgh: Oliver and Boyd.

Gitlow, H. S. and S. J. Gitlow, 1987. *The Deming Guide to Quality and Competitive Position*. Englewood Cliffs, NJ: Prentice-Hall, Inc.

Gorman, J. W., 1966. Discussion of "Extreme Vertices Design of Mixture Experiments" by R. A. McLean and V. L. Anderson. *Technometrics* 8(3):455–456.

Grant, E. L. and R. S. Leavenworth, 1988. *Statistical Quality Control*, 6th ed. New York: McGraw-Hill, Inc.

Grayson, Jr., C. J. and C. O'Dell, 1988. *American Business: A Two-Minute Warning*. New York: The Free Press.

Guaspari, J., 1985. *I Know It When I See It*. New York: AMACOM.

Haber, A. and P. Bergeron, 1989. Organizing formulations using mixture design techniques. *Plastics Engineering* August: 37–40.

Hicks, C. R., 1982. *Fundamental Concepts in the Design of Experiments*, 3rd ed. New York: Holt, Rinehart, and Winston.

Holland, C. W., 1983. *Experimental Design Techniques for Properly Designing and Developing Processes*, Seminar Four. Powell, TN: Qualpro, Inc.

Hunter, J. S., 1968. *Fractional Factorial Designs*. Pittsburgh, PA: Westinghouse Learning Press.

Imai, M., 1986. *Kaizen, the Key to Japan's Competitive Success*. New York: Random House, Inc.

Ishikawa, K., 1984. *Guide to Quality Control*. Tokyo: Asian Productivity Organization

Juran, J. M., ed., 1988. *Juran's Quality Control Handbook*, 4th ed. New York: McGraw-Hill Book Company.

Juran, J. M., 1989. *Juran on Leadership for Quality: An Executive Handbook*. New York: The Free Press.

Juran, J. M. and F. M. Gryna, Jr., 1980. *Quality Planning and Analysis*. New York: McGraw-Hill Book Company.

Keuls, M., 1952. The use of the studentized range in connection with an analysis of variance. *Euphytical* 1:112–22.

Letize, L. and M. Donovan, 1990. The supervisor's changing role in high involvement organizations. *The Journal for Quality and Participation* March:62–65.

Lin, P. K. H., L. P. Sullivan, and G. Taguchi, 1990.

Using Taguchi methods in quality engineering. *Quality Progress* September:55–59.

Lochner, R. H. and J. E. Matar, 1990. *Designing for Quality*. White Plains, NY: Quality Resources.

Lundquist, E. D., 1988. Determining an appropriate setup sample size for use in process average setting. Internal Memo, Rogers Corporation, 27 May, South Windham, CT.

Manual on Presentation of Data and Control Chart Analysis, 6th ed., 1991. Philadelphia: American Society for Testing and Materials.

Marquandt, D. W. and R. D. Snee, 1974. Test statistics for mixture models. *Technometrics* 16(4):533–537.

McDermott, R., 1990. An employees idea system to enhance your SPC program. 46th Annual Rochester Section American Society for Quality Control Conference Transactions, March, Rochester Section of American Society for Quality Control, Rochester, NY, pp. 105–119.

McGue, F. and D. S. Ermer, 1988. Rational samples—not random samples. *Quality* December:30–34.

McLean, R. A. and V. L. Anderson, 1966. Extreme vertices design of mixture experiments. *Technometrics* 8(3):447–454.

Measurement Systems Analysis Reference Manual, 1990. Troy, MI: Automotive Industry Action Group.

Mikulak, R. J., 1990. Building intercompany cooperation through design of experiments (DOE). *Business Month Conference on Manufacturing Productivity Proceedings* June:229–236.

Mikulak, R. J., 1990. TQM—How does it work? Vernon, CT: Resource Engineering, Inc.

Mikulak, R. J., R. E. McDermott, and M. R. Beauregard, 1991. *First Class Services*. White Plains, NY: Quality Resources.

Montgomery, D. C., 1991. *Design and Analysis of Experiments*, 3rd. ed. New York: John Wiley & Sons, Inc.

Montgomery, D. C., 1991. *Introduction to Statistical Quality Control*, 2nd ed. New York: John Wiley & Sons, Inc.

Nelson, L. S., 1986. Control chart for multiple stream processes. *Journal of Quality Technology* 18(4):255–256

Newman, D., 1939. The distribution of the range in samples from a normal population expressed in terms of an independent estimate of standard deviation. *Biometrika* 31:20–30.

Parks, C. J., 1983. Statistical quality control: It's time for a fresh approach. *Manufacturing Engineering* November:39–44.

Parks, C. J., 1983. Statistical quality control: Management's role. *Manufacturing Engineering* December:59–62.

Parks, C. J., 1984. Worker response to SQC. *Manufacturing Engineering* January:44–45.

Peters, T., 1987. *Thriving On Chaos: Handbook for a Management Revolution*. New York: Alfred A. Knopf.

Plackett, R. L. and J. P. Burman, 1946. The design of optimum multifactorial experiments. *Biometrika* 33:305–325.

Potential Failure Mode and Effects Analysis, 1988. Plymouth, MI: Ford Motor Company

Process Capability and Continuing Process Control, 1983. Plymouth, MI: Ford Motor Company Statistical Methods Office.

Rickmers, A. D., 1984. Oil pipeline concept. Lecture to Rogers Corporation, Willimantic, CT.

Ross, P. J., 1988. *Taguchi Techniques for Quality Engineering*. New York: McGraw-Hill Book Company.

Scheffé, H., 1958. Experiments with mixtures. *Journal of the Royal Statistical Society* 20(B):344–361.

Scheffé, H., 1963. The simplex-centroid design for experiments with mixtures. *Journal of the Royal Statistical Society* 25(B):235–251.

Schrock, E. M., 1950. *Quality Control and Statistical Methods*. New York: Reinhold Publishing Corporation.

Shainin, D., 1986. The case of the incapable lathe. *Quality* August:21–22.

Shainin, D. and P. D. Shainin, 1989. Precontrol plan vs. X and R charting. 43rd Annual Northeast Quality Control Conference Transactions, October, 1989, pp. 217–222.

Shewhart, W. A., 1931. *Economic Control of Quality of Manufactured Product*. New York: Van Nostrand Company, Inc.

Shewhart, W. A., 1939. *Statistical Method from the Viewpoint of Quality Control*. Washington: The Graduate School, The Department of Agriculture.

Snee, R. D., 1971. Design and analysis of mixture experiments. *Journal of Quality Technology* 3(4):159–169.

Snee, R. D., 1973. Techniques for the analysis of mixture data. *Technometrics* 15(3):517–528.

Snee, R. D., 1979. Experimental designs for mixture systems with multicomponent constraints. *Communications in Statistics* A8(4):303–327.

Snee, R. D. and D. W. Marquandt, 1976. Screening concepts and designs for experiments with mixtures. *Technometrics* 18(1):19–29.

Spiers, B., 1990. Analysis of destructive measuring systems. 46th Annual Rochester Section American Society for Quality Control Quality Conference Transactions, March, Rochester Section American Society for Quality Control, Rochester, NY, pp. 165–170

Statistical Process Control Reference Guide, 2nd ed., 1986. Anderson, IN: Delco Remy.

Statistical Quality Control Handbook, 1956. Indianapolis, IN: AT & T Technologies.

Strategy of Formulations Development, rev. ed., 1982. Wilmington, DE: E. I. duPont de Nemour and Company, Inc.

Supplier Quality Improvement Guidelines for Production Parts, 1989. Plymouth, MI: Ford Motor Company.

Taguchi, G., 1987. *System Experimental Design*, Vols. 1 and 2. White Plains, NY: UNIPUB/Kraus International Publications.

Tao, B. Y., 1988. Optimization via the simplex method. *Chemical Engineering* 15 February:85–89.

Traver, R. W., 1990. Measuring equipment repeatability—The rubber ruler. 46th Annual Rochester Section American Society for Quality Control Quality Control Conference Transactions, March, Rochester Section American Society for Quality Control, Rochester, NY, pp. 197–208.

Veen, B., 1971. Standardization of flow charts for process quality control systems. *Quality* 2:35–39.

Wadsworth, H. M., K. S. Stephens, and A. B. Godfrey, 1986. *Modern Methods for Quality Control and Improvement*. New York: John Wiley & Sons, Inc.

Wheeler, D. J., 1982. *Using Statistics to Set a Process Average*. Knoxville, TN: Self-published.

Wheeler, D. J., 1990. *Advanced Topics in Statistical Process Control*. Knoxville, TN: SPC Press, Inc.

Wheeler, D. J. and D. Chambers, 1984. *Understanding Statistical Process Control*, 2nd ed. Knoxville, TN: Statistical Process Control, Inc.

Wheeler, D. J. and R. W. Lyday, 1989. *Evaluating the Measurement Process*, 2nd ed. Knoxville, TN: SPC Press, Inc.

Index